ARLIS

Alaska Resources
Library & Information Services
Anchorage, Alaska

RANDOMIZATION, BOOTSTRAP AND MONTE CARLO METHODS IN BIOLOGY

Second Edition

CHAPMAN & HALL/CRC
Texts in Statistical Science Series

Series Editors
C. Chatfield, *University of Bath, UK*
J. Zidek, *University of British Columbia, Canada*

RANDOMIZATION, BOOTSTRAP AND MONTE CARLO METHODS IN BIOLOGY

Second Edition

Bryan F. J. Manly

CHAPMAN & HALL/CRC

Boca Raton   London   New York   Washington, D.C.

Library of Congress Cataloging-in-Publication Data

Catalog record is available from the Library of Congress

Visit the CRC Press Web site at www.crcpress.com

© 1991, 1997 Bryan F.J. Manly
First edition 1991
Reprinted 1994
Second edition 1997
First CRC Press reprint 2001
Originally Published by Chapman & Hall

No claim to original U.S. Government works
International Standard Book Number 0-412-72130-9
Library of Congress Number 96-72031
Printed in the United States of America 2 3 4 5 6 7 8 9 0
Printed on acid-free paper

To cast lots puts an end to disputes, and decides between powerful contenders.

Proverbs

for Liliana

Contents

Preface to the second edition

The first edition of this book was written in 1988 and 1989. Since that time there have been important developments in ideas about randomization and bootstrap tests, and I have realized for some time that it was desirable to modify the book taking these ideas into account. It also became more and more apparent as time passed that I should have given more attention to bootstrapping in the book. I have therefore been pleased to have the opportunity to produce an updated second edition.

The major changes in this edition are a new Chapter 3 on bootstrapping, substantial changes to the chapters on analysis of variance and regression (now Chapters 7 and 8), a new Chapter 13 on survival and growth data, a new Chapter 15 on Bayesian methods, and a new appendix on computer software. To accommodate some of the extra material the Fortran subroutines that were included in the first edition have been removed. The word 'bootstrap' has been added to the title to reflect the new emphasis in this area.

I would like to thank several people who pointed out errors in the first edition of this book, and particularly Arthur Pewsey who gave me a long list of minor corrections that needed to be made before I started work on the second edition. I am also pleased to acknowledge the useful comments provided by Cajo ter Braak (Agricultural Mathematics Group, Wageningen, The Netherlands), Brian Cade (National Biological Service, Fort Collins, USA), Liliana Gonzalez (University of Otago, Dunedin, New Zealand), and Peter Kennedy (Simon Fraser University, Burnaby, Canada) on draft versions of various chapters, and two reviewers of the entire manuscript. However, I take full responsibility for the final contents.

Bryan F. J. Manly
Dunedin, July 1996

Preface to the first edition

In 1979, Bradley Efron wrote an article called 'Computers and the theory of statistics: thinking the unthinkable'. The subtitle was to do with the fact that at that time statisticians were prepared to contemplate methods for analysing data that would have been unthinkable in 1950 because of the huge number of calculations required. For example, to analyse 16 numbers a statistician might be prepared to do half a million basic arithmetic operations, while to analyse 500 numbers, 100 million such operations might be worthwhile. Ideas like this would have sounded insane in 1950.

The significance of the year 1950 is that it was near the end of the pre-computer days. At that time a slow, noisy and heavy desk calculator was all that was available to most statisticians. This was still true for most statistics students even in the mid 1960s, and extracting square roots quickly and correctly on a hand-operated machine was a skill that many had to master. It was not until the end of the 1960s that computers began to become generally available.

When Efron was writing he would have been thinking of data analysts carrying out their calculations on the large computers at their places of work. The development that we have seen in the 1980s is that these calculations can be moved down to personal microcomputers. They can now be done on a portable computer that can be taken home.

In the long term these developments are clearly going to have a tremendous impact on the way that the subject of statistics is taught and practised. Because statistics inevitably involves calculations, the cost of these calculations has to be an important factor in deciding what type of analysis is best for some data. In the pre-computer days, when calculations were expensive in terms of time and effort, there was an emphasis on finding methods that obtained the maximum information out of the smallest number of calculations. The advent of the computer has changed the rules, because calculations are now fast and inexpensive.

There are three ways that computers are having an effect. First, they are used to do exactly the same calculations as before, but these are done much more quickly, and on larger sets of data. Second, some classical methods of analysis are being superseded by alternative computer-intensive methods, with advantages that are paid for by a large increase in the number of calculations required. Third, computers are being used to solve problems for which no satisfactory solution has previously been proposed.

In the teaching of statistics it is often the first of these effects that is emphasised. Over the last 25 years there has been a steady development in this respect and now introductory texts often assume that all except the simplest of calculations will be done on the computer. However, it can be argued that it is the second and third effects that will turn out to be more important. It is these last effects that this book is concerned with, most particularly in the context of randomization and Monte Carlo methods of inference, these being two related categories within an area that is sometimes referred to as computer-intensive statistics.

Randomization methods can be used to test hypotheses and (under certain conditions) to determine confidence limits for parameters. Randomization testing involves determining the significance level of a test statistic calculated for an observed set of data by comparing the statistic with the distribution of values that is obtained by randomly reordering the data values in some sense. This is generally a computer-intensive procedure because the distribution has to be determined either by enumerating all possible data orders, or by taking a large random sample from the distribution. It is the sampling approach that is emphasized in this book. Randomization confidence limits for a parameter are given by the range of values of that parameter for which randomization testing gives a non-significant result.

Monte Carlo methods can also be used to test hypotheses and construct confidence intervals. Testing still involves comparing an observed test statistic with values obtained by sampling a distribution. However, in this case the distribution sampled need not arise from simply reordering the observed data values. Rather, the distribution is of the test statistics generated by some particular model for how the observed data arose. Confidence intervals can be based directly on the variation in parameter estimates observed in data generated from the model.

It is assumed that readers are familiar with the methods that are usually covered in a first university course in statistics for students who are specializing in other subjects. A somewhat fuller knowledge of analysis of variance, regression and multivariate methods will be helpful in some places.

In writing this book, I hope that I have been able to convey some of the spirit of the new-found freedom that the computer has given us for data analysis. I also hope that I have been able to demonstrate the scope that is available for innovative work in developing new statistical methods in biology using the computer.

Byron Morgan read a draft version of this book and made many helpful suggestions for improvements. Lyman McDonald read much of the text and pointed out a discrepancy in Chapter 6. I thank them both but take full responsibility for any errors that remain.

Bryan F. J. Manly
Dunedin, January 1990

1
Randomization

1.1 THE IDEA OF A RANDOMIZATION TEST

Many hypotheses of interest in science can be regarded as alternatives to null hypotheses of randomness. That is to say, the hypothesis under investigation suggests that there will be a tendency for a certain type of pattern to appear in data, whereas the null hypothesis says that if this pattern is present then this is a purely chance effect of observations in a random order.

Randomization testing is a way of determining whether the null hypothesis is reasonable in this type of situation. A statistic S is chosen to measure the extent to which data show the pattern in question. The value s of S for the observed data is then compared with the distribution of S that is obtained by randomly reordering the data. The argument made is that if the null hypothesis is true then all possible orders for the data were equally likely to have occurred. The observed data order is then just one of the equally likely orders and s should appear as a typical value from the randomization distribution of S. If this does not seem to be the case (so that s is 'significant') then the null hypothesis is discredited to some extent and, by implication, the alternative hypothesis is considered more reasonable.

The significance level of s is the proportion or percentage of values that are as extreme or more extreme than this value in the randomization distribution. This can be interpreted in the same way as for conventional tests of significance: if it is less than 5% this provides some evidence that the null hypothesis is not true; if it is less than 1% it provides strong evidence that the null hypothesis is not true; and if it less than 0.1% it provides very strong evidence that the null hypothesis is not true. To avoid the characterization of belonging to 'that group of people whose aim in life is to be wrong 5% of the time' (Kempthorne and Doerfler, 1969), it is better to regard the level of significance as a measure of the strength of evidence against the null hypothesis rather than showing whether the data are significant or not at a certain level.

In comparison with more standard statistical methods, randomization tests have two main advantages: they are valid even without random samples, and it is relatively easy to take into account the peculiarities of the situation of interest and use non-standard test statistics.

There is a disadvantage with randomization tests that may appear at first sight to be somewhat severe. This is that it is not necessarily possible to generalize the conclusions from a randomization test to a population of interest. What a randomization test tells us is that a certain pattern in data is or is not likely to have arisen by chance. This is completely specific to the data at hand. The concept of a population from which other samples could be taken is not needed, which is the very reason why random sampling is not required.

Some statisticians argue that the lack of a theory for generalizing the results of randomization tests to populations means that these tests have very little value, if any, in comparison with more standard tests for which well-developed methods of statistical inference exist. Others, however, suggest that in reality samples are often not really random at all but simply consist of items that happen to be readily available. The generalization of results then rests on the assumption that the sample obtained is effectively the same as a random sample. This non-statistical judgement is similar to the type of judgement that is made when deciding that the result of a randomization test is what can generally be expected for data collected in a particular way.

As an example, suppose that a physiologist wishes to see whether drinking alcohol in moderation has an effect on reaction times of subjects aged 20. Rather than take a random sample of all possible subjects of this age (which leads to considerable difficulties about the definition of the 'population', and is in any case impossible), he uses all the 20-year-old students in a university class in physiology. These are divided at random into two groups, one of which has reaction times measured after taking a drink with a small amount of alcohol, and the other which has reaction times measured after taking an alcohol-free drink.

Various methods can be used to analyse the results of an experiment of this type. For example, if a mean difference between the test scores for the two groups is of interest, then a conventional *t*-test can be used to determine whether the observed difference is significantly different. However, whatever the outcome of such a test is, using it to draw conclusions about the effect of alcohol on all 20-year-olds is only valid on the assumption that the 20-year-olds in the physiology class are equivalent to a random sample of all 20-year-olds with respect to the measurement of reaction times that is used. Hence any such generalization has to be questionable, and requires a judgement as to whether the same type of result is likely to occur again if a different group of subjects is tested.

It seems clear that one experiment of this type will not give a definitive result, no matter how many subjects are used. However, if the experiment is repeated on other groups (law students, factory workers, office workers etc.) and the results always come out about the same, then most people would believe that the effect (or lack of an effect) seen is common to all 20-year-olds. In other words, in the absence of truly random samples, convincing evidence of an effect requires it to be demonstrated consistently at different times in different places. This is a non-statistical type of inference which works equally well with conventional and randomization tests.

Another relevant point is that in many situations the concept of a population is either irrelevant or the data can be considered as representing the whole population. Thus an example that is considered in the next section concerns the relationship between the world distribution of earwigs and the positions of the continents. There is only one distribution of earwigs that can exist, and one set of continents, so the data are not samples from populations except in a most unrealistic sense. Another example in the next section addresses the question of whether there is a cycle in the times of mass extinctions of animals and plants in the geologic past. Here only one extinction record can exist and the concept of this being a random record from a population of possible records is again somewhat artificial.

Although random samples are not necessarily required in order to justify randomization tests, there are times when they do provide the justification. For example, in the reaction time experiment there would be no need to divide the subjects at random into two groups if initially there were two random samples available from the population of 20-year-olds. In that case either group could be the one given the alcohol and a valid comparison between the test scores of the two groups to examine the effect of alcohol could be made using a randomization test or a more conventional alternative.

Randomization tests are most easily justified if either the samples being analysed are random or the experimental design itself justifies randomization testing. This has led some authors (e.g. Kempthorne and Doerfler, 1969) to use the description 'permutation tests' for situations where random samples justify the calculations, and 'randomization tests' for situations where the experimental design provides the justification. Here these descriptions will both be used for any situation where randomly reordering observations is used to determine the significance level of a test statistic.

1.2   EXAMPLES OF RANDOMIZATION TESTS

To clarify the procedures and principles that are used with randomization testing, it will be helpful to consider some examples at this point.

Example 1.1: mandible lengths of male and female golden jackals   The data shown below are mandible lengths in mm for male and female golden jackals (*Canis Aureus*) for 10 of each sex in the collection in the British Museum (Natural History):

Males: 120, 107, 110, 116, 114, 111, 113, 117, 114, 112

Females: 110, 111, 107, 108, 110, 105, 107, 106, 111, 111

The lengths were measured as part of a study by Higham *et al.* (1980) on the relationship between prehistoric canid bones from Thailand and similar bones from modern species. For the present example the question addressed is whether there is any evidence of a difference in the mean lengths for the two sexes.

Data like the above are often collected in the belief that there will be a difference in the results for the two groups. In fact, it is a reasonable supposition that male jackals will tend to be larger than females. The result expected before collecting the data was therefore that the male mean would be higher than the female mean. This can be tested indirectly by setting up a null hypothesis which says that any difference between the two sample means is purely due to chance. If this null hypothesis is consistent with the data then there is no reason to reject this in favour of the alternative hypothesis of that males have a higher mean.

It may seem strange to test the hypothesis of interest by setting up a null hypothesis and seeing how the data compare with this. However, there is frequently little choice in the matter. It is possible to work out probabilities of different sample results or to generate possible sample results using the null hypothesis. To do this for the hypothesis that is of real interest is not possible without specifying the magnitude of any effect. This magnitude is, of course, not known, because if it were known then there would be no reason for carrying out a test in the first place.

Before describing a randomization test for a difference in the male and female means, it is interesting to look at how the data might be analysed with a more conventional approach. There are a number of standard tests that could be used. The *t*-test with a pooled estimate of the within-group standard deviation is one possibility, and that is the one that will be considered here.

Rather than restrict attention only to the data in hand, suppose that there are sample sizes of n_1 for one group and n_2 for a second group, with corresponding sample means and standard deviations of $\bar{X}_1$ and $\bar{X}_2$, and s_1 and s_2, respectively. Assume that the test scores for the first group are a random sample from a normal distribution with mean μ_1 and standard deviation σ, and the test scores for the second group are a random sample from a normal distribution with mean μ_2 and standard deviation σ. Then the hypothesis of interest is that $\mu_1 > \mu_2$ while the null hypothesis is that $\mu_1 = \mu_2$. The null hypothesis is tested by first obtaining a pooled estimate of the common within-group standard deviation,

$$s = \sqrt{[\{(n_1 - 1)s_1^2 + (n_2 - 1)s_2^2\}/\{n_1 + n_2 - 2\}]},$$

and then calculating the statistic

$$T = (\bar{X}_1 - \bar{X}_2)/\{s\sqrt{(1/n_1 + 1/n_2)}\}.$$

If the null hypothesis is true then T will be a random value from the t-distribution with $n_1 + n_2 - 2$ degrees of freedom.

The t-test for the golden jackal data consists, then, of calculating T and seeing if it is within the range of values that can reasonably be expected to occur by chance alone if the null hypothesis is true. If it is within this range then the conclusion is that there is no evidence against the null hypothesis (of no sex difference) and hence no reason to accept the alternative hypothesis (males tend to be larger than females). On the other hand, if the T value is not within the reasonable range for the null hypothesis then this casts doubt on the null hypothesis and the alternative hypothesis of a sex difference may be considered more plausible.

Only large positive values of T favour the hypothesis of males tending to be larger than females. Any negative values (whatever their magnitude) are therefore considered to support the null hypothesis. This is then a one-sided test as distinct from a two-sided test that regards both large positive and large negative values of T as evidence against the null hypothesis.

For the given data the means and standard deviations for the two samples are $\bar{X}_1 = 113.4\,\text{mm}$, $\bar{X}_2 = 108.6\,\text{mm}$, $s_1 = 3.72\,\text{mm}$, and $s_2 = 2.27\,\text{mm}$. The pooled estimate of the common standard deviation is $s = 3.08$, and $T = 3.48$ with 18 degrees of freedom. The probability of a value this large is 0.0013 if the null hypothesis is true. The sample result is therefore nearly significant at the 0.1% level, which would generally be regarded as strong evidence against the null hypothesis, and consequently for the alternative hypothesis.

The assumptions being made in this analysis are:

1. random sampling of individuals from the populations of interest;
2. equal population standard deviations for males and females; and
3. normal distributions for mandible length scores within groups.

Assumption 1 is certainly questionable because the data are from museum specimens collected in some unknown manner. Assumption 2 may be true. At least, the sample standard deviations do not differ significantly on an F-test $(F = s_1^2/s_2^2 = 2.68$ with 9 and 9 degrees of freedom is not significantly different from 1.0 in comparison with the critical value in the F table, using a 5% level of significance). Assumption 3 may be approximately true but this cannot be seriously checked with samples as small as 10.

In fact, the t-test considered here is known to be quite robust to deviations from assumptions 2 and 3, particularly if the two-sample sizes are equal, or nearly so. Nevertheless, these assumptions are used in deriving the test, and violations can be important, particularly when sample sizes are small.

Consider now a randomization test for a sex difference in the mean mandible lengths of golden jackals. This can be based on the idea that if there is no difference then the distribution of lengths seen in the two samples will just be a typical result of allocating the 20 lengths at random into two groups of size 10. The test therefore involves comparing the observed mean difference between the groups with the distribution of differences found with random allocation. The choice of measuring the difference here by the sample mean is arbitrary. A t-statistic could be calculated rather than a simple mean difference, or the median difference could be used equally well. The randomization test procedure is described by the following steps.

A randomization test procedure

(a) Find the mean scores for males and females and the difference D_1 (male − female) between these.
(b) Randomly reallocate 10 of the sample lengths to a 'male' group and the remaining 10 scores to a 'females' group and determine the difference D_2 between the means thus obtained.
(c) Repeat step (b) a large number of times to find a sample of values from the distribution of D that occurs by randomly allocating the scores actually observed to two groups of 10. This estimates what will be called the randomization distribution.
(d) If D_1 looks like a typical value from the randomization distribution then conclude that the allocation of lengths to the males and females that occurred in reality seems to be random. In other words, conclude that there is no sex difference in the distribution of lengths of mandibles. On the other hand, if D_1 is unusually large then the data are unlikely to have arisen if the null model is true and it can be concluded that the alternative hypothesis (males tend to be larger than females) is more plausible.

At step (d) in the above procedure it is a question of whether D_1 is unusually large in comparison with the randomization distribution. If D_1

is in the bottom 95% part of the randomization distribution we can say that the test result is 'not significant', but if D_1 is amongst the values in the top 5% tail of the distribution then the result is 'significant at the 5% level'. Similarly, a value in the top 1% tail is significant at the 1% level, and a value in the top 0.1% tail is significant at the 0.1% level, these representing increasing levels of significance and increasing evidence against the null model.

A convenient way to summarize the randomization results involves calculating the proportion of all the observed D values that are greater than or equal to D_1, counting the occurrence of D_1 in the numerator and denominator. This is an estimate of the probability of such a large value, which is the significance level of D_1. Including D_1 in the numerator and denominator is justified because if the null hypothesis is true then D_1 is just another value from the randomization distribution.

The obvious way to carry out the randomizations is by computer. A pseudo-random number generator is available on most computers, or can be programmed easily enough, although there are pitfalls for the novice in this area. It is fairly straightforward to use random numbers to determine random assignments of 20 test scores to two groups of 10, and the randomizations can be made quite quickly even on a small computer. The procedure is obviously computer-intensive because instead of calculating a single test statistic and using a table to determine whether this is significant, the data are randomized and the test statistic calculated hundreds or even thousands of times.

Computer software for randomization testing and the other methods that are discussed in this book is reviewed in the appendix. The calculations for this example and the other examples in this chapter were carried out using the package RT (Manly, 1996a).

Because the mean of the observed lengths for the males is 113.4 mm and the mean for the females is 108.6 mm, there is a difference of $D_1 = 4.8$ mm. Figure 1.1 shows the distribution obtained for the mean difference from 4999 randomizations plus D_1. Only nine of these differences were 4.8 or more, these being D_1 itself, six other values of 4.8, and the two larger values 5.0 and 5.6. Clearly 4.8 is a most unlikely difference to arise if the null model is true, the significance level being estimated as $9/5000 = 0.0018$ (0.18%). There is strong evidence against the null model, and therefore in favour of the alternative. Note that the estimated significance level here is quite close to the level of 0.0013 for the t-test on the same data discussed before.

It has been noted earlier that in a one-sided test like this it is only large positive values of the test statistic that give evidence against the null hypothesis and in favour of the alternative hypothesis that males tend to be larger than females. Negative values of test statistics, however large,

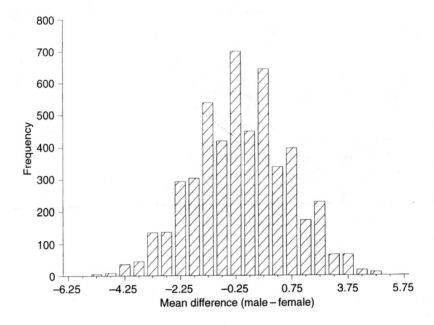

Figure 1.1 The distribution of the differences observed between the mean for males and the mean for females when 20 measurements of mandible lengths are randomly allocated, 10 to each sex.

are regarded as being consistent with the null hypothesis. In practice, of course, common sense must prevail here because a large negative test statistic that has a low probability of occurring is in a sense contradicting both the alternative and the null hypotheses. The appropriate reaction to such an occurrence will depend on the particular circumstances. In the case of the present example it would indicate that females tend to have larger mandibles than males, which may or may not be considered possible. If this was considered to be possible before collecting the data then the test should have been two-sided in the first place.

In this example it is possible to determine the randomization distribution of the male–female mean differences exactly by considering every possible allocation of the 20 test scores to two samples of size 10. This is the number of combinations of 20 things taken 10 at a time to be labelled 'males', $^{20}C_{10} = 184\,756$. Although this is a large number, the allocations could be made on a computer in a reasonable amount of time. With larger samples such a direct enumeration of possibilities may not be practical.

Problems involving the comparison of two samples are considered further in Chapter 6.

Example 1.2: comparing distance matrices Some time ago it was suggested (Popham and Manly, 1969) that there are two plausible explanations for the present distribution of earwigs. They may have evolved in the northern hemisphere and subsequently spread into the southern continents or, alternatively, they may have evolved throughout the southern proto-continent of Gondwanaland as it was disintegrating about 150 million years ago into South America, Africa, India, Australasia and the Antarctic, and spread from there into the northern hemisphere. If the first explanation is correct then the similarities between species in different parts of the world should roughly reflect their present distances apart. In particular, the southern continents should show the maximum species differences. On the other hand, if the second explanation is correct then the southern continents should have rather similar species.

In recent years the theory of continental drift has become widely accepted by scientists. However, this was not the case in 1969. The main purpose of the study by Popham and Manly was therefore to consider some biological data that could lend credibility to this theory. This second example of a randomization test considers the question of which of the two explanations for the distribution of earwigs best matches the available data when these data are summarized in terms of a matrix of geographical distances between continents and a matrix of species closeness between continents.

It is surprising how many problems can be looked at in terms of whether or not there is a relationship between two matrices that measure the similarities or differences between a set of objects. With this approach, each of the matrices is an array in which the element in row i and column j is a measure of the closeness or distance between object i and object j on the basis of some particular criterion. A table of road distances between towns will be a familiar example. Often it is interesting to test whether the elements of the first matrix are significantly related to the elements of the second matrix.

For the example being considered, the 'objects' are eight areas in the world: Europe and Asia, Africa, Madagascar, the Orient, Australia, New Zealand, South America and North America. The first of the two matrices that are to be compared contains measures of the similarity between these areas in terms of genera of earwigs, as shown in Table 1.1. These are coefficients of association defined by Yule and Kendall (1965), based on the distribution of genera in two land areas. A value of +1 occurs if exactly the same genera are present in two areas, a value of 0 is what is expected from a random allocation of genera, and a value of −1 occurs if the distribution of genera is mutually exclusive in two areas. Only the lower

Table 1.1 Coefficients of association for earwig species in different areas of the world

		E&A	AFR	MAD	ORE	AUS	NZ	SA	NA
Europe & Asia	(E&A)	—							
Africa	(AFR)	0.30	—						
Madagascar	(MAD)	0.14	0.50	—					
The Orient	(ORE)	0.23	0.50	0.54	—				
Australia	(AUS)	0.30	0.40	0.50	0.61	—			
New Zealand	(NZ)	−0.04	0.04	0.11	0.03	0.15	—		
South Africa	(SA)	0.02	0.09	0.14	−0.16	0.11	0.14	—	
North America	(NA)	−0.09	−0.06	0.05	−0.16	0.03	−0.06	0.36	—

Table 1.2 Distances between land masses in terms of 'jumps' assuming their present positions

		E&A	AFR	MAD	ORE	AUS	NZ	SA	NA
Europe & Asia	(E&A)	—							
Africa	(AFR)	1	—						
Madagascar	(MAD)	2	1	—					
The Orient	(ORE)	1	2	3	—				
Australia	(AUS)	2	3	4	1	—			
New Zealand	(NZ)	3	4	5	2	1	—		
South Africa	(SA)	2	3	4	3	4	5	—	
North America	(NA)	1	2	3	2	3	4	1	—

Table 1.3 Distances between land masses in terms of 'jumps' assuming the positions before continental drift

		E&A	AFR	MAD	ORE	AUS	NZ	SA	NA
Europe & Asia	(E&A)	—							
Africa	(AFR)	1	—						
Madagascar	(MAD)	2	1	—					
The Orient	(ORE)	1	1	1	—				
Australia	(AUS)	2	1	1	1	—			
New Zealand	(NZ)	3	2	2	2	1	—		
South Africa	(SA)	2	1	2	2	2	3	—	
North America	(NA)	1	2	3	2	3	4	1	—

triangular part of Table 1.1 is given because the table is symmetric and, for example, the similarity between the Orient and Africa and the similarity between Africa and the Orient are the same, both being 0.50.

The second matrix to be considered contains the distance between the eight areas in terms of the number of 'steps' required to get from one to another, which depends upon the relative positions of the areas when these steps are made. The step distances shown in Table 1.2 are based on the present positions of the continents, assuming one step from Asia to North America by a land bridge across the Bering Strait. For example, the distance from Africa to South America is 3 (Africa–Eurasia–North America–South America). The alternative distances shown in Table 1.3 are based on the assumption that there were direct links between South America and Africa, Africa and India, and India and Australia while earwigs were evolving in Gondwanaland. Obviously, these 'step' distance matrices are no more than crude first approximations to the distances that earwigs had to cover when they were evolving.

It is interesting to look at the relationship between the species similarity matrix and the present time 'step' distance matrix, and see how this compares with the relationship between the species similarity matrix and the Gondwanaland 'step' distance matrix. If the first of these relationships is strongest then this suggests that evolution occurred with the continents close to their present positions. Conversely, if the second relationship is strongest then this suggests that evolution occurred in Gondwanaland.

Before making this comparison it is necessary to decide how to measure the relationship between two distance matrices. One obvious possibility is to use the Pearson correlation coefficient between corresponding elements in the two matrices. Thus in the matrix in Table 1.1 there are 28 distances $0.30, 0.14, 0.50, \ldots, 0.36$. The corresponding distances in Table 1.2 are $1, 2, 1, \ldots, 1$. The correlation between these distances is -0.217, which can be taken as the measure of the strength of the relationship between the two matrices. This has an appropriate negative sign because large values in Table 1.1 indicate similar species distributions, which should be associated with small distances in Table 1.2. However, the correlation between the species similarities and the Gondwanaland distances is much more negative at -0.605. It seems, therefore, that the present-day distribution of earwigs is explained best in terms of evolution in Gondwanaland.

The standard theory of correlation does not provide any way to decide whether either of the observed levels of correlation is more than can be explained by chance because the distances within a table are not independent of each other. (For example, A close to B and A close to C imply that B is close to C.) This deficiency can be overcome by using a Mantel (1967) matrix randomization test. The principle behind such a test is that if the species distances are completely unrelated to the geographical distances then the labelling of the continents for one of the matrices will look random with respect to the labelling on the other matrix. In other words, the observed correlation will look like a typical value from the distribution of correlations obtained by randomly shuffling the continent labels in one of the matrices. The test therefore involves carrying out the random shuffling to find this randomization distribution and then obtaining the significance of the observed correlation with respect to this. A one-sided test is needed because only negative correlations are expected for a relationship generated by evolution.

When the randomizations are carried out using the distance matrices of Tables 1.1 to 1.3 a rather definite result is obtained. Comparing the species distances matrix (Table 1.1) and the geographical distance matrix for the present distribution of the continents (Table 1.2), 4999 randomizations produced 913 cases of a correlation of -0.217 or less. The significance level of the observed correlation is therefore $914/5000 = 0.183$ (18.3%), counting the observed correlation as another randomization value. The observed correlation is therefore a fairly typical value to obtain by chance and there is no real evidence of a relationship between species distances and the present-day distances between the continents.

The situation is very different when the species similarity matrix is compared with the Gondwanaland geographical distances (Table 1.3). Here only 6 out of 4999 randomizations produced correlations as negative as the observed -0.605. The observed result therefore has an estimated significance level of $7/5000$, or 0.14%. There is strong evidence of the type of relationship expected if earwigs evolved in Gondwanaland as it was breaking up. It seems, therefore, that the data favour the idea that earwigs evolved in Gondwanaland rather than in the northern hemisphere with continents in their present positions.

An alternative way to determine the randomization distribution of the matrix correlation would be to enumerate the correlation for every possible one of the $8! = 40\,320$ possible orders of the eight continents for one of the distance matrices. This is a feasible undertaking, but it would obviously take much more computer time than the generation of 5000 random orders.

Further details of the use of the Mantel (1967) matrix randomization test are provided in Chapter 9.

Example 1.3: periodicity in the extinctions of animals   Virtually all species of animals and plants that have ever existed are now extinct, with fossil records suggesting that many extinctions occurred over fairly short periods of time. In some cases it is thought that 95% of species may have disappeared in one of these periods of mass extinctions. All this has been known for some time. Indeed, the boundaries between geological 'ages' reflect to some extent the times of mass extinctions. However Raup and Sepkoski (1984) caused something of a stir in the scientific world when they produced evidence that there is periodicity in extinction rates, with about 26 million years between times of high rates. It had previously been assumed that extinction rates were generally at a fairly low level and that the mass extinctions were completely unpredictable random occurrences caused by unlikely events such as the impact of comets. A regular cycle implies some other mechanism at work, possibly related to the position of the Solar System in the galaxy. One suggestion that was taken seriously was that the Sun has a companion star 'Nemesis' which comes close to the Solar System about every 26 million years. (Nemesis was the Greek goddess of retributive justice.)

There are considerable problems with fossil data caused by uncertainties about the dates at which different geological ages began and the fact that fossil records are not complete. These matters are discussed by Raup and Sepkoski who decided to use two time-scales to measure extinctions in terms of families of species, and to go back 253 million years into the past. They used randomization testing as the main basis of their statistical analyses because standard methods for analysing time series are not easily applicable to this example.

What they did was to try extinction cycle lengths between 12 and 60 million years, in one year increments, finding the starting point for each cycle length that was most in agreement with the data in terms of matching the peak extinctions. 'Agreement' was measured by the standard deviation of the difference between the data peak and the closest predicted peak. To determine whether the agreement that was observed was greater than could be expected by chance for any particular cycle length, they randomized the order of the data thousands of times and ran the same analysis as before on each shuffled set. This indicated that the fit of a 26 million year cycle to the original data is such as would only be obtained by chance somewhere between 1 in 10 000 and 5 in 1000 times, depending upon how the randomization of data is done. Somewhat significant results were also found for cycles with lengths of about 19 million years and 30 million years.

Following Raup and Sepkowski's paper similar types of randomization testing have been used to find evidence of 30 million year cycles in the record of reversals of the Earth's magnetic field and in numbers of comet/asteroid impacts. These studies have been controversial, with claims that significant results are due to either errors in the data or flaws in the procedures used. They are considered in more detail in Chapter 11.

1.3 ASPECTS OF RANDOMIZATION TESTING RAISED BY THE EXAMPLES

The tests considered in the previous section are typical of randomization tests in general. In each case there is a particular hypothesis that is of interest (males are larger than females, species similarities are negatively related to geographical distances between continents, there is a cycle in the times of mass extinctions), which is tested by comparing the data with what is expected from a null model which says that the data are in a random order. The meaning of 'random order' is different in each case. Data are summarized by test statistics which also depend very much on the circumstances.

There are several matters related to randomization testing in general that are raised by the examples. It is worth discussing these further at this point.

1.3.1 Sampling the randomization distribution or systematic enumeration

In the first two examples, mention was made of determining the randomization distribution exactly by enumerating all of the equally likely possible orders for the data. This is certainly a practical proposition with some tests and overcomes the problem that different random samples from this distribution will give slightly different significance levels for the observed data. However, in many cases the large number of possible data permutations makes a complete enumeration extremely difficult (if not impossible) with present-day computers.

One way around the problem of enumerating all possibilities in order to analyse one set of data involves choosing a test statistic which is only a function of the relative order of the data values. In that case, if there are no ties then the randomization distribution is the same for all sets of data with the same sample sizes and tables of critical values of test statistics can be constructed. This is in fact the basis of the standard non-parametric tests such as the Mann–Whitney U-test to compare two samples. An important difference between standard non-parametric tests and randomization tests is therefore that the data values are taken into account with randomization tests and not just the order of the values.

In fact, there is usually nothing much to be gained from a complete enumeration. Taking a large sample from the randomization distribution can be expected to give essentially the same result, and avoids the problem of finding a systematic way of working through all possible orders. Furthermore, as noted by Dwass (1957), the significance level of a randomization test is in a sense 'exact' even when the randomization distribution is only sampled.

The sense in which the significance level is exact can be seen by way of an example. Thus suppose for simplicity that a one-sided test at the 5% level of significance is required, with large values of the test statistic providing evidence against the null hypothesis. Assume that 99 randomized sets of data are generated so that 100 sets are available, including the real set. Each of the 100 sets then provides a value for the test statistic and the observed test statistic is declared significant at the 5% level if it is one of the largest five values in the set. This procedure gives a probability of 0.05 of a significant result if there are no tied values for the test statistic and the null hypothesis is true (i.e. the observed test statistic is a random value from the randomization distribution) because the probability of the observed test statistic falling in any position in the list is 0.01. Consequently, the probability of it being one of the five at the top of the list is 0.05.

If there are ties then the situation is a little more complicated. However, defining the significance level of the observed data as the proportion of the 100 test statistics that are greater than or equal to the one observed with the real data is equivalent to ranking all the values from smallest to largest and putting the observed value as low as possible. The probability of it being one of the five values at the top of the list is therefore 0.05 or less, depending on where ties occur. The probability of a significant result by chance alone is therefore 0.05 or less.

This argument can be generalized easily enough for a one-sided test with any number of randomizations and any significance level. With a two-sided test significant values can be either high or low, but again the argument generalizes in a straightforward way.

1.3.2 Equivalent test statistics

The randomization test in the first example on mandible lengths used the difference between two sample means as the test statistic. An obvious question then concerns whether the same level of significance is obtained if another statistic is used. In particular, because a t-statistic would commonly be used for the sample comparison one might wonder what results would be obtained by using this in the randomizations.

In fact, it rather conveniently turns out that using a t-statistic must give the same result as using the difference between the two means. This is because these two summary statistics are exactly equivalent for randomization testing in the sense that if randomized sets of data are ordered according to the mean difference between samples 1 and 2 then this will be the same order as that based on t-values.

Obviously not all sensible test statistics are equivalent in this sense. For example, a comparison of two samples based on the difference between the sample medians may give a quite different result from a comparison based on means. Part of the business of randomization testing is to decide what test statistic is appropriate in a given situation, which would be unnecessary if every statistic gave the same result.

What is sometimes important is to recognize the equivalence of several alternative statistics and use for a randomization test the one that can be calculated most quickly. Minor differences such as the multiplication by a constant may become important when the statistic used has to be evaluated thousands of times. Thus it is sensible to use the difference between the sample means in the first example, rather than the t-statistic. Actually, as will be shown in Chapter 6, the sum of the sample 1 values is an even simpler equivalent test statistic.

1.3.3 Significance levels for classical and randomization tests

The first example on mandible lengths also brings up the question of how outcomes compare when a classical test and a randomization test are both used on the same data. In that example two alternative tests were discussed. The first was a t-test using a pooled sample estimate of variance. The second was a randomization test. The outcome was the same in both cases: a significant difference between the experimental and control group means at nearly the 0.1% level. More precisely, it can be noted that the calculated t-value of 3.48 with 18 degrees of freedom is equalled or exceeded with probability 0.0013 according to the t-distribution, whereas the estimated significance from randomization is 0.0018.

A good agreement between a conventional test and a randomization test is not unusual when both tests are using equivalent test statistics. As a general rule it seems that if the data being considered are such that the assumptions required for a classical test appear to be reasonable (e.g. the observations look as though they could have come from a normal distribution because there are no obvious extreme values), then a classical test and a randomization test will give more or less the same results. However, if the data contain one or two anomalous values then the two tests may not agree.

An interesting aspect of the general agreement between classical and randomization tests with 'good' data is that it gives some justification for using classical statistical methods on non-random samples because the classical tests can be thought of as approximations for randomization tests. Thus Romesburg (1985) argues that

> ...scientists can use classical methods with nonrandom samples to good advantage, both descriptively and, in a qualified sense, inferentially. In other words, scientists can perform, say, a paired t-test on a nonrandom sample of data and use this to help confirm knowledge, in spite of warnings against such practice in statistics texts.

Edgington (1995, p. 10) quotes several statisticians who appear to be extending this argument to the extent of saying that the only justification for classical methods is that they usually give the same results as randomization tests, although this seems a rather extreme point of view.

1.3.4   Limitations of randomization tests

There is an obvious limitation with randomization tests in the strictest sense in that they can only be used to test hypotheses involving comparisons between two or more groups (where randomization involves swapping observations between groups), or hypotheses that say that the observations for one group are in a random order (where randomization involves generating alternative random orders). By their very nature they are not applicable to testing hypotheses concerning absolute values of parameters of populations.

For example, consider again the comparison of mandible lengths for male and female golden jackals (Example 1.1). Suppose that it is somehow known that the average length of mandibles for wild males is 115.3 mm. Then it might be interesting to see whether the observed sample mean score for males is significantly different from 115.3. However, on the face of it, a randomization test cannot do this as there is nothing to randomize. This is hardly surprising because the hypothesis of interest concerns a population parameter, and the concept of a population is not needed for randomization tests.

There is a way round this particular limitation if it is reasonable to assume that the male observations are a random sample from a distribution that is symmetric about the population mean. In that case it is possible to subtract the hypothetical mean of 115.3 from each male observation and regard the signs of the resulting values as being equally likely to be positive or negative. The observed sample means of the modified observations can then be compared with the distributions of means that are obtained by randomly assigning positive and negative signs to the data values. This test, which is commonly referred to as Fisher's one-sample randomization test, is discussed in more detail in Chapter 6. Here it will merely be noted that it does not fall within the definition of a randomization test adopted for most of this book, which requires that observations themselves can be randomly reordered.

1.4   CONFIDENCE LIMITS BY RANDOMIZATION

Under certain circumstances, the concept of a randomization test allows confidence intervals to be placed on treatment effects. The principle involved was first noted by Pitman (1937a). To explain the idea, it is again helpful to consider the example of the comparison of mandible lengths for male and female golden jackals. As has been seen before, the sample mean length for the males is $\bar{X}_1 = 113.4$ mm and the sample mean length for the females is $\bar{X}_2 = 108.6$ mm, so that the mean difference is $D = 4.8$ mm. Randomization has shown that the difference is significantly larger than zero at about the 0.18% level on a one-sided test.

Suppose that the effect of being female is to reduce a golden jackal's mandible length by an amount μ_D, with this being the same for all females. Then the difference between the male and female distributions of mandible length can be removed by simply subtracting μ_D from the length for each male, and when that is done, the probability of a significant result on a randomization test at the 5% level of significance will be 0.05.

Now, whatever the value of μ_D might be, the probability of the data giving a significant result is still 0.05. Hence the assertion that a randomization test does not give a significant result can be made with 95% confidence. Thus it is possible to feel 95% confident that the true value of μ_D lies within the range of values for which the randomization test is not significant.

The 95% confidence limit defined in this way can be calculated in a straightforward way, at the expense of some considerable computing. What have to be found are the two constants L and U which, when subtracted from the male measurements, just avoid giving a significant difference between the two sample means. The first constant L will be a low value: subtracting this will result in the male–female mean difference being positive and just at the top 2.5% point of the randomization distribution. The second constant U will be a high value: subtracting this will result in the male–female mean difference being negative and just at the bottom 2.5% point of the randomization distribution. The values L and U can be positive or negative providing that $L < U$. The 95% confidence interval is $L < \mu_D < U$.

It is possible to calculate limits with any level of confidence in exactly the same way. For 99% confidence, L and U must give mean differences between the two groups at the 0.5% and 99.5% points on the randomization distribution; for 99.9% confidence the equivalent percentage points must be 0.05% and 99.95%.

The L and U values can be determined by trial and error or by some more systematic search technique. For example, the calculations to determine a 95% confidence interval for the golden jackal male–female difference can be done as follows. First, the upper percentage points (percentages of randomization differences greater than or equal to the observed difference between male and female means) can be determined for a range of trial values for L. One choice of trial values gives the results shown below:

Trial L: 4.8   2.8   2.3   2.0   1.9   1.8

Upper % point: 50.3   8.0   4.2   3.2   2.3   1.7

Here linear interpolation suggests that subtracting $L = 1.92$ from the male measurements leaves a difference between male and female means that is on the borderline of being significantly large at about the 2.5% level. Next the lower percentage points (percentages of randomization differences less than or equal to the observed difference between male and female means) can be determined for some trial values of U. For example:

Trial U: 4.8   6.8   7.7   7.8

Lower % point: 51.8   3.8   2.7   1.9

Here interpolation suggests that subtracting $U = 7.72$ from the male measurements leaves a difference between the male and female means that is on the borderline of being significantly small at about the 2.5% level. Combining the results for L and U now gives an approximately 95% confidence interval for the male–female difference of 1.92 to 7.72.

The upper and lower trial percentage points for evaluating L and U just reported were all determined using the same 5000 randomizations of observations to males and females as were used for the randomization test on the unmodified data because keeping the randomizations constant avoids a certain amount of variation in the results. Note, however, that using a different set of 5000 randomizations would result in values slightly different from those shown here.

A 95% confidence interval for the male–female difference that is determined using the t-distribution has the form

$$(\bar{X}_1 - \bar{X}_2) \pm 2.10s\sqrt{(1/n_1 + 1/n_2)},$$

where 2.10 is the appropriate critical value for the t-distribution with 18 degrees of freedom, $s = 3.08$ is the pooled sample estimate of standard deviation, and $n_1 = n_2 = 10$ are the sample sizes. It is interesting to note that this yields limits (1.91 to 7.69) that are virtually the same as the randomization limits.

1.5   APPLICATIONS OF RANDOMIZATION IN BIOLOGY

The range of situations where randomization methods have been applied in biology is very wide, as is shown by the following references, which have been placed into broad classifications according to their nature. Some of these applications are discussed further in the chapters that follow.

1.5.1 Single species ecology

Applications of randomization that involve data from a single species include searching for spatial patterns in vegetation or animal distributions (Mead, 1974; Goodhall, 1974; Ludwig and Goodhall, 1978; Galiano et al., 1987; Perry and Hewitt, 1991; Perry, 1995a,b); testing for a relationship between the spatial closeness of nests and the laying dates of birds (Besag and Diggle, 1977); testing whether biological differences between colonies of animals and plants are related to geographical and environmental differences (Douglas and Endler, 1982; Dillon, 1984; Manly, 1983, 1986; Legendre and Troussellier, 1988; Tilley et al., 1990; Brown and Thorpe, 1991; Brown et al., 1991, 1993; Rejmánkova et al., 1991; Thorpe and Brown, 1991; Somers and Green, 1993; Somers and Jackson, 1993; Castellano et al., 1994; Pierce et al., 1994; Britten et al., 1995); testing whether behavioural differences and interactions between animals are related to other types of difference (Schnell et al., 1985; Hemelrijk, 1990; Dagosto, 1994); testing for differences in prey utilization by different groups of animals (Linton et al., 1989; Luo and Fox, 1994, 1996); comparison of the dispersion of birds banded at different sites (Lokki and Saurola, 1987; Barker, 1990); detection of density dependence and delayed density dependence in time series of animal abundance, including those for stage-structured populations (Pollard et al., 1987; den Boer and Reddingius, 1989; Reddingius and den Boer, 1989; den Boer, 1990; Manly, 1990a, Chapter 7; Vickery, 1991; Crowley, 1992; Holyoak and Lawton, 1992; Holyoak, 1993, 1994; Holyoak and Crowley, 1993); testing relationships between genetic and biometric measurements on human populations and language groupings in Europe (Sokal et al., 1988, 1989, 1990; Legendre et al., 1990; Barbujani and Sokal, 1991; Livshits et al., 1991); testing for the independence of successive animal relocations in space (Solow, 1989b); finding confidence limits for the size of an animal population (Buckland and Garthwaite, 1991); assessing resource selection by animals by comparing measurements on used locations with randomly selected alternative locations (Alldredge et al., 1991; Caton et al., 1992; Squires et al., 1992); and testing the significance of the effects of explanatory variables in generalized additive models for fisheries survey data (Swartzman et al., 1992).

1.5.2 Genetics, evolution and natural selection

Applications of randomization in the study of genetics, evolution and natural selection include testing for parallel genetic variation in two closely related species (Borowsky, 1977); testing whether the basal metabolic rates of mammals are related to diet and habitat when taxonomic affinity is allowed for, and other analyses of data involving comparisons across taxonomic groups (Elgar and Harvey, 1987; Pagel and Harvey, 1988); testing for non-random information in area cladograms and phylogenetic trees (Simberloff, 1987; Page, 1988; Archie, 1989a,b; Bryant, 1992; Faith, 1991; Faith and Cranston, 1991, 1992; Carpenter, 1992; Källersjö et al., 1992; Faith and Ballard, 1994; Farris et al., 1994); comparing genetic and phenotypic correlations (Cheverud, 1988); small sample tests on genetic data (Roff and Bentzen, 1989); for the comparison of taxonomic classifications based on different methods of construction and different variables and the comparison of these classifications with other characteristics of species (Lapointe and Legendre, 1990, 1991, 1992a,b; Legendre et al., 1994); and for the evaluation of competing hypotheses concerning human origins (Waddle, 1994, 1996; Konigsberg, 1994; Cole, 1996).

1.5.3 Community ecology

Community ecology is the area where randomization methods appear to have received most use. Some applications are for testing whether the correlations between 48 mosquito species based on 71 larval characters are related to the correlations between the same species based on 77 adult characters (Rohlf, 1965); testing whether species cluster significantly into guilds on the basis of their use of resources (Sale, 1974; Joern and Lawlor, 1980, 1981; Jaksić and Medel, 1990); testing 'null models' for species presences and absences or species abundances (Harper, 1978; Strauss, 1982; Wilson, 1987, 1988, 1989, 1995; Wilson et al., 1987, 1992a, 1992b, 1994, 1995a, 1995b, 1995c, 1995d; Clarke and Green, 1988; Vassiliou et al., 1989; Roberts and Stone, 1990; Stone and Roberts, 1990, 1992; Jackson et al., 1992; Watkins and Wilson, 1992; Bycroft et al., 1993; Kadmon and Pulliam, 1993; Wilson and Roxburgh, 1994; Wilson and Watkins, 1994; Kadmon, 1995; Manly, 1995a; van der Maarel et al., 1995; Wilson and Gitay, 1995; Wilson and Whittaker, 1995; Winston, 1995); testing whether the stability levels in natural communities differ significantly from the levels obtained from randomly constructed communities (Lawlor, 1980; Hallett, 1991; Crowley and Johnson, 1992); testing for differences in prey utilization by different species (Patterson, 1986, 1992; Green, 1989; Dawson et al., 1992); testing for relationships between community structure and environmental variation (ter Braak, 1988; Faith and Norris, 1989; Borcard et al., 1992; Cumming et al., 1992; Nantel and Neumann, 1992; Dixit et al., 1993; Myklestad and Birks, 1993; Borcard and Legendre, 1994; Hinch et al., 1994); environmental impact assessment (Carpenter et al., 1989; Clarke, 1993; Lotter and Birks, 1993; Solow, 1993; Verdonschot and ter Braak, 1994); testing for seasonal effects in the number of species shared between two populations using an analysis of variance type of approach (Smith,

1989); detecting clusters in the spatial distribution of the species in a community from quadrat samples (Solow and Smith, 1991); comparing rates of change for plankton and water quality variables from limnological field experiments (France *et al.*, 1992); detecting edges in two-dimensional data on vegetation composition using 'wombling' and other methods (Fortin, 1994; Williams, 1996); studying patterns of inter-island plant and bird species movements in the Galápagos Islands (Harvey, 1994); testing for significant consistency in species abundance ranks for replicate samples from plant communities (Watkins and Wilson, 1994); and testing the correlation between the abundance of fruits and frugivorous birds (Verdú and García-Fayos, 1994).

1.6 RANDOMIZATION AND OBSERVATIONAL STUDIES

The advantage that randomization has over other methods for testing hypotheses is that it is conditional upon the observed data values only, with the null hypothesis being that in some sense the different sample units could have occurred in any of the possible orders. The justification for this approach is clearly strongest in an experimental situation, where the experimenter really does carry out a randomization before the sample units receive different treatments. It is weakest for observational studies where there is a suspicion that all possible orderings of the sample units may not have been equally likely.

For example, Rosenbaum (1994) discusses the situation where a group of 23 subjects who had consistently eaten fish contaminated with methylmercury for three years is compared with 16 subjects who seldom ate fish and had no known history of consuming contaminated fish. The comparison was made in terms of the level of mercury in the blood, the percentage of cells with structural abnormalities, and the percentage of cells with a particular type of abnormality. Rosenbaum developed a test statistic that was designed to be powerful against what he called a coherent alternative (a tendency for higher levels for all the measured variables in the exposed group) and found strong evidence of a group difference in the expected direction. However, he noted that this evidence has reduced impact because of the possibility of a hidden bias in the way that subjects ended up in the two groups, i.e. some subjects may have been more likely than others to have ended up in the exposed group.

If there was information on variables that could be used to model the probability of being in the exposed group then this could be used in a modified analysis. In the absence of such information, Rosenbaum suggests that it can be assumed that there is an unobserved covariate v_i for the ith subject and the probability of a particular assignment of subjects to the exposed and unexposed groups is then a function of the values of this covariate. This then leads to a sensitivity analysis so that the significance level for the difference between the groups can be calculated for different levels of the hidden bias. In this way it is possible to determine how much bias is required in order to explain away the observed difference.

See Rosenbaum and Krieger (1990) and Rosenbaum (1991, 1994) for more details about this type of sensitivity analysis. It does not overcome all of the problems of dealing with non-random observational data. Nevertheless it does at least allow some quantification of the possible effects of bias on the outcome of tests.

1.7 CHAPTER SUMMARY

- A randomization test is defined as a procedure that involves comparing an observed test statistic with a distribution that is generated by randomly reordering the data values in some sense.
- Randomization tests have the advantages of being valid with non-random samples and allowing the user to choose a test statistic that is appropriate for the particular situation being considered. A disadvantage, that results cannot necessarily be generalized to a population of interest, is less serious than it might seem at first sight because the generalization that is often made with conventional statistical procedures is based on the unverifiable assumption that non-random samples are equivalent to random samples, or the data are for the entire population of interest.
- Randomization tests are exact, i.e. they give a level of significance that is less than or equal to the nominal level.
- Two test statistics are said to be equivalent if they give the same significance level on a randomization test.
- Randomization and conventional tests usually give similar significance levels if the assumptions of the conventional test hold.
- Although randomization tests usually involve the comparison between parameters for different groups, this is not always the case. Fisher's one-sample randomization test is an example where a single parameter is tested.
- Confidence limits by randomization can be obtained as the range of values for a parameter such that a randomization test fails to give a significant result.
- Numerous examples of randomization tests used in biology are described briefly.
- Methods are being developed for assessing the sensitivity of group comparisons from observational studies to unequal probabilities of sample units being assigned to groups.

2

The jackknife

2.1 THE JACKKNIFE ESTIMATOR

In ordinary usage the word 'jackknife' describes a large pocket knife with a multitude of pull-out tools such as screwdrivers and scissors. The value of this device is that it is easy to carry around and it can be used for a large variety of tasks, although it is seldom ideal for any one. By analogy, the word 'jackknife' was proposed by Tukey (1958) for use in statistics to describe a general approach for testing hypotheses and calculating confidence intervals in situations where no better methods are easily used. This method is not really computer-intensive by present-day standards. However, it is appropriate to give it some consideration because it is sometimes used as an alternative to the other methods that are discussed in this book.

One way of justifying the idea of jackknifing is to think in terms of what is usually done when estimating a mean value, but from an unusual point of view. Suppose, then, that a random sample of n values $X_1, X_2, \ldots, X_n$ is taken and the sample mean

$$\bar{X} = \sum_{i=1}^{n} X_i/n$$

is used to estimate the mean of the population. Next, suppose that the sample mean is calculated with the jth observation missed out, to give

$$\bar{X}_{-j} = (\sum_{i=1}^{n} X_i - X_j)/(n-1). \tag{2.1}$$

Then the last two equations can be solved for X_j to give

$$X_j = n\bar{X} - (n-1)\bar{X}_{-j}.$$

This shows that it is possible to determine the sample value X_j from the overall mean and the mean with X_j removed.

This is not a useful result if the sample values are known in the first place. However, it is potentially useful in situations where the parameter being estimated is something other than a sample mean. Thus, in a general situation, suppose that a parameter θ is estimated by some function of the n sample values. This estimate can be denoted by $\hat{E}(X_1, X_2, \ldots, X_n)$, and written briefly as $\hat{E}$. With X_j removed this becomes the 'partial estimate' $\hat{E}_{-j}$. By analogy with equation (2.1) there are then the set of 'pseudo-values'

$$\hat{E}_j^* = n\hat{E} - (n-1)\hat{E}_{-j}, \tag{2.2}$$

for $j = 1, 2, \ldots, n$. These pseudo-values play the same role as the values X_j in estimating a mean, and the average of the pseudo-values

$$\hat{E}^* = \sum_{j=1}^{n} \hat{E}_j^*/n, \tag{2.3}$$

is the jackknife estimate of θ. Treating the pseudo-values as a random sample of independent estimates then suggests that the variance of this jackknife estimate can be estimated by s^2/n, where s^2 is the sample variance of the pseudo-values. Going one step further suggests that an approximate $100(1 - \alpha)\%$ confidence interval for θ is given by $\hat{E}^* \pm t_{\alpha/2, n-1} s/\sqrt{n}$, where $t_{\alpha/2, n-1}$ is the value that is exceeded with probability $\alpha/2$ for the t-distribution with $n - 1$ degrees of freedom.

On the face of it this is a way of transforming many estimation problems into the problem of estimating a sample mean. All that is necessary is that the usual estimator of the parameter of interest is some function of n sample values. In reality, as might be expected, things are not quite as simple as that. In particular, the pseudo-values are likely to be correlated to some extent, so that the estimated variance of the jackknife estimator is biased upwards or downwards. Whether or not this is the case is difficult to predict theoretically. Usually, therefore, the jackknife variance needs to be justified by numerical studies of a range of situations before it can be relied on.

Before the jackknife was suggested as a general tool for inference by Tukey (1958), Quenouille (1956) had already pointed out that replacing an estimator by its jackknifed version has the effect of removing bias of order $1/n$. To see this, suppose that the expected (mean) value of the estimator $\hat{E}$ of θ based on a full set of n observations is $\theta(1 + A/n)$. Then the expected value of the partial estimator $\hat{E}_{-j}$ with the jth observation removed is $\theta\{1 + A/(n-1)\}$. It follows that the expected value of the jth pseudo-value as defined by equation (2.2) is $n[\theta(1 + A/n)] - (n-1)[\theta\{1 + A/(n-1)\}] = \theta$, with the bias factor A cancelling out. This holds for each value of j and consequently the expected value of $\hat{E}^*$, the average of the pseudo-values, is also θ.

These ideas can be generalized so that a pth-order jackknife estimator can be used to remove bias of order $1/n^p$, for $p = 2, 3$, etc., where this involves the concept of comparing the full sample estimate with the different possible estimates that are obtained when p observations are excluded (Gray and Schucany, 1972). Another generalization is to the two-sample or multi-sample jackknife where the statistic of interest is some form of comparison between the samples or is a combined estimate of a parameter. However, these generalizations will not be considered further here.

Tukey's paper led to a keen interest in developing the uses of jackknife estimation from the mid 1960s through to the late 1970s. Review of early developments are provided by Gray and Schucany (1972) and Miller (1974), and a bibliography of 162 references has been produced by Parr and Schucany (1980), with an update by Frangos (1987). A more recent review is provided by Hinkley (1983).

Example 2.1: estimating a standard deviation   Suppose that a random sample of size 20 from a certain population yields the observations 3.56, 0.69, 0.10, 1.84, 3.93, 1.25, 0.18, 1.13, 0.27, 0.50, 0.67, 0.01, 0.61, 0.82, 1.70, 0.39, 0.11, 1.20, 1.21 and 0.72, from which it is necessary to estimate the population standard deviation σ. Suppose also that it is decided that the estimator will be the sample standard deviation $\hat{\sigma} = \sqrt{\{\sum(x_i - \bar{x})^2/20\}}$, where x_i is the ith observation and $\bar{x} = \sum x_i/20$ is the sample mean, the summations being for i from 1 to 20. (The division by 20 in the equation for $\hat{\sigma}$ is chosen on purpose here, although it might be thought that a division by 19 is more conventional.)

Three questions might now be asked:

1. Can any bias in the estimator be removed by modifying the estimator?
2. Can the standard error of the estimator be estimated to indicate the level of sampling errors?
3. Can a 95% confidence interval for σ be constructed?

In all cases it is desirable to assume that the population being sampled does not necessarily have a normal distribution.

Jackknifing provides answers to these three questions, with the calculations involved being shown in Table 2.1. In fact, two jackknife answers are provided, with one being based on the direct calculation of standard deviations and the other on log-transformed values.

Consider first the results without a transformation. In this case the first row of the column headed SD is the standard deviation $\hat{\sigma} = 1.03$ of the 20 data values. The estimates with one value at a time removed are shown below this value. These are the partial estimates. The corresponding pseudo-values from equation (2.2), $\hat{\sigma}_1^*$ to $\hat{\sigma}_{20}^*$, are shown on the right of the partial estimates. The average of the pseudo-values is the jackknife estimate of σ, $\hat{\sigma}^* = 1.10$. The estimated standard error is $s/\sqrt{20} = 0.27$, where $s = 1.22$ is the standard deviation of the 20 pseudo-values, and the nominal confidence limits 0.53 to 1.67 are calculated as $1.10 \pm 2.09 \times 0.27$, where

Table 2.1 Jackknife calculations with the estimation of the population standard deviation from a sample of size 20. Row 1 shows the sample values, their standard deviation of 1.03 and the natural logarithm of the standard deviation of 0.03. The following rows are the partial samples with single values removed, followed in the right-hand columns by the partial estimate of the standard deviation, the corresponding pseudo-value (PV), the partial estimate of the logarithm of the standard deviation and the corresponding pseudo-value

Data	1	2	3	4	5	6	7	8	9	10	11	12	13	14	15	16	17	18	19	20	SD	PV	Log SD	PV
	3.56	0.69	0.10	1.84	3.93	1.25	0.18	1.13	0.27	0.50	0.67	0.01	0.61	0.82	1.70	0.39	0.11	1.20	1.21	0.72	1.03		0.03	
1		0.69	0.10	1.84	3.93	1.25	0.18	1.13	0.27	0.50	0.67	0.01	0.61	0.82	1.70	0.39	0.11	1.20	1.21	0.72	0.87	3.96	−0.13	3.10
2	3.56		0.10	1.84	3.93	1.25	0.18	1.13	0.27	0.50	0.67	0.01	0.61	0.82	1.70	0.39	0.11	1.20	1.21	0.72	1.05	0.58	0.06	−0.39
3	3.56	0.69		1.84	3.93	1.25	0.18	1.13	0.27	0.50	0.67	0.01	0.61	0.82	1.70	0.39	0.11	1.20	1.21	0.72	1.03	0.97	0.04	−0.02
4	3.56	0.69	0.10		3.93	1.25	0.18	1.13	0.27	0.50	0.67	0.01	0.61	0.82	1.70	0.39	0.11	1.20	1.21	0.72	1.04	0.84	0.04	−0.15
5	3.56	0.69	0.10	1.84		1.25	0.18	1.13	0.27	0.50	0.67	0.01	0.61	0.82	1.70	0.39	0.11	1.20	1.21	0.72	0.81	5.21	−0.21	4.58
6	3.56	0.69	0.10	1.84	3.93		0.18	1.13	0.27	0.50	0.67	0.01	0.61	0.82	1.70	0.39	0.11	1.20	1.21	0.72	1.05	0.54	0.06	−0.43
7	3.56	0.69	0.10	1.84	3.93	1.25		1.13	0.27	0.50	0.67	0.01	0.61	0.82	1.70	0.39	0.11	1.20	1.21	0.72	1.04	0.89	0.04	−0.10
8	3.56	0.69	0.10	1.84	3.93	1.25	0.18		0.27	0.50	0.67	0.01	0.61	0.82	1.70	0.39	0.11	1.20	1.21	0.72	1.06	0.52	0.06	−0.45
9	3.56	0.69	0.10	1.84	3.93	1.25	0.18	1.13		0.50	0.67	0.01	0.61	0.82	1.70	0.39	0.11	1.20	1.21	0.72	1.04	0.82	0.04	−0.17
10	3.56	0.69	0.10	1.84	3.93	1.25	0.18	1.13	0.27		0.67	0.01	0.61	0.82	1.70	0.39	0.11	1.20	1.21	0.72	1.05	0.66	0.05	−0.32
11	3.56	0.69	0.10	1.84	3.93	1.25	0.18	1.13	0.27	0.50		0.01	0.61	0.82	1.70	0.39	0.11	1.20	1.21	0.72	1.05	0.59	0.06	−0.39
12	3.56	0.69	0.10	1.84	3.93	1.25	0.18	1.13	0.27	0.50	0.67		0.61	0.82	1.70	0.39	0.11	1.20	1.21	0.72	1.03	1.06	0.03	0.06
13	3.56	0.69	0.10	1.84	3.93	1.25	0.18	1.13	0.27	0.50	0.67	0.01		0.82	1.70	0.39	0.11	1.20	1.21	0.72	1.05	0.61	0.05	−0.36
14	3.56	0.69	0.10	1.84	3.93	1.25	0.18	1.13	0.27	0.50	0.67	0.01	0.61		1.70	0.39	0.11	1.20	1.21	0.72	1.05	0.54	0.06	−0.43
15	3.56	0.69	0.10	1.84	3.93	1.25	0.18	1.13	0.27	0.50	0.67	0.01	0.61	0.82		0.39	0.11	1.20	1.21	0.72	1.04	0.73	0.05	−0.25
16	3.56	0.69	0.10	1.84	3.93	1.25	0.18	1.13	0.27	0.50	0.67	0.01	0.61	0.82	1.70		0.11	1.20	1.21	0.72	1.04	0.73	0.05	−0.25
17	3.56	0.69	0.10	1.84	3.93	1.25	0.18	1.13	0.27	0.50	0.67	0.01	0.61	0.82	1.70	0.39		1.20	1.21	0.72	1.03	0.95	0.04	−0.04
18	3.56	0.69	0.10	1.84	3.93	1.25	0.18	1.13	0.27	0.50	0.67	0.01	0.61	0.82	1.70	0.39	0.11		1.21	0.72	1.06	0.53	0.06	−0.44
19	3.56	0.69	0.10	1.84	3.93	1.25	0.18	1.13	0.27	0.50	0.67	0.01	0.61	0.82	1.70	0.39	0.11	1.20		0.72	1.06	0.53	0.06	−0.44
20	3.56	0.69	0.10	1.84	3.93	1.25	0.18	1.13	0.27	0.50	0.67	0.01	0.61	0.82	1.70	0.39	0.11	1.20	1.21		1.05	0.57	0.06	−0.41

Jackknife estimate (mean)	1.10	0.13
Jacknife standard error	0.27	0.29
Lower 95% confidence limit	0.53	−0.47
Upper 95% confidence limit	1.67	0.74

2.09 is the upper 2.5% point for the t-distribution with 19 degrees of freedom. Because the pseudo-values are being treated as if they were independent values from a normal distribution their standard deviation is calculated as $s = \sqrt{\{\sum(\hat{\sigma}_j^* - \hat{\sigma}^*)^2/19\}}$, i.e. with a division by 19 rather than 20 as has been used with $\hat{\sigma}$.

It is obviously of interest to know how well this procedure works. This can be ascertained with the example being considered because the 20 data values are actually independent computer-generated pseudo-random numbers from an exponential distribution with a mean and a standard deviation both equal to 1. Therefore the jackknife confidence limits of 0.53 to 1.67 do contain the true population standard deviation. In this sense the procedure has been successful.

Unfortunately, however, when the process of generating a set of 20 values and producing jackknife confidence intervals was repeated 1000 times in a Lotus 1-2-3 spreadsheet (Lotus Development Corporation, 1991) the results were not very good. The population standard deviation of 1 was only inside the nominal 95% confidence limits 81% of the time, with the top limit being too low 19% of the time. Two factors contributed to this poor performance. First, the estimated standard error of the jackknife estimate tended to be too low. Second, the distribution of the jackknife estimate was not very close to a normal distribution, as is required for the use of the t-distribution in producing confidence limits. On the other hand, jackknifing was successful at removing bias. The mean value of $\hat{\sigma}$ from the 1000 sets of generated data was too low at 0.93, while the mean of the jackknife estimates was 0.99, which is nearly equal to the correct value of 1.

When the jackknife fails because the estimator being used is not normally distributed it is sometimes possible to improve the situation by applying a normalizing transformation. In the present case this suggests that working with $\log_e(\hat{\sigma})$ might help, on the grounds that logarithms of estimated standard deviations tend to be more normally distributed than the standard deviations themselves. The calculations with this modification are shown in the last two columns of Table 2.1, where it can be seen that the nominal 95% confidence limits for $\log_e(\sigma)$ are -0.47 to 0.74. The limits for σ are then $\exp(-0.47) = 0.63$ to $\exp(0.74) = 2.10$.

When the jackknife procedure was carried out on the 1000 computer-generated sets of data it was found that the confidence limits contained the true population standard deviation 87% of the time, with the top limit being too low 12% of the time. This is a marked improvement on what was found without the logarithmic transformation, but it is still hardly satisfactory. The problems of non-normality and underestimation of the standard error are still present.

Of course, in real life the form of the true population distribution will not be known, let alone the value of the parameter being estimated. Nevertheless, simulation studies of the type described here can be used to assess whether or not jackknifing is likely to be satisfactory for a particular application. In the situation being considered in this example there clearly are some difficulties. But it must be stressed that this is not necessarily the case, and often jackknifing works much better than it does for this example. Indeed, the problem with this example appears to be simply that the sample size is not large enough, because Miller (1968) has shown that jackknifing $\log(\hat{\sigma})$ does work for large enough samples.

2.2 APPLICATIONS OF JACKKNIFING IN BIOLOGY

The jackknife has been widely used in biology. Some applications are shown below under three broad headings.

2.2.1 Single species analyses

Applications of jackknifing involving data from single species include the analysis of dose–response and other types of proportion data (Salsburg, 1971; Frawley, 1974; Gladen, 1979; Does et al., 1988; Cyr et al., 1990, 1993; Carr and Portier, 1993; Frangos and Schucany, 1995); the estimation of population size from mark–recapture data (Manly, 1977; Burnham and Overton, 1978, 1979; Pollock and Otto, 1983; Chao, 1984; Rosenberg et al., 1995); the estimation of population density from line–transect data (Seber, 1979; Buckland, 1982); the estimation of catch per unit effort in fisheries (Smith, 1980); the estimation of population growth rates (Mueller and Ayala, 1981; Lenski and Service, 1982; Meyer et al., 1986); the analysis of experiments on the time at which tumours occur as the result of different treatments (Mantel et al., 1982); assessing the accuracy of principal components analysis of morphometric variation (Gibson et al., 1984); the comparison of plant frequency distributions (Higgins et al., 1984); and testing for differences between correlation matrices for samples of aphids from different locations (Riska, 1985).

2.2.2 Genetics, evolution and natural selection

Applications of jackknifing under the above heading include the study of genetic variation and selection in natural and artificial populations (Arvesen and Schmitz, 1970; Johnson and Johnston, 1989; Knapp and Bridges, 1988; Knapp et al., 1989; Mitchell-Olds and Bergelson, 1990; Lo, 1991; Roff and Preziosi, 1994; Roff, 1995); the estimation of genetic distances (Mueller, 1979; Mueller and Ayala, 1982); and the study of phylogenetic trees (Lapointe et al., 1994).

2.2.3 Community ecology

Applications of jackknifing with data from communities include the estimation of indices of diversity, aggregation and size hierarchy (Zahl, 1977; Fagen, 1978; Routledge, 1980, 1984; Reed, 1983; Heltshe and Forrester, 1983b, 1985; Lyons and Hutcheson, 1986; Dixon, 1993); the estimation of species richness (Heltshe and Forrester, 1983a; Smith and van Belle, 1984; Palmer, 1990, 1991; Mingoti and Meeden, 1992); and the estimation of measures of the species similarity between two communities and niche overlap (Smith *et al.*, 1979; Smith, 1985; Mueller and Altenberg, 1985; Heltshe, 1988; Dixon, 1993).

Example 2.2: estimating species richness One of the important parameters in many ecological studies is the species richness, which is the number of species in a community. This can be estimated by the number of species recorded in one or more samples from the community in question, but this will clearly often be an underestimate as it cannot be expected that all species will be seen. Therefore, a variety of methods have been proposed for estimating the number of species present but not recorded. In this example the situation is considered where n quadrat samples are taken and the species present in each sample are recorded. The question of interest is then: how many more species would be found if the number of samples taken was increased without limit?

Burnham and Overton (1978, 1979) suggested using jackknifing to answer this question. They showed that the first-order jackknife produces the estimator

$$\hat{S}^* = S + \{(n - 1)/n\}f_1, \tag{2.4}$$

where S is the total number of species seen in all samples and f_1 is the number of species recorded in just one of the n samples. They also gave explicit equations for the jackknife estimators of order 2 to 5, and a selection procedure for choosing which of these to apply with a particular set of data. The use of the first-order jackknife was also proposed independently by Heltshe and Forrester (1983a), and this was compared to the bootstrap by Smith and van Belle (1984). An empirical study of the performance of alternative methods was carried out by Palmer (1990, 1991), who concluded that the first-order jackknife estimator was the best out of the eight methods used. Rosenberg *et al.* (1995) also present results that favour the first- and second-order jackknife estimator.

On the other hand, Mingoti and Meeden (1992) have provided evidence that an empirical Bayes estimator is better than the jackknife estimator in cases where the quadrats sampled are a random sample from a known finite population of N possible quadrats in the study area. Their estimator is an increasing function of the population size N for a fixed sample size n, which means that there will be difficulty in using it when it is not possible to decide exactly how many quadrat samples could be taken. The approach seems promising, but it will not be considered further here. Instead, it will be assumed that the sampled quadrats are from an effectively infinite population of possible quadrats, so that Mingoti and Meeden's method cannot be used.

The first-order jackknife estimate can either be calculated using the general equation (2.3) or using the specific equation (2.4). In the first case, the full sample estimate is the number of different species seen in all n quadrats, and the jth partial estimate is the number of different species seen when the jth quadrat is ignored. The n pseudo-values are then found using equation (2.2) and the jackknife estimate is the mean of these pseudo-values. The standard error of the mean can be estimated in the usual way on the assumption that the pseudo-values behave as independent estimators of the species richness. Alternatively, an equation for the standard error is provided by Heltshe and Forrester (1983a) and Smith and van Belle (1984).

To illustrate the calculations, consider the data in Table 2.2, which were collected by J. Hyland in 1978 (Heltshe and Forrester, 1983a). Here there

Table 2.2 Species presences (1) and absences (0) in 10 quadrat samples taken from a subtidal marsh creek in the Pettaquamscutt River, southern Rhode Island

Species	1	2	3	4	5	6	7	8	9	10	Total
1	0	1	1	1	1	1	1	1	1	1	9
2	1	1	1	1	1	1	1	0	1	1	9
3	0	1	0	0	0	0	0	1	0	0	2
4	1	0	1	1	0	1	0	0	1	1	6
5	0	0	1	1	0	0	1	0	0	1	4
6	1	1	1	1	0	1	0	0	1	1	7
7	1	0	0	0	0	0	0	0	0	0	1
8	1	0	0	0	0	0	0	0	0	0	1
9	0	1	0	0	0	0	0	0	0	0	1
10	0	0	1	0	0	0	0	0	0	0	1
11	0	0	1	0	0	0	0	0	0	1	2
12	0	0	1	1	0	1	0	0	0	1	4
13	0	0	0	0	0	0	0	1	0	0	1
14	1	1	1	1	1	1	1	1	1	1	10

Table 2.3 Jackknife calculations for estimating species richness from the data in Table 2.2

	Sample										Mean	*SD*	*SE*
	1	2	3	4	5	6	7	8	9	10			
Partial estimate	12	13	13	14	14	14	14	13	14	14			
Pseudo-value	32	23	23	14	14	14	14	23	14	14	18.5	6.4	2.0

are 10 quadrat samples of benthic fauna from a subtidal marsh creek in the Pettaquamscutt River in southern Rhode Island, with a total of 14 species seen. Individual samples contained from three to nine species. Five of the species were seen in only one of the samples, but one species was seen in all samples.

The jackknife calculations based on the general equations (2.2) and (2.3) are shown in Table 2.3. It is estimated that the total number of species is 18.5, with a standard error (SE) of 2.0, where this is the standard deviation (SD) of the pseudo-values divided by $\sqrt{10}$. A jackknife 95% confidence interval is therefore $18.5 \pm 2.26 \times 2.0$, or 14.0 to 23.0, where 2.26 is the upper 2.5% point of the t-distribution with 9 degrees of freedom.

Having carried out these calculations it is interesting to know how reliable the results are. To this end, a small simulation experiment was carried out using the Lotus 1-2-3 spreadsheet (Lotus Development Corporation, 1991). An imaginary community consisting of 20 species was set up, such that the probabilities of species 1, 2, 3, ..., 20 being present in a sample are 0.05, 0.10, 0.15, ..., 1.00, respectively. Five hundred trials of jackknife estimation were then carried out, where for each trial 10 independent samples were taken from the population to yield data of the form of Table 2.2 (except that the total number of species seen did not have to be 14). The computer-generated data were then analysed in the same way as the data in Table 2.2.

From this experiment it was found that:

(a) The mean of the 500 simulated jackknife estimates was 20.3, which is close to the actual species richness of 20.

(b) The mean of the 500 jackknife variances was 1.51, which is substantially lower than the observed variance of the 500 species richness estimates of 2.35. Thus jackknife estimates of the variance (and hence jackknife estimates of the standard error) appear to be biased downwards.

(c) In 53 out of the 500 trials the estimated jackknife standard error was zero because all 10 samples from the community had exactly the same species present. As a result, in these 53 cases it was not really sensible to calculate the jackknife 95% confidence limits, because on the face of it the upper and lower limit are both equal to the observed species richness. If these cases are ignored then it is found that the true species richness is within the jackknife limits for 93% of the remaining 447 trials. However, if all 500 trials are considered then the coverage drops to the rather poor figure of 83%.

Essentially, it appears that jackknifing is effective at removing the bias, but it is not so effective for estimating the accuracy of estimation. More extensive simulations indicate that the properties of jackknifing depend on the distribution of the probabilities of species being found in a random sample, that the method may or may not give approximately unbiased estimators, and that the coverage of confidence intervals may be much less than the nominal amount (Burnham and Overton, 1979; Heltshe and Forrester, 1983a). Still, it should be remembered that the jackknife has performed well in comparison with alternative methods for estimating species richness (Palmer, 1990, 1991). Therefore this example should not be thought of as a case where the jackknife fails completely.

2.3   CHAPTER SUMMARY

- The jackknife can be thought of as a method for converting the problem of estimating any population parameter into the problem of estimating a population mean.
- The usual jackknife estimator of a parameter removes bias of order $1/n$, where n is the sample size. Similarly, a pth-order jackknife removes bias of order $1/n^p$.
- An illustration of the use of the jackknife for estimating the standard deviation for an exponential distribution using a random sample of size 20 demonstrates the poor performance that may occur with small sample sizes.
- Numerous applications of the jackknife in biology are described briefly and the application to estimating species richness is described in some detail. For estimating species richness the jackknife is effective for bias removal, but not for estimating the standard error. Nevertheless, previous studies have shown that the jackknife compares well with its competitors for this application.

3

The bootstrap

3.1 RESAMPLING WITH REPLACEMENT

The technique of bootstrapping was first considered in a systematic manner by Efron (1979b), although the generality of the method means that it was used in some particular circumstances before that time. The label 'bootstrapping' was used by Efron to follow in the spirit of Tukey's (1958) use of 'jackknifing' to describe his earlier resampling method. Thus, whereas Tukey's technique is supposed to be analogous to the use of the scout's jackknife, Efron's technique is supposed to be analogous to someone pulling themselves out of mud with their bootstraps.

The essence of bootstrapping is the idea that, in the absence of any other knowledge about a population, the distribution of values found in a random sample of size n from the population is the best guide to the distribution in the population. Therefore, to approximate what would happen if the population was resampled it is sensible to resample the sample. In other words, the infinite population that consists of the n observed sample values, each with probability $1/n$, is used to model the unknown real population. The sampling is with replacement, which is the only difference in practice between bootstrapping and randomization in many applications.

One of the main themes in research on bootstrapping in the last 15 years has been the development of methods to calculate valid confidence limits for population parameters. The literature in this area is large, and still growing. More recently, bootstrap tests of significance have been attracting increasing attention. The main aim of the present chapter is to introduce some of the basic methods that are used in both of these areas in terms of a few relatively simple examples. More complicated bootstrap methods are considered in the later chapters on particular types of data.

3.2 STANDARD BOOTSTRAP CONFIDENCE LIMITS

The simplest method for obtaining bootstrap confidence limits is called the standard bootstrap method. The principle here is that if an estimator $\hat{\Theta}$ is normally distributed with mean Θ and standard deviation σ, then there is a probability of $1 - \alpha$ that the statement

$$\Theta - z_{\alpha/2}\sigma < \hat{\Theta} < \Theta + z_{\alpha/2}\sigma$$

holds for any random value of $\hat{\Theta}$, where $z_{\alpha/2}$ is the value that is exceeded with probability $\alpha/2$ for the standard normal distribution. This statement is equivalent to

$$\hat{\Theta} - z_{\alpha/2}\sigma < \Theta < \hat{\Theta} + z_{\alpha/2}\sigma,$$

which therefore holds with the same probability. This leads to the following result.

The standard bootstrap confidence interval

With the standard bootstrap confidence interval σ is estimated by the standard deviation of estimates of a parameter Θ that are found by bootstrap resampling of the values in the original sample of data. The interval is then

$$\text{Estimate} \pm z_{\alpha/2}(\text{Bootstrap standard deviation}). \qquad (3.1)$$

For example, using $z_{0.025} = 1.96$ gives the standard 95% bootstrap interval.

The requirements for this method to work are that:

(a) $\hat{\Theta}$ has an approximately normal distribution;
(b) $\hat{\Theta}$ is unbiased so that its mean value for repeated samples from the population of interest is Θ; and
(c) bootstrap resampling gives a good approximation to σ.

Whether or not these conditions hold depends on the particular circumstances. If they do then the method is potentially useful whenever an alternative method for approximating σ is not easily available. In practice it may be possible to avoid requirement (b) by estimating the bias in $\hat{\Theta}$ as part of the bootstrap procedure.

The number of bootstrap samples that need to be taken is discussed in the next chapter. Here it can merely be noted that 100 may be sufficient to get a good estimate of the standard deviation of an estimator. The following example uses 1000 bootstrap samples because the same situation is considered later in this chapter with the application of other types of bootstrap confidence interval that require this larger number.

Example 3.1: confidence limits for a standard deviation   Consider again the situation discussed in Example 2.1, where there is interest in assessing the accuracy of the standard deviation of a random sample of size 20 as an estimator of the standard deviation of the distribution from which the sample was drawn. The distribution was in fact exponential with a mean and standard deviation of 1.

The sample values are 3.56, 0.69, 0.10, 1.84, 3.93, 1.25, 0.18, 1.13, 0.27, 0.50, 0.67, 0.01, 0.61, 0.82, 1.70, 0.39, 0.11, 1.20, 1.21 and 0.72. The estimated standard deviation is $\hat{\sigma} = \sqrt{\{\sum(x_i - \bar{x})^2/20\}} = 1.03$, where x_i is the ith sample value and $\bar{x} = \sum x_i/20$ is the sample mean, the summations being for i from 1 to 20. The division by 20 in the equation for the sample standard deviation is done on purpose, although a division by 19 would be more conventional.

As in Example 2.1, three questions will be considered:

1. Can any bias in the estimator be removed by modifying the estimator?
2. Can the standard error of the estimator be estimated to indicate the level of sampling errors?
3. Can a 95% confidence interval for σ be constructed?

No particular assumptions will be made about the distribution that the observations come from.

It must be said from the start that this is not an easy application for bootstrapping. Indeed, the example of constructing a confidence interval for the variance of a normal distribution on the basis of a sample of size 20 has been used by Schenker (1985) to highlight the need for caution in the use of the bootstrap.

The bootstrap answers to questions 1 and 2 are relatively straightforward to obtain. The bias in the estimator can be approximated by resampling the data a large number of times and calculating the difference between the mean of the bootstrap estimates and the known standard deviation of the bootstrap population of 1.03 (although a better method is described in section 3.9). The standard error of $\hat{\sigma}$ can be approximated by the standard deviation of the bootstrap estimates. Table 3.1 shows part of the calculations when 1000 bootstrap samples were used for this purpose. All of the calculations were done in a Lotus 1-2-3 spreadsheet (Lotus Development Corporation, 1991), which is a convenient tool for small bootstrap analyses (Willemain, 1994).

It can be seen from Table 3.1 that the mean of the 1000 bootstrap estimates of σ was 0.97. The bias in $\hat{\sigma}$ is therefore estimated to be $0.97 - 1.03 = -0.06$. This then suggests that the original sample estimate of 1.03 is too low by about this amount as well, so that a bias-corrected estimate of the standard deviation of the original population sampled is $1.03 - (-0.06) = 1.09$. This is slightly larger than the estimate that is

Table 3.1 Part of the output from taking 1000 bootstrap samples from the data considered in Example 3.1. The first line in the body of the table shows the 20 data values with their standard deviation $\hat{\sigma}(SD)$. This is followed by rows giving the results for the first five and the last five of 1000 bootstrap samples. The bootstrap estimates of the bias in $\hat{\sigma}$ is $0.97 - 1.03 = -0.06$, the difference between the bootstrap mean of $\hat{\sigma}$ and the known standard deviation for the bootstrapped population of 1.03. The bootstrap estimate of the standard error of $\hat{\sigma}$ is 0.25, the standard deviation of the bootstrap estimates

	1	2	3	4	5	6	7	8	9	10	11	12	13	14	15	16	17	18	19	20	SD
Original sample																					
	3.56	0.69	0.10	1.84	3.93	1.25	0.18	1.13	0.27	0.50	0.67	0.01	0.61	0.82	1.70	0.39	0.11	1.20	1.21	0.72	1.03
Bootstrap data																					
1	1.13	0.18	1.84	0.01	0.50	0.72	0.61	0.69	0.39	0.67	1.20	1.25	0.50	0.50	0.01	0.27	0.69	1.25	0.18	0.72	0.46
2	1.13	0.50	0.39	0.27	3.93	0.67	0.27	3.93	0.72	0.82	0.11	0.82	1.20	0.10	1.20	3.93	1.20	1.25	3.93	1.21	1.33
3	0.61	0.67	0.01	1.84	0.10	0.11	0.72	1.20	0.01	0.67	0.82	0.50	1.21	0.39	0.10	0.50	0.10	1.21	1.70	1.21	0.54
4	1.70	1.13	1.20	0.01	0.18	3.93	1.25	1.25	0.82	3.56	0.10	0.50	1.21	1.13	1.84	3.93	0.69	1.21	3.56	1.13	1.21
5	0.69	3.93	1.70	1.25	0.72	0.10	0.72	3.93	1.20	3.93	0.72	1.25	0.69	0.39	1.20	0.01	0.18	0.50	1.84	0.01	1.23
.																					
996	3.56	0.39	0.18	0.01	0.27	0.72	0.18	0.11	0.67	1.70	3.56	1.84	1.25	3.93	1.25	0.61	0.39	1.70	0.10	0.27	1.21
997	1.20	1.70	3.56	1.84	0.67	0.27	0.39	0.72	0.50	1.21	0.50	1.20	1.13	0.01	0.50	0.39	0.72	0.27	0.72	3.56	0.96
998	1.84	0.18	1.21	1.25	1.21	0.69	3.56	0.61	3.93	1.21	0.67	1.13	0.01	1.13	1.13	3.93	0.82	0.82	1.70	1.13	1.09
999	0.18	1.20	0.82	3.56	0.61	1.25	1.25	1.13	0.39	0.82	0.61	0.67	0.61	3.56	0.01	1.21	1.21	0.10	1.70	3.56	1.07
1000	0.82	1.70	0.01	0.50	1.84	1.25	0.50	1.13	1.21	1.70	3.93	0.50	0.50	0.27	0.27	0.69	1.70	0.67	0.18	1.70	0.87

Bootstrap mean 0.97
Bootstrap SD 0.25
Estimated bias −0.06

obtained using the usual unbiased variance estimator of $\sqrt{\{\sum(x_i - \bar{x})^2/}$ $(n-1)\} = 1.06$. An estimate of the standard error of $\hat{\sigma}$ is the bootstrap standard deviation of 0.25.

The standard bootstrap 95% confidence interval is $\hat{\sigma} \pm 1.96$(bootstrap standard deviation), which is $1.03 \pm 1.96(0.25)$, or 0.54 to 1.52. However, because the bootstrap calculations have indicated that the estimator has a bias of -0.06 it appears to be better to centre the limits on the adjusted estimate of 1.09 and make them $1.09 \pm 1.96(0.25)$, or 0.60 to 1.58. These are then quite similar to the jackknife 95% confidence limits of 0.53 to 1.67 that were calculated in Example 2.1.

In fact, it is known that the sample being considered came from an exponential distribution with standard deviation $\sigma = 1$. On the face of it the bootstrap calculations therefore seem reasonable, with the population standard deviation being well within the bootstrap 95% confidence limits with or without a bias adjustment. However, it is interesting to see whether this is a lucky coincidence resulting from the particular sample values used, or is a general property of the bootstrap standard confidence interval. It may be recalled in this connection that simulations discussed in Example 2.1 show that jackknifing does not perform well in the situation being considered.

To test bootstrapping, 1000 further samples of size 20 were taken from the exponential distribution with mean and standard deviation equal to 1, and these were analysed in the same manner as just described. It was found that the mean of the estimates of the standard deviation after the bootstrap bias corrections was 0.97, which indicates that the bootstrap bias correction works reasonably well. It was also found that the mean of the 1000 bootstrap estimates of the standard deviation of $\hat{\sigma}$ was 0.20, which is substantially lower than the actual standard deviation of 0.30.

The bias-corrected standard bootstrap 95% confidence interval for the population standard deviation was also constructed for each set of the 1000 sets of data, and a check was made to see whether this included the true value of 1. Only 72.9% of the intervals included the population standard deviation, which is far from being satisfactory. Furthermore, 25.9% of the time the upper limit of the confidence interval was lower than 1.

With jackknifing it was found that using a logarithmic transformation slightly improved confidence intervals, mainly because logarithms of sample standard deviations are more normally distributed than the sample standard deviations themselves. This was therefore also tried with bootstrapping. Essentially this just involved applying the same methods as used with $\hat{\sigma}$ to $\log_e(\hat{\sigma})$ on the 1000 sets of simulated data. This did lead to some improvement, with 81.8% of the sets of data producing a nominal 95% confidence interval for $\log_e(\sigma)$ that included the true value of zero. But 81.8% coverage for a 95% confidence interval is hardly satisfactory, showing that some alternative method for constructing confidence limits is needed for this application.

3.3 SIMPLE PERCENTILE CONFIDENCE LIMITS

Several approaches to constructing confidence limits that have received considerable attention are based on attempts to approximate the percentiles of the distribution of an estimator using percentiles generated by bootstrapping. In this section, two of these methods are described. Section 3.4 covers some modifications that have been proposed with the idea of making one of these methods more generally applicable.

In an early paper on bootstrapping, Efron (1979b) described what is often called the percentile method. With this, the $100(1 - \alpha)\%$ limits for a parameter are just the two values that contain the central $100(1 - \alpha)\%$ of the estimates obtained from bootstrapping the original sample. This is justified on the basis of the assumption that a transformation exists that will convert the distribution of the estimator being considered into a normal distribution.

Thus, suppose that the problem is to find a $100(1 - \alpha)\%$ confidence interval for a parameter Θ, for which an estimate $\hat{\Theta}$ is available based on a random sample $x_1, x_2, \ldots, x_n$ from the population of interest. Assume also that a monotonic increasing function exists, such that the transformed values $f(\hat{\Theta})$ are normally distributed with a mean of $f(\Theta)$ and standard deviation of 1. Note that only the existence of this transformation is assumed. The mathematical form of the transformation does not need to be known.

In this situation it is clear that there is a probability of $1 - \alpha$ that the statement

$$f(\Theta) - z_{\alpha/2} < f(\hat{\Theta}) < f(\Theta) + z_{\alpha/2},$$

will hold, where, as before, $z_{\alpha/2}$ is the value exceeded with probability $\alpha/2$ for the standard normal distribution (Figure 3.1). By rearranging this result it follows with the same probability that

$$f(\hat{\Theta}) - z_{\alpha/2} < f(\Theta) < f(\hat{\Theta}) + z_{\alpha/2}, \tag{3.2}$$

which is a $100(1 - \alpha)\%$ confidence interval for $f(\Theta)$.

If the transformation function f is known then the limits for the interval (3.2) can be back-transformed to give an interval for Θ with the same level of confidence. The problem is that the transformation is usually not known. However, suppose that the distribution of the transformed estimator is generated by bootstrapping the original data. That is to say, let $f(\hat{\Theta}_B)$ be a transformed estimate obtained by bootstrap resampling of the original

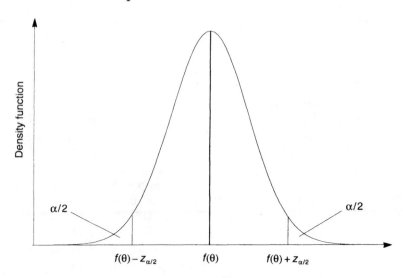

Figure 3.1 The normal distribution of the transformed estimator $f(\hat{\Theta})$ for the percentile confidence interval. There is a probability $\alpha/2$ of a value less than $f(\Theta) - z_{\alpha/2}$ and the same probability of a value greater than $f(\Theta) + z_{\alpha/2}$.

data. A large number of such values of $f(\hat{\Theta}_B)$ can be generated, and hence their distribution obtained to any required level of accuracy. This distribution should then be similar to the one that is shown in Figure 3.1, except that the mean will be $f(\hat{\Theta})$ instead of $f(\Theta)$. The interesting thing about this distribution is that the two values that encompass the central $100(1 - \alpha)\%$ of the distribution are the two end-points $f(\hat{\Theta}) - z_{\alpha/2}$ and $f(\hat{\Theta}) + z_{\alpha/2}$ of the confidence interval (3.2). Therefore, one way to find a confidence interval for $f(\Theta)$ involves bootstrap resampling the original data and finding the value that exceeds a fraction $\alpha/2$ of the generated values (the lower confidence limit) and the value that exceeds $1 - \alpha/2$ of the generated values (the upper confidence limit).

Furthermore – and this is the crucial part – because of the monotonic nature of the transformation being considered, the ordering of the transformed bootstrap estimates $f(\hat{\Theta}_B)$ from smallest to largest must correspond to the ordering of the untransformed bootstrap estimates $\hat{\Theta}_B$. Therefore, the limits for Θ corresponding to the confidence interval (3.2) are simply the value that exceeds a fraction $\alpha/2$ of the bootstrap estimates of this parameter, and the value that exceeds $1 - \alpha/2$ of the bootstrap estimates of this parameter. Thus the transformation function does not need to be known to find these limits. In practice, therefore, the first percentile confidence interval is obtained as follows.

The first percentile method (Efron)

Bootstrap resampling of the original data is used to generate the bootstrap distribution of the parameter of interest. The $100(1 - \alpha)\%$ confidence interval for the true value of the parameter is then given by the two values that encompass the central $100(1 - \alpha)\%$ of this distribution. For example, a 95% confidence interval is given by the value that exceeds 2.5% of the generated distribution and the value that exceeds 97.5% of the generated distribution.

Hall (1992a, p. 36) called the percentile confidence interval just described the 'other percentile method' and suggested that it is analogous to looking up the wrong statistical table backwards. The reasoning behind this suggestion was based on the concept that a bootstrap distribution should mimic the particular distribution of interest. This might then be taken to imply that the distribution of the error in $\hat{\Theta}$, $\varepsilon = \hat{\Theta} - \Theta$, should be approximated by the error in the bootstrap distribution, $\varepsilon_B = \hat{\Theta}_B - \hat{\Theta}$. On this basis, a bootstrap distribution of ε_B can be generated to find two errors ε_L and ε_H such that

$$\text{Prob}(\varepsilon_L < \hat{\Theta}_B - \hat{\Theta} < \varepsilon_H) = 1 - \alpha,$$

in such a way that the probability of an error of less than ε_L is $\alpha/2$ and the probability of an error of more than ε_H is also $\alpha/2$. Then on the assumption that the bootstrap distribution of errors is a good approximation for the real distribution of errors it can be asserted that

$$\text{Prob}(\varepsilon_L < \hat{\Theta} - \Theta < \varepsilon_H) = 1 - \alpha,$$

so that

$$\text{Prob}(\hat{\Theta} - \varepsilon_H < \Theta < \hat{\Theta} - \varepsilon_L) = 1 - \alpha.$$

This is then a second type of percentile confidence interval, which can be summarized as follows.

The second percentile method (Hall)

Bootstrap resampling is used to generate a distribution of estimates $\hat{\theta}_B$ for a parameter θ of interest. The bootstrap distribution of difference between the bootstrap estimate and the estimate of θ in the original sample $\varepsilon_B = \hat{\theta}_B - \hat{\theta}$ is then assumed to approximate the distribution of errors for $\hat{\theta}$ itself. On this basis the bootstrap distribution of ε_B is used to find limits ε_L and ε_H for the sampling error such that $100(1 - \alpha)\%$ of errors are between these limits. The $100(1 - \alpha)\%$ confidence limits for θ are then $\hat{\Theta} - \varepsilon_H < \Theta < \hat{\Theta} - \varepsilon_L$. For example, to obtain a 95% confidence interval ε_L and ε_H should be chosen as the two values that define the central 95% part of the distribution of the bootstrap sampling errors ε_B.

From the definition of ε_L and ε_H it can be seen that these are calculated as $\varepsilon_L = \hat{\Theta}_L - \hat{\Theta}$ and $\varepsilon_H = \hat{\Theta}_H - \hat{\Theta}$, where $\hat{\Theta}_L$ is the value in the bootstrap distribution that is exceeded with probability $1 - \alpha/2$ and $\hat{\Theta}_H$ is the value that is exceeded with probability $\alpha/2$. The first type of percentile confidence interval is then given by the equation

$$\text{Prob}(\hat{\Theta}_L < \Theta < \hat{\Theta}_H) = \alpha \tag{3.3}$$

while the second type is given by

$$\text{Prob}(2\hat{\Theta} - \hat{\Theta}_H < \Theta < 2\hat{\Theta} - \hat{\Theta}_L) = \alpha. \tag{3.4}$$

The role of the bootstrap distribution is quite different for the two percentile intervals that have just been described. In particular, if the bootstrap distribution is skewed then the intervals will not agree well at all. For example, if the bootstrap distribution is highly skewed to the right then $\hat{\Theta}_L$ will be much closer to $\hat{\Theta}$ than $\hat{\Theta}_H$ and the first type of percentile confidence interval will indicate that high values of Θ are possible. On the other hand, for the second type of interval the large value for $\hat{\Theta}_H$ will translate into a lower confidence limit that is some distance from $\hat{\Theta}$ and an upper confidence limit that is much less extreme.

Potential users of bootstrap methods may well be concerned that there are two percentile methods that are both based on arguments that seem sensible, and yet these two methods work quite differently. It might be thought that it is possible to decide which one is 'correct'. However, unfortunately this is not possible. The assumptions are different, and sometimes one is best and at other times the other is best. Sometimes both are unsatisfactory but may possibly be improved by suitable modifications, as discussed in the following sections of this chapter.

The calculation of percentile confidence limits requires more bootstrap samples than does the calculation of the standard confidence limits because of the need to estimate percentage points accurately for the bootstrap distribution, instead of just estimating the mean and standard deviation of this distribution. Results presented by Efron (1987), which are discussed further in the next chapter, suggest that there may be little point in taking more than 100 bootstrap samples in order to estimate the bootstrap mean and standard deviation, but that 1000 bootstrap samples or more may be needed for percentile methods. A table presented by Buckland (1984) is useful in this context because it indicates the range of coverage that will be obtained with different numbers of bootstrap samples for the first percentile method (assuming that the assumptions of this method are met). This table shows, for example, that if 1000 bootstrap samples are used then the calculated 95% percentile confidence interval will have an actual confidence level of between 93.6% and 96.4% with probability 0.95. With 200 bootstrap samples the range of actual confidence levels widens to between 92% and 98% with 0.95 probability. With 10 000 bootstrap samples it narrows to between 94.6% and 95.4% with 0.95 probability.

Example 3.2: percentile confidence limits for a standard deviation   It is interesting to compare the two percentile methods just described in terms of the problem addressed in Example 3.1. That is to say, given the 20 observations in a random sample 3.56, 0.69, 0.10, 1.84, 3.93, 1.25, 0.18, 1.13, 0.27, 0.50, 0.67, 0.01, 0.61, 0.82, 1.70, 0.39, 0.11, 1.20, 1.21 and 0.72, which provide an estimated population standard deviation of $\hat{\sigma} = \sqrt{\{\sum(x_i - \bar{x})^2/20\}} = 1.03$, what is a satisfactory 95% confidence interval for the population standard deviation?

According to the first percentile method that was described, the 95% confidence limits should be obtained by generating the bootstrap distribution for the estimate of the standard deviation and finding the two values that include the central 95% of this distribution, as in equation (3.3). These calculations were carried out in a Lotus 1-2-3 spreadsheet (Lotus Development Corporation, 1991), and the limits were found to be 0.44 to 1.40. The limits for the second percentile method were also calculated, using equation (3.4). These second limits are $2 \times 1.03 - 1.40$ to $2 \times 1.03 - 0.44$, or 0.66 to 1.62. The first type of percentile limits are therefore substantially lower than the second type, although both include the true population value of 1.

To assess the accuracy of the limits more fully, 1000 samples of 20 observations were taken from the exponential distribution with a mean and a standard deviation of 1, and the two percentile methods were then used to obtain confidence limits for the population standard deviation. The results were rather unsatisfactory. The first percentile method of equation (3.3) gave nominal 95% confidence limits that included the true standard deviation only 65.9% of the time, with the lower limit being too high on 0.1% of occasions and the upper limit being too low on 34.0% of occasions. The second method, using equation (3.4), was slightly better. The nominal 95% confidence limits included the true population standard deviation 72.7% of the time, with the lower limit being too high on 2.6% of occasions and the upper limit being too low on 24.7% of occasions.

Comparison of these results with those of Examples 2.1 and 3.1 shows that the percentile method of equation (3.3) has given a worse performance than the jackknife (81% coverage), the jackknife on the logarithms of the sample standard deviations (87% coverage), the standard bootstrap interval (73% coverage), and the standard bootstrap interval on logarithms of the sample standard deviations (81% coverage). On the other hand, the percentile method of equation (3.4) has given about the same performance as the standard bootstrap method.

There is one further matter that can be mentioned before leaving this example. The first percentile method of equation (3.3) is transformation-respecting in the sense that if it is applied with any monotonic transformation of the parameter of interest and the resulting confidence limits are untransformed, then the final result is the same as if there were no transformation. For example, if this method is applied with $\log(\hat{\sigma})$ instead of $\hat{\sigma}$ then the limits found for $\log(\sigma)$ are just the logarithms of the limits that would be found for σ.

This property is sometimes thought of as being an advantage that the first percentile method has over other methods (Efron and Tibshirani, 1993, p. 175). However, it does mean that some transformations that are used in a routine way in statistics cannot possibly improve the method. This is not the case with the second percentile defined by equation (3.4), for which an appropriate transformation may give some improvement. In the example being considered this suggests trying the second percentile method on logarithms of sample standard deviations because this transformation is often used in this context to produce a more normal distribution.

This works quite well. For the 1000 simulated sets of data already considered the second percentile method was applied to logarithms of sample standard deviations, and the resulting limits were then untransformed. The coverage obtained from nominal 95% limits was 79.0%, with the lower limit being too high 7.0% of the time and the upper limit too high 14.0% of the time. Although this cannot be described as satisfactory, the coverage is about as good as any of the other methods considered so far.

3.4 BIAS-CORRECTED PERCENTILE CONFIDENCE LIMITS

One of the problems with the first percentile method when it is applied as in the last example is that it does not seem to be the case that a transformation exists such that the transformed variable $f(\hat{\Theta})$ is normally distributed with

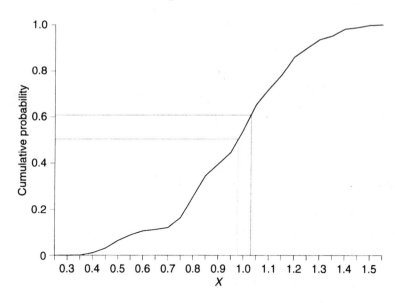

Figure 3.2 The cumulative probability distribution obtained for standard deviations from bootstrap resampling of a sample of size 20 from an exponential distribution with mean and standard deviation of 1. Slightly less than 40% of the bootstrap estimates exceeded 1.03 although this was the standard deviation for the distribution that was bootstrapped. The median of the distribution is approximately 0.98.

mean $f(\Theta)$, as shown in Figure 3.1. This can be seen because if the situation is as shown in Figure 3.1 then

$$\text{Prob}\{f(\hat{\Theta}) > f(\Theta)\} = \text{Prob}(\hat{\Theta} > \Theta) = 0.5,$$

with the two probabilities being equal because of the assumption that the transformation is monotonic and increasing. However, if the bootstrap distribution of sample standard deviations is considered for the data used in the last example then it gives the cumulative frequency polygon shown in Figure 3.2, for which this property does not apply: the bootstrap samples are taken from a distribution with a true standard deviation of $\hat{\sigma} = 1.03$, but the probability of exceeding this value for bootstrap estimates is about 0.4 instead of 0.5.

If 1000 bootstrap samples are adequate to represent the true bootstrap distribution, it appears that it may be useful to modify the first percentile method to remove any bias that arises because the true parameter value is not the median of the distribution of estimates. If this is done then the resulting limits are called bias-corrected percentile limits (Efron, 1981a).

The assumption that is made to construct bias-corrected confidence limits for a parameter Θ is that a monotonic increasing transformation of an estimator $\hat{\Theta}$ exists such that the transformed values $f(\hat{\Theta})$ are normally distributed with a mean of $f(\Theta) - z_0$ and a standard deviation of 1. In that case, there is a probability of $1 - \alpha$ that the statement

$$-z_{\alpha/2} < f(\hat{\Theta}) - f(\Theta) + z_0 < +z_{\alpha/2}$$

will hold where, as before, $z_{\alpha/2}$ is the value exceeded with probability $\alpha/2$ for the standard normal distribution. Rearrangement of this statement gives the required confidence interval

$$f(\hat{\Theta}) + z_0 - z_{\alpha/2} < f(\Theta) < f(\hat{\Theta}) + z_0 + z_{\alpha/2} \qquad (3.5)$$

for $f(\Theta)$, and untransforming these limits gives a confidence interval for Θ.

In order to use this idea it is necessary to estimate z_0. This can be done by noting that for any value s:

$$\begin{aligned} \text{Prob}\{f(\hat{\Theta}) > s\} &= \text{Prob}\{f(\hat{\Theta}) - f(\Theta) + z_0 > s - f(\Theta) + z_0\} \\ &= \text{Prob}\{Z > s - f(\Theta) + z_0\} \end{aligned}$$

where Z has a standard normal distribution. In particular, taking $s = f(\Theta)$ gives:

$$\text{Prob}\{f(\hat{\Theta}) > f(\Theta)\} = \text{Prob}(Z > z_0),$$

from which it also follows that

$$\text{Prob}(\hat{\Theta} > \Theta) = \text{Prob}\{Z > z_0\}.$$

At this point it is assumed that the probability on the left-hand side of the last equation can be estimated by p, the proportion of times that the bootstrap estimate $\hat{\Theta}_B$ exceeds Θ, the true value of Θ for the bootstrapped population. Then z_0 can be approximated by solving the equation

$$z_0 = z_p \qquad (3.6)$$

where z_p is the value for the standard normal distribution that is exceeded with probability p.

To make this clearer, suppose that 1000 bootstrap estimates $\hat{\Theta}_B$ are obtained, and that 400 of these estimates are greater than the original sample estimate $\hat{\Theta}$. Then $p = 0.4$ is the estimate of the probability that $\hat{\Theta}$ will be greater than Θ, and tables of the standard normal distribution show that $z_0 = 0.25$ is the value that is exceeded with probability 0.4.

Having found z_0 it is necessary to use bootstrapping to find the limits for Θ that correspond to the limits for $f(\Theta)$ from equation (3.5). Consider first the upper limit, $f(\hat{\Theta}) + z_0 + z_{\alpha/2}$. Let p_U be the probability of being less than this value for the bootstrap distribution of $f(\hat{\Theta}_B)$. Then

$$
\begin{aligned}
p_U &= \text{Prob}\{f(\hat{\Theta}_B) < f(\hat{\Theta}) + z_0 + z_{\alpha/2}\} \\
&= \text{Prob}\{f(\hat{\Theta}_B) - f(\hat{\Theta}) + z_0 < z_0 + z_{\alpha/2} + z_0\} \\
&= \text{Prob}\{Z < 2z_0 + z_{\alpha/2}\}.
\end{aligned}
$$

where Z has a standard normal distribution. Thus the bootstrap upper confidence limit for $f(\Theta)$ is the value that just exceeds a proportion p_U of the bootstrap distribution generated for $f(\hat{\Theta}_B)$, and the corresponding upper confidence limit for Θ is the value that just exceeds a proportion p_U of the bootstrap distribution generated for $\hat{\Theta}_B$.

The lower confidence limit for Θ is found in a similar way. For $f(\Theta)$ the limit is $f(\hat{\Theta}) + z_0 - z_{\alpha/2}$ and the probability of being less than this value for the bootstrap distribution of $f(\hat{\Theta}_B)$ is

$$
\begin{aligned}
p_L &= \text{Prob}\{f(\hat{\Theta}_B) < f(\hat{\Theta}) + z_0 - z_{\alpha/2}\} \\
&= \text{Prob}\{f(\hat{\Theta}_B) - f(\hat{\Theta}) + z_0 < z_0 - z_{\alpha/2} + z_0\} \\
&= \text{Prob}\{Z < 2z_0 - z_{\alpha/2}\}.
\end{aligned}
$$

Thus p_L is the probability of being less than $2z_0 - z_{\alpha/2}$ for the standard normal distribution, the bootstrap lower confidence limit for $f(\Theta)$ is the value that just exceeds a proportion p_L of the bootstrap distribution generated for $f(\hat{\Theta}_B)$, and the corresponding bootstrap lower confidence limit for Θ is the value that just exceeds a proportion p_L of the bootstrap distribution generated for $\hat{\Theta}_B$.

From these results, the bias-corrected percentile confidence limits that has just been justified at some length can be written as

$$
\text{INVCDF}\{\phi(2z_0 - z_{\alpha/2})\} \quad \text{and} \quad \text{INVCDF}\{\phi(2z_0 + z_{\alpha/2})\}, \quad (3.7)
$$

where $\text{INVCDF}(p)$ means the value in the bootstrap distribution corresponding to a cumulative probability of p, $\phi(z)$ is the proportion of the standard normal distribution that is less than the value z, and z_0 is as defined by equation (3.6). If the bootstrap distribution of $\hat{\Theta}_B$ has the median value $\hat{\Theta}$ then $z_0 = 0$ and the limits (3.7) become

$$
\text{INVCDF}(\alpha/2) \quad \text{and} \quad \text{INVCDF}(1 - \alpha/2).
$$

These are then the values that cut off a fraction $\alpha/2$ of the bottom tail and a fraction $\alpha/2$ of the top tail of the bootstrap distribution for $\hat{\Theta}$. That is to say, they are just the simple percentile confidence limits for Θ as defined by equation (3.3).

The calculations for bias-corrected percentile limits are slightly complicated, but can be summarized as follows.

Bias-corrected percentile $100(1 - \alpha)\%$ confidence limits

(a) Generate values $\hat{\theta}_B$ from the bootstrap distribution for estimates of the parameter θ of interest. Find the proportion of times p that $\hat{\theta}_B$ exceeds $\hat{\theta}$, the estimate of θ from the original sample. Hence calculate z_0, the value from the standard normal distribution that is exceeded with probability p. (This is $z_0 = 0$ if $p = 0.5$.)

(b) Calculate $\phi(2z_0 - z_{\alpha/2})$, which is the proportion of the standard normal distribution that is less than $2z_0 - z_{\alpha/2}$, where $z_{\alpha/2}$ is the value that is exceeded with probability $\alpha/2$ for the standard normal distribution. For example, $z_{0.025} = 1.96$ is the value to use for a 95% confidence interval.

(c) The lower confidence limit for θ is the value that just exceeds a proportion $\phi(2z_0 - z_{\alpha/2})$ of all values in the bootstrap distribution of estimates $\hat{\theta}_B$.

(d) Calculate $\phi(2z_0 + z_{\alpha/2})$, which is the proportion of the standard normal distribution that is less than $2z_0 + z_{\alpha/2}$. Use $z_{\alpha/2} = z_{0.025} = 1.96$ for a 95% confidence interval.

(e) The upper confidence limit for θ is the value that just exceeds a proportion $\phi(2z_0 + z_{\alpha/2})$ of all values in the bootstrap distribution of estimates $\hat{\theta}_B$.

In order to calculate the bias-corrected percentile confidence limits it is useful to have equations available for calculating the value z_p that is exceeded by a proportion p of the standard normal distribution, and $\phi(z)$, the proportion of the standard normal distribution that is less than the value z. There are various approximations available for this purpose, including

$$z_p \approx -0.862 + 1.202\sqrt{\{-0.639 - 1.664 \log_e(p)\}}$$

where $0.0005 \leq p \leq 0.5$, and

$$\phi(z) \approx 1 - 0.5\exp(-0.717z - 0.416z^2)$$

for $0 \leq z \leq 3.29$ (Lin, 1989). The first of these approximations has an error of less than 0.018, and the second an error of less than 0.006. If $p > 0.5$ then it is possible to use the relationship $z_{1-p} = -z_p$, while if $z < 0$ then it is possible to use $\phi(z) = 1 - \phi(-z)$.

3.5 ACCELERATED BIAS-CORRECTED PERCENTILE LIMITS

There is a third type of percentile confidence interval introduced by Efron and Tibshirani (1986) and discussed more fully by Efron (1987), which is called the accelerated bias-corrected percentile method. This relies on the assumption that a transformation $f(\hat{\Theta})$ of the estimator of interest $\hat{\Theta}$ exists such that $f(\hat{\Theta})$ has a normal distribution with mean $f(\Theta) - z_0\{1 + af(\Theta)\}$ and standard deviation $1 + af(\Theta)$, where z_0 and a are constants. This is a less restrictive assumption than that required for the bias-corrected percentile method, which effectively sets $a = 0$. The introduction of the constant a allows the standard deviation of $f(\hat{\Theta})$ to vary linearly with $f(\Theta)$, which may be a reasonable approximation to the true situation.

Assuming for the moment that the constants z_0 and a, and the transformation $f(\Theta)$ are all known, it is possible to derive a confidence interval for $f(\Theta)$. To begin with it can be noted that if $f(\hat{\Theta})$ has the normal distribution as claimed then this implies that

$$\text{Prob}[-z_{\alpha/2} < [f(\hat{\Theta}) - f(\Theta) + z_0\{1 + af(\Theta)\}]/[(1 + af(\Theta)] < +z_{\alpha/2}] = 1 - \alpha,$$

because the random variable in the centre of the inequalities in this statement has a standard normal distribution. Rearranging the inequalities shows that it will also be the case that

$$\text{Prob}\left[\frac{f(\hat{\Theta}) + z_0 - z_{\alpha/2}}{1 - a(z_0 - z_{\alpha/2})} < f(\Theta) < \frac{f(\hat{\Theta}) + z_0 + z_{\alpha/2}}{1 - a(z_0 + z_{\alpha/2})}\right] = 1 - \alpha,$$

say

$$\text{Prob}[L < f(\Theta) < U] = 1 - \alpha,$$

where L and U are $100(1 - \alpha)\%$ confidence limits for the true value of $f(\Theta)$.

To calculate the limit L by bootstrapping, the following results can be used, where $f(\hat{\Theta}_B)$ denotes a transformed value obtained by bootstrap resampling of the original data:

$$\text{Prob}[f(\hat{\Theta}_B) < L] = \text{Prob}[f(\hat{\Theta}_B) < \{f(\hat{\Theta}) + z_0 - z_{\alpha/2}\}/\{1 - a(z_0 - z_{\alpha/2})\}].$$

Therefore,

$$\text{Prob}[f(\hat{\Theta}_B) < L] = \text{Prob}\left[\frac{f(\hat{\Theta}_B) - f(\hat{\Theta})}{1 + af(\hat{\Theta})} + z_0 < \frac{z_0 - z_{\alpha/2}}{1 - a(z_0 - z_{\alpha/2})} + z_0\right]$$

$$= \text{Prob}[Z < (z_0 - z_{\alpha/2})/\{1 - a(z_0 - z_{\alpha/2})\} + z_0], \tag{3.8}$$

where $Z = \{f(\hat{\Theta}_B) - f(\hat{\Theta})\}/\{1 + af(\hat{\Theta})\} + z_0$ is a value from the standard normal distribution. It is being assumed here that the bootstrap distribution of $f(\hat{\Theta}_B)$ mirrors the distribution of $f(\hat{\Theta})$ and, in particular, $f(\hat{\Theta}_B)$ is normally distributed with mean $f(\hat{\Theta}) - z_0\{1 + af(\hat{\Theta})\}$ and standard deviation $1 + af(\hat{\Theta})$.

What equation (3.8) says is that the probability of a bootstrap estimate $f(\hat{\Theta}_B)$ being less than the lower confidence limit for $f(\Theta)$ is the probability of a standard normal variable being less than

$$z_L = (z_0 - z_{\alpha/2})/\{1 - a(z_0 - z_{\alpha/2})\} + z_0,$$

which is $\phi(z_L)$ with the notation used before. This lower confidence limit can therefore be estimated by generating a large number of values for $f(\hat{\Theta}_B)$ and finding the one that is just greater than a fraction $\phi(z_L)$ of these values. Actually, this is not possible without knowing the transformation function $f(\Theta)$. However, this does not matter because it is the lower confidence limit for Θ that is really required. Because the transformation is monotonic and increasing this lower limit must be the value that exceeds a fraction $\phi(z_L)$ of the values in the bootstrap distribution for $\hat{\Theta}$.

The argument that has just been given produces a lower confidence limit for the true value of Θ. A similar argument leads to the conclusion that an upper confidence limit for Θ is given by the value in the bootstrap distribution for $\hat{\Theta}$ that exceeds a fraction $\phi(z_U)$ of this distribution, where

$$z_U = (z_0 + z_{\alpha/2})/\{1 - a(z_0 + z_{\alpha/2})\} + z_0.$$

Taking the lower and upper limits together shows that the $100(1 - \alpha)\%$ confidence limits for Θ given by the accelerated bias-corrected method are

$$\text{INVCDF}\{\phi(z_L)\} \quad \text{and} \quad \text{INVCDF}\{\phi(z_U)\}, \tag{3.9}$$

where $\text{INVCDF}(p)$ is the value in the cumulative bootstrap distribution for $\hat{\Theta}$ that just exceeds a proportion p of the distribution.

The constant z_0 can be estimated from the bootstrap distribution by assuming, as before, that $f(\hat{\Theta}_B)$ is normally distributed with mean $f(\hat{\Theta}) - z_0\{1 + af(\hat{\Theta})\}$ and standard deviation $1 + af(\hat{\Theta})$. Then

$$\text{Prob}[f(\hat{\Theta}_B) > f(\hat{\Theta})] = \text{Prob}[\{f(\hat{\Theta}_B) - f(\hat{\Theta})\}/\{1 + af(\hat{\Theta})\} + z_0 > z_0]$$
$$= \text{Prob}[Z > z_0],$$

where Z has a standard normal distribution. But it also follows that

$$\text{Prob}[\hat{\Theta}_B > \hat{\Theta}] = \text{Prob}[Z > z_0]$$

because of the monotonic increasing nature of the transformation $f(\Theta)$. Therefore, if p is the proportion of values in the bootstrap distribution for $\hat{\Theta}_B$ that are greater than $\hat{\Theta}$, then z_0 can be estimated as

$$z_0 = z_p, \tag{3.10}$$

where z_p is the value for the standard normal distribution that is exceeded with probability p. This is the same as equation (3.6) for the bias-corrected percentile method.

Unfortunately, an equation for the constant a involved in the limits (3.9) cannot be derived in a simple way. One approach that is justified by Efron and Tibshirani (1993, p. 186) is as follows, for the case where no assumptions are to be made about the nature of the population that the original sample came from. Let $\hat{\Theta}_{-i}$ denote the estimate of Θ that is obtained when the ith of the original sample values $x_1, x_2, \ldots, x_n$ is removed from the data, and let $\hat{\Theta} = \sum \hat{\Theta}_{-i}/n$ be the average of these partial estimates. The constant a can then be approximated using the equation

$$a \approx \sum_{i=1}^{n}(\hat{\Theta} - \hat{\Theta}_{-i})^3 / [6\{\sum_{i=1}^{n}(\hat{\Theta} - \hat{\Theta}_{-i})^2\}^{1.5}]. \qquad (3.11)$$

Because the calculation of accelerated bias-corrected confidence limits involves several steps it is useful to list these, as follows.

Accelerated bias-corrected $100(1 - \alpha)\%$ confidence limits

(a) Generate values $\hat{\Theta}_B$ from the bootstrap distribution for estimates of a parameter Θ by resampling the original sample.
(b) Determine z_0 using equation (3.10).
(c) Determine a using equation (3.11).
(d) The confidence limits for Θ are set at INVCDF$\{\phi(z_L)\}$ and INVCDF$\{\phi(z_U)\}$ as in equation (3.9). To do this, the bootstrap estimates $\hat{\Theta}_B$ are ordered from the smallest to the largest. The confidence limits for Θ then correspond to the values that are at the positions up the list corresponding to fractions $\phi(z_L)$ and $\phi(z_U)$ of all the values. For example, if $\phi(z_L) = 0.1$ then the lower confidence limit is the value in the ordered list that just exceeds 10% of all bootstrap estimates.

Example 3.3: better percentile confidence limits for a standard deviation Consider again the problem of constructing a confidence interval for a population standard deviation that has been discussed in Examples 3.1 and 3.2. It may be recalled that the values 3.56, 0.69, 0.10, 1.84, 3.93, 1.25, 0.18, 1.13, 0.27, 0.50, 0.67, 0.01, 0.61, 0.82, 1.70, 0.39, 0.11, 1.20, 1.21 and 0.72 provide an estimated population standard

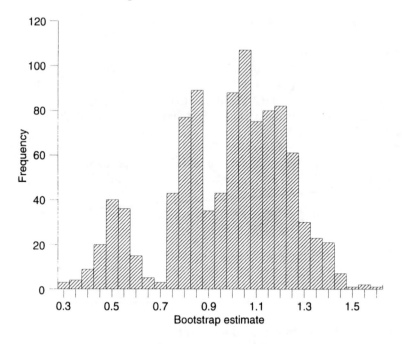

Figure 3.3 The distribution of 1000 bootstrap estimates of standard deviation taken from an original sample of size 20. The standard deviation of the original sample was 1.03.

deviation of $\hat{\sigma} = \sqrt{\{\sum(x_i - \bar{x})^2/20\}} = 1.03$. The problem is to find a satisfactory 95% confidence interval for the true population standard deviation.

The distribution of 1000 bootstrap estimates of standard deviation is shown in Figure 3.3. The distribution is not symmetric about the original sample value of 1.03. In fact, a proportion $p = 0.463$ of the estimates are above that value. This leads to the estimate $z_0 = z_{0.463} = 0.10$ from equation (3.6). Using equation (3.7) the bias-corrected 95% percentile confidence limits are then found to be

$$\text{INVCDF}\{\phi(2 \times 0.10 - 1.96)\} \qquad \text{to} \qquad \text{INVCDF}\{\phi(2 \times 0.10 + 1.96)\},$$

i.e.

$$\text{INVCDF}\{\phi(-1.76)\} \qquad \text{to} \qquad \text{INVCDF}\{\phi(2.16)\},$$

i.e.

$$\text{INVCDF}(0.039) \qquad \text{to} \qquad \text{INVCDF}(0.985),$$

because the area under the standard normal distribution from minus infinity to -1.76 is 0.039, while the area from minus infinity to 2.16 is 0.985. The bias-corrected confidence limits are therefore the values corresponding to cumulative proportions of 0.039 and 0.985 for the bootstrap distribution of $\hat{\sigma}$. These limits are 0.49 and 1.41, so that the true population standard deviation of 1 is well within them.

For the accelerated bias-corrected 95% confidence interval it is necessary to determine the constant a using equation (3.11), with $\hat{\Theta}$ equal to the estimated standard deviation $\hat{\sigma}$. The calculations are shown in Table 3.2, giving $\sum(\hat{\Theta} - \hat{\Theta}_{-i})^3 = 0.013\,28$, and $\sum(\hat{\Theta} - \hat{\Theta}_{-i})^2 = 0.078\,34$. Hence $a = 0.013\,28/(6 \times 0.078\,34^{1.5}) = 0.101$. Using this value and $z_0 = 0.10$ in equation (3.9) shows that the accelerated bias-corrected 95% percentile limits for the population standard deviation are

$$\text{INVCDF}\{\phi(-1.46)\} \quad \text{to} \quad \text{INVCDF}\{\phi(2.70)\},$$

i.e.

$$\text{INVCDF}(0.072) \quad \text{to} \quad \text{INVCDF}(0.997),$$

because the area under the standard normal distribution from minus infinity to -1.46 is 0.072 while the area from minus infinity to 2.70 is 0.997. The required confidence limits are therefore the values from the bootstrap distribution for the standard deviation corresponding to cumulative percentages of 7.2% and 99.7%. From the 1000 bootstrap samples that were generated these limits are found to be 0.54 to 1.50.

To examine the performance of the bias-corrected and accelerated bias-corrected percentile limits more generally, the process of calculating 95% confidence limits was repeated on 1000 samples drawn independently from the exponential distribution with a mean and standard deviation of 1. It was found that the bias-corrected limits included the population standard deviation 68.2% of the time and the accelerated bias-corrected limits included the population standard deviation 72.1% of the time. Thus these methods have failed to give a satisfactory performance for this application, just like the other methods that have been considered before.

It is interesting to conclude this example by comparing the results for all the methods that have been considered for this problem of finding confidence limits for the standard deviation of the exponential distribution. A summary is shown in Table 3.3. The different methods vary considerably both in terms of the limits that they produce for one set of data and also the coverage that they give for nominal 95% limits. The lower confidence limit for the original set of data varies from 0.44 to 0.77, while the upper limit varies from 1.40 to 2.32. For the 1000 sets of simulated data all of the methods have produced poor results. The best performance has been given by the standard confidence limits applied to logarithms of standard

Table 3.2 Calculations needed to evaluate the acceleration constant a using equation (3.11). The columns headed 1 to 20 contain the values in the original sample, with the ith removed in the ith row. The column headed $\hat{\Theta}_{-i}$ contains the partial estimates of the standard deviation, with mean $\hat{\Theta}$. The last three columns are needed to calculate the sums in equation (3.11)

	1	2	3	4	5	6	7	8	9	10	11	12	13	14	15	16	17	18	19	20	$\hat{\Theta}_{-i}$	$\hat{\Theta}_{\cdot}-\hat{\Theta}_{-i}$	$(\hat{\Theta}_{\cdot}-\hat{\Theta}_{-i})^2$	$(\hat{\Theta}_{\cdot}-\hat{\Theta}_{-i})^3$
1		0.69	0.10	1.84	3.93	1.25	0.18	1.13	0.27	0.50	0.67	0.01	0.61	0.82	1.70	0.39	0.11	1.20	1.21	0.72	0.8788	0.1507	0.02270	0.00342
2	3.56		0.10	1.84	3.93	1.25	0.18	1.13	0.27	0.50	0.67	0.01	0.61	0.82	1.70	0.39	0.11	1.20	1.21	0.72	1.0564	-0.0269	0.00072	-0.00002
3	3.56	0.69		1.84	3.93	1.25	0.18	1.13	0.27	0.50	0.67	0.01	0.61	0.82	1.70	0.39	0.11	1.20	1.21	0.72	1.0361	-0.0066	0.00004	-0.00000
4	3.56	0.69	0.10		3.93	1.25	0.18	1.13	0.27	0.50	0.67	0.01	0.61	0.82	1.70	0.39	0.11	1.20	1.21	0.72	1.0430	-0.0135	0.00018	-0.00000
5	3.56	0.69	0.10	1.84		1.25	0.18	1.13	0.27	0.50	0.67	0.01	0.61	0.82	1.70	0.39	0.11	1.20	1.21	0.72	0.8134	0.2161	0.04670	0.01009
6	3.56	0.69	0.10	1.84	3.93		0.18	1.13	0.27	0.50	0.67	0.01	0.61	0.82	1.70	0.39	0.11	1.20	1.21	0.72	1.0586	-0.0291	0.00084	-0.00002
7	3.56	0.69	0.10	1.84	3.93	1.25		1.13	0.27	0.50	0.67	0.01	0.61	0.82	1.70	0.39	0.11	1.20	1.21	0.72	1.0400	-0.0104	0.00011	-0.00000
8	3.56	0.69	0.10	1.84	3.93	1.25	0.18		0.27	0.50	0.67	0.01	0.61	0.82	1.70	0.39	0.11	1.20	1.21	0.72	1.0595	-0.0300	0.00090	-0.00003
9	3.56	0.69	0.10	1.84	3.93	1.25	0.18	1.13		0.50	0.67	0.01	0.61	0.82	1.70	0.39	0.11	1.20	1.21	0.72	1.0439	-0.0144	0.00021	-0.00000
10	3.56	0.69	0.10	1.84	3.93	1.25	0.18	1.13	0.27		0.67	0.01	0.61	0.82	1.70	0.39	0.11	1.20	1.21	0.72	1.0519	-0.0224	0.00050	-0.00001
11	3.56	0.69	0.10	1.84	3.93	1.25	0.18	1.13	0.27	0.50		0.01	0.61	0.82	1.70	0.39	0.11	1.20	1.21	0.72	1.0560	-0.0265	0.00070	-0.00002
12	3.56	0.69	0.10	1.84	3.93	1.25	0.18	1.13	0.27	0.50	0.67		0.61	0.82	1.70	0.39	0.11	1.20	1.21	0.72	1.0313	-0.0018	0.00000	-0.00000
13	3.56	0.69	0.10	1.84	3.93	1.25	0.18	1.13	0.27	0.50	0.67	0.01		0.82	1.70	0.39	0.11	1.20	1.21	0.72	1.0547	-0.0252	0.00064	-0.00002
14	3.56	0.69	0.10	1.84	3.93	1.25	0.18	1.13	0.27	0.50	0.67	0.01	0.61		1.70	0.39	0.11	1.20	1.21	0.72	1.0584	-0.0288	0.00083	-0.00002
15	3.56	0.69	0.10	1.84	3.93	1.25	0.18	1.13	0.27	0.50	0.67	0.01	0.61	0.82		0.39	0.11	1.20	1.21	0.72	1.0484	-0.0189	0.00036	-0.00001
16	3.56	0.69	0.10	1.84	3.93	1.25	0.18	1.13	0.27	0.50	0.67	0.01	0.61	0.82	1.70		0.11	1.20	1.21	0.72	1.0484	-0.0189	0.00036	-0.00001
17	3.56	0.69	0.10	1.84	3.93	1.25	0.18	1.13	0.27	0.50	0.67	0.01	0.61	0.82	1.70	0.39		1.20	1.21	0.72	1.0366	-0.0071	0.00005	-0.00000
18	3.56	0.69	0.10	1.84	3.93	1.25	0.18	1.13	0.27	0.50	0.67	0.01	0.61	0.82	1.70	0.39	0.11		1.21	0.72	1.0590	-0.0295	0.00087	-0.00003
19	3.56	0.69	0.10	1.84	3.93	1.25	0.18	1.13	0.27	0.50	0.67	0.01	0.61	0.82	1.70	0.39	0.11	1.20		0.72	1.0590	-0.0294	0.00087	-0.00003
20	3.56	0.69	0.10	1.84	3.93	1.25	0.18	1.13	0.27	0.50	0.67	0.01	0.61	0.82	1.70	0.39	0.11	1.20	1.21		1.0569	-0.0274	0.00075	-0.00002

$$\hat{\Theta}_{\cdot} = 1.0295$$

Sum=0.0000 0.07834 0.01328

Table 3.3 Confidence limits obtained by various methods for the standard deviation of an exponential distribution, based on random samples of 20 observations. The left-hand side of the table shows how 95% confidence limits compare for the set of data used in Examples 3.1–3.3. The right-hand side of the table shows the percentages of times that the population standard deviation was above the top confidence limit or below the bottom confidence limit. These percentages are based on 1000 sets of simulated data, with errors of estimation (1.96 standard errors of the estimated percentages) being approximately 1%

Method	Results for one set of data		Outside limits (%)	
	Lower limit	Upper limit	Lower limit too high	Upper limit too low
1. Standard	0.60	1.58	1.4	25.9
2. Standard on logarithms	0.65	1.96	2.3	16.6
3. Efron's percentile	0.44	1.40	0.1	34.0
4. Hall's percentile	0.66	1.62	2.6	24.7
5. Hall's percentile on logarithms	0.77	2.32	7.0	14.0
6. Bias corrected percentile	0.49	1.41	0.8	31.0
7. Accelerated bias corrected	0.54	1.50	1.9	26.0
		Desired	2.5%	2.5

deviations, but even for this method the upper limit was too low nearly 17% of the time, instead of the desired 2.5%. The worst performance has been given by Efron's original percentile method, with the upper limit too low 34% of the time. The bias-corrected and accelerated bias-corrected methods are slightly better, but far from satisfactory.

The difficulties that bootstrapping has with this example are basically due to the fact that a sample of size 20 from an exponential distribution cannot be expected to represent the distribution very well. Indeed, the theory of bootstrapping shows that it will work on standard deviations for large enough samples from distributions with a finite fourth moment. This follows, for example, as a special case of the results of Behran and Srivastava (1985) concerning the validity of bootstrapping of functions of a covariance matrix. The magnitude of 'large' in this context is indicated to some extent by the need for a sample size of more than 100 in order for the percentile and bias-corrected percentile methods to give good results when sampling from a normal distribution (Schenker, 1985). It seems likely that an even larger sample size is needed with the exponential distribution.

3.6 OTHER METHODS FOR CONSTRUCTING CONFIDENCE INTERVALS

Several other methods for constructing confidence intervals have been proposed in recent years (DiCiccio and Romano, 1988, 1990). These include non-parametric tilting (Efron, 1981a), the automatic percentile method (DiCiccio and Romano, 1988), non-parametric likelihood methods (Owen, 1988) and the bootstrap-t method (Efron, 1981a).

All of these are more or less difficult to apply, except bootstrap-t, which just consists of using bootstrapping to approximate the distribution of a statistic of the form

$$T = (\hat{\Theta} - \Theta)/\widehat{SE}(\hat{\Theta})$$

where $\hat{\Theta}$ is an estimate of the parameter Θ of interest, with estimated standard error $\widehat{SE}(\hat{\Theta})$. The bootstrap approximation in this case is obtained by taking bootstrap samples from the original data values, calculating the corresponding estimates $\hat{\Theta}_B$ and their estimated standard errors, and hence finding the bootstrapped T-values

$$T_B = (\hat{\Theta}_B - \hat{\Theta})/\widehat{SE}(\hat{\Theta}_B)$$

The hope is then that the generated distribution will mimic the distribution of T.

The assumption behind bootstrap-t is that T is a pivotal statistic, which means that the distribution is the same for all values of Θ. If this is true, then for all Θ, the statement

$$t_{1-\alpha/2} < (\hat{\Theta} - \Theta)/\widehat{SE}(\hat{\Theta}) < t_{\alpha/2}$$

will hold with probability $1 - \alpha$, where $t_{1-\alpha/2}$ and $t_{\alpha/2}$ are chosen so that

$$\text{Prob}(T > t_{\alpha/2}) = \text{Prob}(T < t_{1-\alpha/2}) = \alpha/2.$$

A little algebra then shows that the statement

$$\hat{\Theta} - t_{\alpha/2}\widehat{SE}(\hat{\Theta}) < \Theta < \hat{\Theta} - t_{1-\alpha/2}\widehat{SE}(\hat{\Theta}) \qquad (3.12)$$

will hold with the same probability, which constitutes a $100(1 - \alpha)\%$ confidence interval for Θ. The procedure for bootstrap-t is therefore as follows.

Bootstrap-t $100(1 - \alpha)\%$ confidence limits

(a) Approximate $t_{\alpha/2}$ and $t_{1-\alpha/2}$ using the bootstrap t-distribution, i.e. by finding the values that satisfy the two equations

$$\text{Prob}[(\hat{\Theta}_B - \hat{\Theta})/\widehat{SE}(\hat{\Theta}_B) > t_{\alpha/2}] = \alpha/2$$

and

$$\text{Prob}[(\hat{\Theta}_B - \hat{\Theta})/\widehat{SE}(\hat{\Theta}_B) > t_{1-\alpha/2}] = 1 - \alpha/2,$$

for the generated bootstrap estimates. For example, for a 95% confidence interval the two values of t will encompass the central 95% of the bootstrap-t distribution.

(b) The confidence interval is given by equation (3.12) with the two values of t defined in (a).

The idea behind the bootstrap-t method can be applied whenever there is a suitable pivotal statistic available. In reality this means that an estimate of $\widehat{SE}(\hat{\Theta})$ can be obtained as well as $\hat{\Theta}$, and there is some hope that the distribution of $T = (\hat{\Theta} - \Theta)/\widehat{SE}(\hat{\Theta})$ will be approximately constant. If necessary, $\widehat{SE}(\hat{\Theta})$ can be estimated by jackknifing.

Example 3.4: confidence intervals for a mean value   Consider yet again the sample data from an exponential distribution with mean and standard deviation 1 that have already been considered in Examples 3.1 to 3.3, i.e. 3.56, 0.69, 0.10, 1.84, 3.93, 1.25, 0.18, 1.13, 0.27, 0.50, 0.67, 0.01, 0.61, 0.82, 1.70, 0.39, 0.11, 1.20, 1.21 and 0.72. Suppose that it is required to construct a 95% confidence interval for the population mean. It has been found already that calculating a valid confidence interval for the population standard deviation is not straightforward using bootstrapping. It can be hoped that calculating a valid confidence interval for the mean is easier.

So far in this chapter, six bootstrap procedures that can be used for finding confidence limits for the mean have been described: the standard method using $\bar{x} \pm z_{\alpha/2}$(bootstrap standard deviation), Efron's percentile method, Hall's percentile method, the bias-corrected percentile method, the accelerated bias-corrected percentile method and bootstrap-t. In the present example these methods will all be compared in terms of the results obtained on the particular set of data given above, and also in terms of the coverage obtained from nominal 95% confidence intervals for 1000 independently generated sets of data from the same exponential distribution. All of the calculations were done using Lotus 1-2-3 (Lotus Development Corporation, 1991).

In fact, it makes little sense to use the standard method in this situation because it is known that the bootstrap standard deviation of estimated means should equal $s/\sqrt{20}$, where $s^2 = \sum(x_i - \bar{x})^2/19$, the usual unbiased estimator of the sample variance. Confidence intervals can therefore be calculated using $s/\sqrt{20}$ rather than the bootstrap standard deviation. Nevertheless, the bootstrap standard method has been used here so that it can be compared with its competitors.

Consider first the bootstrap-t interval applied to the particular sample of 20 observations given above. The natural estimate of the population mean is the sample mean $\bar{x}$, leading to the t-statistic

$$T = (\bar{x} - \mu)/(s/\sqrt{20})$$

because $\widehat{SE}(\bar{x}) = s/\sqrt{20}$. The bootstrap version of this statistic is then

$$T_B = (\bar{x}_B - \bar{x})/(s_B/\sqrt{20}),$$

where $\bar{x}_B$ and s_B are calculated from a bootstrap sample.

When 1000 bootstrap values of t_B were generated it was found that these varied from a low of -7.50 to a high of 2.69. For constructing a 95% confidence interval using equation (3.12) it is necessary to find the value of T_B that is less than 97.5% of the distribution, which is $t_{0.975} = -3.76$, and the value that is less than 2.5% of the distribution, which is $t_{0.025} = 1.72$. The sample mean is 1.045, the sample standard deviation is 1.060, and the estimated standard error of the mean is 0.237. The confidence limits are therefore

$$1.045 - 1.72 \times 0.237 < \mu < 1.045 + 3.76 \times 0.237$$

or

$$0.64 < \mu < 1.93.$$

These limits and the limits obtained from the other five methods are shown in the first two columns of Table 3.4. The limits are reasonably close for the different methods except that the upper limit for the bootstrap-t method is substantially higher than the upper limits for the other methods.

The last two columns of Table 3.4 show coverage information for the 1000 generated sets of data. The best method is clearly bootstrap-t. For this method the coverage of the confidence limits was 95.2%, with the lower limit too high 1.4% of the time and the upper limit too low 3.4% of the time. The performance of the accelerated bias-corrected method is also quite reasonable, with 92.4% coverage.

Although too much should not be made of special cases, this example, together with the one on the obtaining confidence limits for the population standard deviation, does suggest that bootstrap methods should not necessarily be assumed to work with small samples. The lesson to be learned is that bootstrap methods should be tested out before they are relied upon for a new application.

Table 3.4 Confidence limits obtained by various methods for the mean of an exponential distribution, based on random samples of 20 observations. The left-hand side of the table shows how 95% confidence limits compare for the set of data used in Examples 3.4. The right-hand side of the table shows the percentages of times that the population standard deviation was above the top confidence limit or below the bottom confidence limit. These percentages are based on 1000 sets of simulated data, with errors of estimation (1.96 standard errors of the estimated percentages) being approximately 1%

Method	Results for one set of data		Outside limits (%)	
	Lower limit	Upper limit	Lower limit too high	Upper limit too low
1. Standard	0.58	1.50	1.1	9.4
2. Efron's percentile	0.63	1.57	1.7	8.2
3. Hall's percentile	0.52	1.46	0.7	10.5
4. Bias corrected percentile	0.64	1.58	2.1	7.0
5. Accelerated bias corrected percentile	0.68	1.63	2.5	5.1
6. Bootstrap-*t*	0.64	1.93	1.4	3.4
		Desired	2.5	2.5

3.7 TRANSFORMATIONS TO IMPROVE BOOTSTRAP-*t* INTERVALS

Recently there has been interest in the possibility of improving bootstrap-*t* types of confidence interval by the use of transformations (Abramovitch and Singh, 1985; Bose and Babu, 1991; Hall, 1992b). Here, only one of the methods that were discussed by Hall (1992b) for confidence limits for a mean value will be described. The transformation in this case can be extended to differences between two means and linear regression problems quite easily. See Hall's paper for more details about these extensions.

Suppose that a random sample of size n is taken from a distribution with mean μ, and that the sample mean $\bar{x}$, variance $\hat{\sigma}^2 = \sum(x_i - \bar{x})^2/n$ and skewness $\hat{v} = \sum(x_i - \bar{x})^3/\hat{\sigma}^3$ are calculated, where the summations are for i from 1 to n. Then it can be shown that the statistic

$$Q(W) = W + \hat{v}W^2/3 + \hat{v}^2W^3/27 + \hat{v}/(6n), \qquad (3.13)$$

where $W = (\bar{x} - \mu)/\hat{\sigma}$, will be approximately normally distributed with mean zero and variance $1/n$. It can also be shown that $Q(W)$ is a strictly increasing function of W, with inverse

$$W(Q) = 3([1 + \hat{v}\{Q - \hat{v}/(6n)\}]^{1/3} - 1)/\hat{v}. \qquad (3.14)$$

On this basis, Hall suggests that Q should be bootstrapped rather than the pivotal statistic $T = (\bar{x} - \Theta)/SE(\bar{x})$. The procedure is as follows.

Hall's bootstrap-t transformation

(a) Generate bootstrap samples and calculate values for the statistic

$$W_B = (\bar{x}_B - \bar{x})/\hat{\sigma}_B,$$

where $\bar{x}_B$ and $\hat{\sigma}_B$ denote estimates from a bootstrap sample.

(b) Transform the statistics to $Q(W_B)$ using equation (3.13).

(c) The bootstrap distribution of $Q(W_B)$ (with $\hat{v}$ recalculated for each bootstrap sample) is assumed to approximate the distribution of $Q(W)$ that would be obtained from resampling the original population. On this basis two values $q_{\alpha/2}$ and $q_{1-\alpha/2}$ are estimated such that

$$\mathrm{Prob}\{q_{\alpha/2} < Q(W)\} = \mathrm{Prob}\{q_{1-\alpha/2} > Q(W)\} = \alpha/2.$$

(d) Because $\mathrm{Prob}\{q_{1-\alpha/2} < Q(W) < q_{\alpha/2}\} = 1 - \alpha$, it follows that

$$\mathrm{Prob}\{W(q_{1-\alpha/2}) < (\hat{\Theta} - \Theta)/\hat{\sigma} < W(q_{\alpha/2})\} = 1 - \alpha,$$

and

$$\mathrm{Prob}\{\bar{x} - W(q_{\alpha/2})\hat{\sigma} < \mu < \bar{x} - W(q_{1-\alpha/2})\hat{\sigma}\} = 1 - \alpha,$$

where $W(q_{\alpha/2})$ and $W(q_{1-\alpha/2})$ are evaluated using equation (3.14). Hence

$$\bar{x} - W(q_{\alpha/2})\hat{\sigma} < \mu < \bar{x} - W(q_{1-\alpha/2})\hat{\sigma} \qquad (3.15)$$

is an approximately $100(1 - \alpha)\%$ confidence interval for μ.

One point to note is that $1 + \hat{v}\{Q - \hat{v}/(6n)\}$ in equation (3.14) may be negative. In that case the cube root must be interpreted as minus the cube root of the absolute value of the quantity because this is what gives the correct answer when raised to the third power.

3.8   PARAMETRIC CONFIDENCE INTERVALS

So far this chapter has been concerned with what can be called non-parametric bootstrapping techniques, in the sense that they are intended to be valid for samples from any distribution. However, parametric bootstrapping is also possible, where this involves fitting a particular model to the data and then using this model to produce bootstrap samples. It is in principle possible to construct exact confidence limits for a single unknown parameter in this setting (Buckland and Garthwaite, 1990; Garthwaite and Buckland, 1992).

3.9   A BETTER ESTIMATE OF BIAS

In Example 3.1 it was suggested that the bias in an estimator can be approximated by the difference between the mean of the estimates found from a large number of bootstrap samples and the known value for the 'population' being bootstrap resampled. It is possible to improve on this approximation by taking into account the characteristics of the particular bootstrap samples taken.

To see how this can be done, suppose that an initial sample of values $x_1, x_2, \ldots, x_n$ is available, from which B bootstrap samples are taken. There are then a total of nB values randomly selected from the initial sample, of which a certain proportion p_i are x_i. Assume that it is possible to calculate the value of the parameter of interest for the distribution that consists of the values x_1 with probability p_1, x_2 with probability $p_2, \ldots, x_n$ with probability p_n, and let this value be Θ^*. Then the better bias approximation is

$$\text{Bias}(\hat{\Theta}) \approx \bar{\Theta}_B - \Theta^*, \tag{3.16}$$

where $\bar{\Theta}_B$ is the mean of the bootstrap estimates from the B samples.

Efron and Tibshirani (1993, p. 342) give a theoretical justification for why this gives a better bias approximation than the difference between $\bar{\Theta}_B$ and the value of Θ for the original sample of values. One intuitive justification is that the approximation (3.16) compares the bootstrap mean of Θ with the value of this parameter in the population that is exactly represented by the bootstrap samples as a whole. Also, if it can be shown that if $\hat{\Theta} = \bar{x}$, so that it is just a mean value that is being estimated, then equation (3.16) always gives the correct bias of zero. This is not the case for the difference between $\bar{\Theta}_B$ and the original sample mean.

The calculation of Θ for a distribution that has the value x_i with probability p_i will usually be quite straightforward. For example, if the parameter of interest is a population mean value then

$$\Theta^* = \sum_{i=1}^{n} p_i x_i$$

in equation (3.16).

3.10   BOOTSTRAP TESTS OF SIGNIFICANCE

Bootstrap tests of significance have not been as well studied as bootstrap confidence intervals, although they represent an obvious application of the basic idea. Indeed, one way to carry out a bootstrap test of the hypothesis that the parameter Θ takes the particular value Θ_0 involves simply calculating a bootstrap confidence interval for Θ and seeing whether this includes Θ_0. This appears to indicate that the extension of the theory of bootstrapping to tests of significance is trivial. However, as noted by Fisher and Hall (1990) and Hall and Wilson (1991), there is an important difference between the two applications that needs to be recognized. This is that when carrying out a test of significance it is important to obtain accurate estimates of critical values of the test statistic even if the null hypothesis is not true for the population from which the sample being bootstrapped came from. Basically it is a question of deciding exactly how the null hypothesis being tested should influence the choice of the statistic being bootstrapped. Example 3.5, below, should clarify what this means.

Whatever test statistic S is used, a bootstrap test involves seeing whether the value of S for the available data is sufficiently extreme, in comparison with the bootstrap distribution of S, to warrant rejecting the null hypothesis. Generally the test can operate in a similar way to a randomization test. Thus suppose that large values of S provide evidence against the null hypothesis. Then the observed data provide a value S_1 of S, and bootstrap resampling of the data produces another $B - 1$ values $S_2, S_3, \ldots, S_B$. All B test statistics are from the same distribution if the null hypothesis is true. Hence S_1 is significantly large at the $100\alpha\%$ level if it is one of the largest $100\alpha\%$ of the B test statistics. Or, to put it a different way, the significance level for the data is $p = m/B$, where m is the number of the statistics S_1 to S_B that are greater than or equal to S_1.

This argument applies with any number of bootstrap samples, but generally a large number is better than a small number in order to reduce the effect of the random sampling from the bootstrap distribution. The key requirement for the test to be valid is that the bootstrap distribution really does mimic the distribution of S when the null hypothesis is true. This may or may not be true in practice.

Example 3.5: testing a sample mean   One of the examples that was discussed by Hall and Wilson (1991) concerned testing whether the mean of a random sample of 20 observations is significantly different from a specified value. The data are temperature readings in degrees Celsius, as follows: 431, 450, 431, 453, 481, 449, 441, 476, 460, 482, 472, 465, 421, 452, 451, 430, 458, 446, 466 and 476. The original source is Cox and Snell (1981). The sample mean and standard deviation are $\bar{x} = \sum x_i/20 = 454.6$ and $s = \sqrt{[\sum(x_i - \bar{x})^2/19]} = 18.04$, so that an estimate of the standard error of the mean is $18.04/\sqrt{20} = 4.02$. The hypothetical population mean is $\mu_0 = 440$. A two-sided test is required.

At first sight it might seem appropriate to use

$$D_{B1} = |\bar{x}_B - 440|$$

as a bootstrap test statistic, where $\bar{x}_B$ is the mean of a bootstrap sample of size 20 taken from the original data. The distribution of this statistic should mimic the distribution of $|\bar{x} - 440|$ when the null hypothesis is true. However, a moment's reflection will indicate that this is not sensible, on the grounds that any difference between the sample mean $\bar{x}$ and 440 will be built in to the bootstrap distribution: if $|\bar{x} - 440|$ is large then all the values of D_{B1} will tend to be large as well. In fact, the bootstrap distribution will in this case be reflecting the distribution of the statistic when the null hypothesis is not true rather than the distribution when the null hypothesis is true.

This consideration leads to the idea of adjusting the data in order to make the null hypothesis true before bootstrapping is carried out. The simplest way to do this in the present case involves changing the ith data value from x_i to $x'_i = x_i - \bar{x} + 440$, and finding the bootstrap distribution of

$$D_{B2} = |\bar{x}'_B - 440| = |\bar{x}_B - \bar{x}|,$$

where $\bar{x}'_B$ is the mean of the adjusted values for a bootstrap sample. As can be seen from the last equation, this amounts to using the distribution of absolute differences between bootstrap sample means and the original sample mean to mimic the distribution of the difference between the original sample mean and the population mean.

Although D_{B2} is a sensible statistic it does suffer from being heavily dependent on the variation observed in the original sample, although this is not relevant to the null hypothesis of interest. This suggests that it may be better to use a statistic that is less influenced by this variation, such as the t-statistic:

$$T_B = D_{B2}/(s_B/\sqrt{20}) = |\bar{x}_B - \bar{x}|/(s_B/\sqrt{20}),$$

where s_B is the sample standard deviation for a bootstrap sample. This now amounts to attempting to reconstruct the distribution of the usual t-statistic for samples from the population of interest using the distribution of this same statistic as obtained from the population that consists of the original sample values with equal probability. In practice, the factor $\sqrt{20}$ can be omitted from the test statistic because this is the same for all bootstrap samples.

It seems likely that using T_B may well work better than using D_{B2} because it can be seen that if the null hypothesis is true but the original sample happens by chance to display either a very high or a very low variance then the distribution of D_{B2} will not reflect the distribution of $|\bar{x} - 440|$ very well. However, the distribution of T_B may still be a close approximation to the distribution of $|\bar{x} - 440|/(s/\sqrt{20})$.

Bootstrap tests were carried out using all three test statistics to see how results compared, using the same 999 bootstrap samples in each case. The observed values of the test statistics are $D_1 = 14.5$, $D_2 = 14.5$ and $T = 3.62$ and the significance levels were found to be 0.51, 0.001 and 0.005, respectively. As expected, the bootstrap distribution of D_1 was more or less centred on 14.5. Hence, rather than looking like an unusual value if the null hypothesis is true (which it clearly is), D_1 looks as typical as it could be. Both of the other statistics are highly significant, but D_2 is more so. From what is known about bootstrapping in general it can be argued that the test based on t is the one that can be trusted most.

3.11   BALANCED BOOTSTRAP SAMPLING

Hall and Wilson (1991) used balanced bootstrap resampling for the test discussed in the last example. This is one of several methods that have been proposed for increasing the efficiency of bootstrap sampling that has the advantage of being relatively simple to apply (Davison *et al.*, 1986; Gleason, 1988).

Balanced bootstrap sampling operates as follows. Suppose that B bootstrap samples are to be taken from a sample of size n. To begin with, B copies of the integers 1 to n are made up to form a list of length nB. Then the items in this list are randomly permuted, and the first bootstrap sample uses the first n observation numbers in the permuted list, the second bootstrap sample uses the observation numbers in positions $n + 1$ to $2n$ in the permuted list, and so on. In this way it is ensured that each observation in the original sample is used the same number of times. This can considerably reduce the amount of resampling needed in order to estimate the moments of a distribution with a specified accuracy, but is not so effective with the estimation of percentage points (Hinkley, 1988).

Some other methods for improving the efficiency of bootstrap resampling are discussed in Chapter 5.

3.12 APPLICATIONS OF BOOTSTRAPPING IN BIOLOGY

Bootstrapping has been widely used to estimate standard errors, construct confidence intervals and carry out tests of significance. Some applications in different biological areas are summarized below.

3.12.1 Single species ecology

Applications of bootstrapping with single species include the estimation of the size of an animal population from mark–recapture data (Buckland, 1980, 1984; Chao, 1984; Huggins, 1989; Buckland and Garthwaite, 1991; Garthwaite and Buckland, 1992; Anderson, 1995); the analysis of line–transect data (Buckland, 1982; Quang, 1990; Quang and Lanctot, 1991; Buckland et al., 1993); the comparison of principal components from real and randomly constructed sets of ecological data (Stauffer et al., 1985); the estimation of the production and biomass for marine animals (Haslett and Wear, 1985; Brey, 1990); the estimation of population growth rates (Meyer et al., 1986; Alvarez-Buylla and Slatkin, 1994; Eberhardt, 1995); deciding on the sample sizes required for study designs (Bros and Cowell, 1987; Hamilton and Collings, 1991; Manly, 1992; Roff, 1995); the estimation of the Gini coefficient as a measure of the inequality in a distribution of plant sizes (Dixon et al., 1987; Dixon, 1993); spatial data analysis (Solow, 1989a); the estimation of models for dose–response and other types of proportion data (Muller and Wang, 1990; Carr and Portier, 1993; Frangos and Schucany, 1995); testing the null hypothesis that a fossil collection represents a single hominid species (Cope and Lacy, 1992; Grine et al., 1993; Kramer et al., 1995); the analysis of fisheries and bacteria survey data using generalized additive models and other methods (Swartzman et al., 1992; Solow and Gaines, 1993; Perry and Smith, 1994); assessing the accuracy of models for the distribution of wildlife populations using information of explanatory variables recorded at different levels of resolution in geographical information systems (Buckland and Elston, 1993); testing for density-dependence in time series of the abundance of animals (Kemp and Dennis, 1993; Wolda and Dennis, 1993; Dennis and Taper, 1994); determining the accuracy of waterbird population indices (Howe et al., 1989; Underhill and Prŷs-Jones, 1994); and seeing whether there is evidence that time series of the abundance of Holarctic microtine rodents are chaotic (Falck et al., 1995).

3.12.2 Genetics, evolution and natural selection

Applications of bootstrapping in the areas of genetics, evolution and natural selection include the estimation of phylogenies (Ward, 1985;

Bermingham and Avise, 1986; Felsenstein, 1986, 1992; Gilinsky and Bambach, 1986; Sanderson, 1989, 1995; Hillis and Bull, 1993; Lecointre *et al.*, 1993; Rodrigo *et al.*, 1993; Dopazo, 1994; Harshman, 1994; Hasegawa and Kishino, 1994; Navidi, 1995; Sitnikova *et al.*, 1995; Lake, 1995; Van Dongen, 1995) and the estimation of various genetic parameters (Schluter and Smith, 1986; Schluter, 1988; Abrahamson *et al.*, 1989; Aastveit, 1990; Lo, 1991; Skibinski *et al.*, 1993; Smith, 1993; Roff, 1995; Long and Singh, 1995; Van Dongen and Backeljau, 1995).

3.12.3 Community ecology

Applications of bootstrapping in community ecology include the estimation of species richness (Smith and van Belle, 1984; Palmer, 1990; Mingoti and Meeden, 1992); the estimation of niche overlap (Mueller and Altenberg, 1985); assessing the statistical significance of clusters obtained from species abundance data and comparing two dendrograms obtained on the same entities using different variables (Nemec and Brinkhurst, 1988a,b); studying the structure of plant communities (Zobel and Zobel, 1988; Zobel *et al.*, 1993); the assessment of the stability of vegetation ordination methods (Knox, 1989; Knox and Peet, 1989); the reconstruction of series of past values of limnological variables for lake water (Birks *et al.*, 1990; Cumming *et al.*, 1992; Dixit *et al.*, 1993); the comparison of niche overlap indices for different groups of animals (Manly, 1990b); assessing the properties of an index of biological integrity used to evaluate water resources (Fore *et al.*, 1994); assessing the sampling variation of principal components obtained from data on moth abundance counts from light traps (Milan and Whittaker, 1995); assessing the accuracy of summary statistics for trends in population numbers for groups of bird species (Link and Sauer, 1995); and assessing the evidence for multimodality of animal size distributions using kernel estimated distribution functions (Manly, 1996b).

3.13 FURTHER READING

It is possible to approximate likelihood functions using Owen's (1988, 1990) empirical likelihood method or Davison *et al.*'s (1992) bootstrap likelihood method. These and other approaches are reviewed by Efron and Tibshirani (1993, Chapter 24). They have not yet received much use, but hold considerable promise. See also Hall and Owen's (1993) development of empirical likelihood for finding confidence bands for estimates of density functions.

Donegani and Unternährer (1991) and Donegani (1992) discuss the use of bootstrapping to find the variance of adaptive estimators. In this situation there are two or more potential methods available to estimate some parameter of interest from a random sample, and the method used is chosen after the sample values are known. For example, a robust estimator of a mean might be chosen for use if the sample appears to contain outliers. Donegani and Unternhrer (1991) and Donegani (1992) report good results for an example of this type and with two-factor analysis of variance. Related problems are the choice of the trimming proportion in a trimmed mean, the choice of the optimal weight for a weighted combination of two estimators and smoothing parameter selection with curve fitting (Léger *et al.*, 1992; Efron and Tibshirani, 1993, Chapter 18).

Sitter (1992), Chen and Sitter (1993) and Booth *et al.* (1994) discuss various bootstrap methods for estimating the variance of estimators for complex sample surveys of finite stratified populations.

Bello (1994) proposes a method for determining the standard error of a statistic when there are complications due to missing data. In order that bootstrap sets of data are a good representation of the original data, he suggests resampling the complete observations and the incomplete observations separately.

Other recent applications of bootstrapping that are worth mentioning are with logistic regression (Swanepoel and Frangos, 1994) and for allowing for errors in predictor variables with generalized linear models (Haukka, 1995).

3.14 CHAPTER SUMMARY

- The essence of bootstrapping is that in the absence of any other information about a population, the values in a random sample are the best guide to the distribution, and resampling the sample is the best guide to what can be expected from resampling the population.
- Much of the research on bootstrapping has been aimed at developing reliable methods for constructing confidence limits for population parameters, but recently bootstrap tests of significance have been attracting more interest.
- The standard bootstrap confidence limits are calculated as the estimate $\pm z_{\alpha/2}$(bootstrap standard deviation), where the standard deviation is estimated by resampling the original data values. A bias correction can be made if necessary.
- Efron's percentile confidence limits are the two values that cut off fixed percentages in the tails of the bootstrap distribution of an estimate. For example, the bootstrap 95% limits are the two values that include 95% of the bootstrap distribution of an estimate between them. Hall's percentile method gives different limits for estimates with non-symmetric distributions. These limits are based on approximating the distribution of sampling errors by bootstrapping.
- Two methods are available that may improve Efron's percentile limits.

The bias-corrected percentile method adjusts for the median of the distribution of an estimate not being equal to the mean. The accelerated bias-corrected method also attempts to adjust for the standard deviation of the distribution varying with the mean of the distribution.

- Other methods for constructing confidence intervals include bootstrap-t, which attempts to approximate the distribution of a statistic of the form $T = $ (Estimate − Mean)/(Standard Error) using equivalent values constructed by resampling the original data. Transformations have been proposed to improve bootstrap-t confidence intervals by making the distribution of T more like a normal distribution.

- A simple approximation for the bias of an estimator is the difference between the mean of this estimator from bootstrapping and the estimate from the sample that is bootstrapped. A better bias approximation is the difference between the bootstrap mean and the value of the parameter of interest for a population consisting of the original sample values with relative frequencies equal to those that are found for the bootstrap samples as a whole.

- With a bootstrap test of significance a statistic calculated from the original sample data is compared with a distribution found by bootstrap resampling. It is important to ensure that the bootstrap distribution reflects the situation where the null hypothesis is true, which may require the original data to be adjusted before being resampled.

- Balanced bootstrap resampling is an improvement on simple bootstrap resampling where it is ensured that each of the original sample values is reused the same number of times.

- Numerous applications of bootstrapping in biology are briefly described.

4

Monte Carlo methods

4.1 MONTE CARLO TESTS

With a Monte Carlo test the significance of an observed test statistic is assessed by comparing it with a sample of test statistics obtained by generating random samples using some assumed model. If the assumed model implies that all data orderings are equally likely then this amounts to a randomization test with random sampling of the randomization distribution. Hence randomization tests can be thought of as special cases within a broader category of Monte Carlo tests. Bootstrapping can also be thought of as the Monte Carlo method applied in a particular manner. At least, this is the point of view adopted in this book.

One example of a Monte Carlo test that cannot obviously be thought of as a randomization or a bootstrap test is discussed by Besag and Diggle (1977). This is a test to see whether a set of points appear to be distributed randomly in space within a given region. Here one reasonable test statistic is the sum of the distances from each point to its nearest-neighbour, with high values indicating a tendency for points to be regularly spaced and low values indicating a certain amount of clustering. The significance of an observed statistic can then be evaluated by comparing it with the distribution of values obtained when the points are placed in random positions over the study region.

The generation of one set of data for a Monte Carlo test may need considerably more calculations than are involved in randomly reordering the observed values. For this reason it is not uncommon to find that the number of random test statistics generated for a Monte Carlo test is rather small. In an extreme case, a test at the 5% level of significance might only involve comparing the observed test statistic with 19 other values. This low number of randomly generated values can be justified on the grounds that the argument used in section 1.3.1 for randomization tests with random data permutations being exact applies equally well with Monte Carlo tests in general (Barnard, 1963). Nevertheless, a large number of random values of the test statistic is always desirable to avoid inferences being strongly dependent on the properties of a small sample.

Monte Carlo methods can also be used to calculate confidence limits for

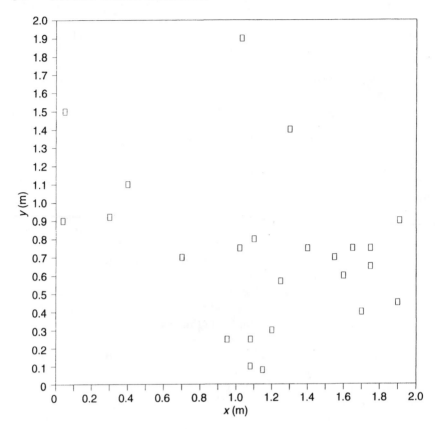

Figure 4.1 The positions of 24 plants in a square with 2 m sides.

population parameters. Essentially the idea is to use computer-generated data to determine the amount of variation to be expected in sample statistics using, for example, one of the percentile methods that have been discussed for bootstrapping in Chapter 3.

Example 4.1: testing for spatial randomness   The principle behind Monte Carlo testing is quite straightforward. Nevertheless, an example may be helpful at this point. Suppose therefore that Figure 4.1 shows the positions of 24 plants in a square plot of land with sides of length 2 m. The question to be considered is whether the plants are in random positions. To be more specific, the null model to be tested states that each of the plants was equally likely to occur anywhere within the square, independent of all the other plants.

There are many different test statistics that could be used to summarize

Table 4.1 Nearest-neighbour statistics for 24 plants

	g_1	g_2	g_3	g_4	g_5	g_6	g_7	g_8	g_9	g_{10}
Observed	0.217	0.293	0.353	0.419	0.500	0.559	0.606	0.646	0.698	0.739
Mean	0.227	0.354	0.455	0.544	0.625	0.708	0.780	0.846	0.909	0.968
Significance	0.729	0.080	0.009	0.003	0.009	0.005	0.008	0.004	0.004	0.002

the data, and there is no reason why more than one should not be used at the same time. These matters are discussed more fully in Chapter 10, but here what will be considered is a set of nearest-neighbour statistics. The first, g_1, is the mean of the distances between plants and their nearest-neighbours. As there are 24 plants there will be 24 distances to be averaged. The second statistic, g_2, is the mean of the distances between plants and their second nearest-neighbours, which is again an average of 24 distances, one for each plant. More generally, the ith statistic to be considered is the mean distance between plants and their ith nearest-neighbours, for i from 1 to 10.

The statistics g_1 to g_{10} are shown in the first row of Table 4.1. To assess their significance, 999 random sets of data were generated. For each of these sets, 24 points were allocated to random positions within 2 m by 2 m squares and g statistics calculated. The significance level for g_i is the estimated probability of getting a value as extreme as the one observed, which has been interpreted as the probability of a value as far or further from the mean of the simulated distribution. Both the mean and the probability can be estimated from the generated data, including the observed g_i as a 1000th value. The second and third rows in Table 4.1 show these estimates.

It can be seen that the distances between plants and their third- and higher order nearest-neighbours are smaller than can reasonably be expected from randomness. For example, the observed value of g_4 is 0.419, whereas the randomization mean is 0.544. The difference is 0.125. The probability of a value as far (or further) than 0.419 is from 0.544 is estimated as the proportion of the 1000 values that are 0.419 or less, or 0.669 or more. There are three values this extreme, so the significance of the result is estimated as 0.003. The conclusion must be that there is clear evidence that the plants shown in Figure 4.1 are not randomly placed within the square area.

At first sight the definition of the significance level of a test statistic as the probability of a value as far or further from the mean as that observed seems reasonable. However, because the test statistics may not have symmetric distributions about the mean this definition is not altogether satisfactory. It is conceivable, for instance, that when a statistic is observed at a certain distance above the mean it is not even possible to obtain a value that far below the mean. This has led to the suggestion that the significance level should be calculated as the minimum of $2p_L$ and $2p_U$, where P_L is the probability of a value as small or smaller than that observed and p_U is the probability of a value as large or larger than that observed. Other definitions are also possible, and some authors suggest just reporting the minimum of p_L and p_U with the direction in which the test statistic is extreme because of the difficulty of deciding which definition is 'correct' (George and Mudholkar, 1990).

4.2 GENERALIZED MONTE CARLO TESTS

Besag and Clifford (1989) have defined a generalized Monte Carlo test as one where (i) an observed set of data is one of many sets that could have occurred, (ii) all possible sets of data can be generated by a series of one-step changes to the data, (iii) the null hypothesis of interest states that all the possible sets of data were equally likely to occur, and (iv) each possible set of data is summarized by a test statistic S. Given these conditions, Besag and Clifford proposed two algorithms for tests of the null hypothesis based on the idea of comparing an observed set of data with alternative sets of data generated by making stepwise changes to this observed set.

With the serial method, the observed set of data is made to be the kth set in a series of N sets of data by choosing k as a random number between 1 and N. The remaining sets of data are then determined by moving backwards $k - 1$ steps from the observed set of data by making this number of random changes to the data, and then moving forward $N - k$ steps from the observed data, again by making this number of random changes to the data. The nature of the random changes to the data depends on the particular circumstances of the situation being considered. Often it will just amount to randomly switching two data values.

For each of the sets of data in the series of length N, a test statistic of interest is calculated. Besag and Clifford (1989) have then shown that if the null hypothesis is true (the observed set of data was equally likely to be any one of the possible sets of data that could occur) then this procedure gives a probability of m/N that the observed test statistic is one of the m largest statistics when the N statistics for all the sets of data are placed in order of magnitude, assuming that no ties occur. Therefore, a one-sided test at the $100\alpha\%$ level of significance consists of seeing whether the observed test statistic is among the largest $100\alpha\%$ of the N values. If tied values of the test statistic occur then a conservative test is obtained by requiring the observed statistic to be larger than $100(1 - \alpha)\%$ of all N statistics. Two-sided tests can be handled by testing the absolute value of the statistic of interest.

With the parallel method, r backward steps are made from the observed set of data to produce a base set of data. Then r forward steps are made to produce a randomized set of data with which to compare the observed set. The forward steps from the base configuration are repeated $M - 1$ times, to produce $M - 1$ randomized sets of data. Test statistics are then calculated for the original set of data and the $M - 1$ randomized sets, and the observed statistic is significantly large at the $100\alpha\%$ level if it exceeds $100(1 - \alpha)\%$ of the set of all M statistics. This test is valid, because if the null hypothesis is true then the observed set of data is equally likely to be any one of the possible sets of data that can be produced from the base set of data by making r random changes.

The serial and parallel algorithms have been described in terms of 'forwards' and 'backwards' steps, but there is no need for these to be different in any way. In practice the steps will often just consist of random interchanges of data that are in no way directional, but the algorithms will still be valid.

Although it may not appear to be the case at first sight, it turns out that the framework for a generalized Monte Carlo test is able to include many other types of computer-intensive test as special cases. One of these special cases is the usual randomization test to compare two samples, as described in the next example, and, indeed, any randomization test can be carried out using Besag and Clifford's algorithms providing that the randomization can be thought of in terms of a series of single random switches of data values. However, the real value of the concept of a generalized Monte Carlo test is that it can be used in situations where alternative methods are either not available or are not easy to use. One such situation is testing for randomness in the co-occurrences of species on islands, which is discussed at some length in section 14.3.

Example 4.2: mean mandible lengths for male and female golden jackals Example 1.1 was concerned with the comparison of mean mandible lengths for male and female golden jackals (*Canis Aureus*) for 10 of each sex in the collection in the British Museum (Natural History). In that example a randomization test for a significant difference was carried out using the usual method of comparing the observed mean difference with the distribution of the mean difference found by randomly allocating the data values to two samples of size 10. Now essentially the same test will be carried out using Besag and Clifford's serial algorithm.

It must be stressed from the start that using the serial algorithm in this situation is not meant to suggest that this algorithm is recommended for general use with this test. Rather, the purpose of the present example is just to demonstrate how the algorithm works in a relatively simple setting. In fact, some simulation results presented by Manly (1993) show

Table 4.2 The serial test to see whether the sample mean mandible length is significantly larger for male golden jackals than it is for female golden jackals. The actual set of data is put into random position 78 in a series of 100 sets. Random swaps of sample 1 values with sample 2 values are made to produce the other 99 sets of data from the original set, with observation R1 in the first sample swapped with observation R2 in the second sample. Swapped values are shown in italic type. 'Mean diff' is the sample mean difference and Ind is 1 if the absolute sample mean difference is greater than or equal to the observed mean difference of 4.8 mm. The average of Ind is 0.01, showing that none of the modified sets of data gave a mean difference as large as 4.8. Parts of the results are omitted to save space

Data	R1	R2	Sample 1 (males)										Sample 2 (females)										Mean diff	Ind
			1	2	3	4	5	6	7	8	9	10	1	2	3	4	5	6	7	8	9	10		
Backwards stepwise changes																								
1	1	6	*110*	106	111	114	110	114	111	113	117	108	111	116	120	110	107	*107*	111	107	105	112	0.8	0
2	1	9	*107*	106	111	114	110	114	111	113	117	108	111	116	120	110	107	110	111	107	*105*	112	0.2	0
.																								
76	10	8	120	107	110	116	114	111	113	117	110	*106*	114	111	107	108	110	105	107	*112*	111	111	2.8	0
77	9	1	120	107	110	116	114	111	113	117	*110*	112	*114*	111	107	108	110	105	107	106	111	111	4.0	0
Original data																								
78			120	107	110	116	114	111	113	117	110	112	110	111	107	108	110	105	107	106	111	111	4.8	1
Forward stepwise changes																								
79	5	2	120	107	110	116	*111*	111	113	117	114	112	110	*114*	107	108	110	105	107	106	111	111	4.2	0
80	10	8	120	107	110	116	111	111	113	117	114	*106*	110	114	107	108	110	105	107	*112*	111	111	3.0	0
.																								
99	5	7	111	112	107	111	*111*	110	120	114	107	106	110	114	110	108	117	105	*111*	116	113	107	−0.2	0
100	2	4	111	*108*	107	111	111	110	120	114	107	106	110	114	110	*112*	117	105	111	116	113	107	−1.0	0

Mean = 0.01

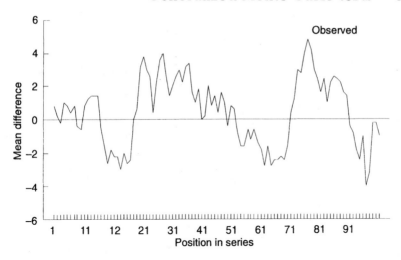

Figure 4.2 Sample mean differences obtained for a series of 100 sets of data obtained by making random interchanges of values in two samples. The original data, at position 78, have a mean difference of 4.8 mm, which is the largest value for the series.

that the usual method of randomization and the serial method are about as efficient for a test of this type, while the parallel method is slightly less efficient.

The mandible length data to be considered are as follows, in mm:

Males: 120, 107, 110, 116, 114, 111, 113, 117, 114, 112

Females: 110, 111, 107, 108, 110, 105, 107, 106, 111, 111

The mean for males is $\bar{x}_1 = 113.4$ mm and the mean for females is $\bar{x}_2 = 108.6$ mm. The observed mean difference is therefore $D = 4.8$ mm. It will be assumed that a one-sided test is required at the 5% level of significance to determine whether this mean difference is significantly large, on the assumption that if a size difference exists then the males will be larger than the females.

Table 4.2 shows the results from using the serial algorithm with a series length of $N = 100$. A random number between 1 and 100 was found to be 78. This was therefore the position in the series that was chosen for the observed data, with the other 99 sets generated by randomly switching observations between the two samples. None of the randomly modified sets of data gave a sample mean difference as large as 4.8, and therefore the observed difference can be declared to be significantly large at the 1% level ($p = 0.01$).

Figure 4.2 gives a plot of the sample mean differences. This is of some interest because it indicates that generally random switching of values between the two samples leads to mean values within a range from about -4 to $+4$. A theoretical analysis shows that with two-sample sizes of 10 it requires about 21 switches to randomize the sample allocation of values effectively, i.e. to do a complete random shuffle (Manly, 1993).

With important sets of data the length of the series would be made more than 100. It also seems worthwhile to repeat the test several times with different random positions for the real data in order to avoid the position having any important influence on the test result. These and other matters are discussed more fully with the example of a generalized Monte Carlo test in section 14.3.

4.3   IMPLICIT STATISTICAL MODELS

An implicit statistical model is one that is defined in terms of a stochastic generating mechanism, rather than in terms of equations for the distributions of random variables. Such a model can in principle be simulated to generate data for specific values of one or several parameters. This then allows the possibility of using simulation to approximate the likelihood function, or some other function that measures the extent to which a given set of data agrees with data simulated with different parameter values.

Based on this concept, Diggle and Gratton (1984) have discussed how maximum likelihood estimation can be used with a model for the response of the flour beetle (*Tribolium castaneum*) to poison. They also discuss an example where a more *ad hoc* method of estimation can be used for parameters that are related to the spatial distribution of misplaced amacrine cells in the retina of a rabbit. Minta and Mangel (1989) used this type of approach with the maximum likelihood estimation of population size from mark–recapture data, while McPeek and Speed (1995) used it for modelling interference in genetic recombination.

The use of implicit statistical models is similar in concept to the methods for approximating likelihood functions using bootstrapping called empirical likelihood and bootstrap likelihood, as referenced in section 3.13. See also Gelman's (1995) discussion of how methods of moments estimation can be applied with data generated from a simulation model.

4.4   APPLICATIONS OF MONTE CARLO METHODS IN BIOLOGY

Some applications of Monte Carlo methods in biology that do not fall obviously within the categories of randomization and bootstrapping, and which have not already been mentioned, are listed below, separately for single-species ecology, genetics and evolution, and community ecology.

4.4.1 Single-species ecology

With single-species ecology, applications of Monte Carlo methods have included testing for spatial randomness and the independence of different types of events in space (Harkness and Isham, 1983; Pearson, 1991, 1992; Larkin 1992; Penttinen et al., 1992); testing for whether habitat utilization by organisms is random (Kincaid and Bryant, 1983); testing for genetic changes across language boundaries in Europe by comparing observed changes with those generated by randomly placing boundaries on the map of Europe (Sokal et al., 1988); testing for a relationship between spatial and temporal patterns in cases of leukaemia by comparing observed cases with cases generated from an artificial control population (Raybould, 1990); testing the broken-stick model for the positions of breaks in fossil crinoid stalks (Baumiller and Ausich, 1992); and for making an allowance for measurement errors in examining the relationship between blood pressure and coronary heart disease (Cook and Stefanski, 1994).

The method described by Cook and Stefanski (1994) seems to be potentially useful in many situations where there is a need to allow for the effects of measurement errors. Essentially what is done is to simulate data similar to that observed but with extra measurement errors added to observations. The effect of different amounts of measurement error on estimators can then be directly observed, and extrapolation indicates what the estimates would have been in the absence of any measurement error at all.

4.4.2 Genetics and evolution

Applications in genetics and evolution have included testing for Hardy–Weinberg equilibrium with multiple alleles (Guo and Thompson, 1992) and searching for patterns in phylogenetic trees (Kirkpatrick and Slatkin, 1993).

4.4.3 Community ecology

In the area of community ecology, applications of Monte Carlo methods have included testing whether the use of resources by organisms along an axis in time or space shows evidence for staggering or clustering (Poole and Rathcke, 1979; Cole, 1981; Tonkyn and Cole, 1986; Fleming and Partridge, 1984; Tokeshi, 1986; Pleasants, 1990, 1994; Arita, 1993; Williams, 1995); testing for non-random communities of species selected from a source pool (Strong et al., 1979; Case, 1983; Case and Sidell, 1983, Fox; 1987, 1989; Patterson and Atmar, 1986; Patterson, 1990; Patterson and Brown, 1991; Fox and Brown, 1993, 1995; Andrén, 1994; Wilson,

1995); the search for linear combination of species abundances that have maximal correlation with an environmental gradient, with significance tested by comparison with the correlations obtained from randomly generated data (Pielou, 1984); testing for interaction between species in space (Andersen, 1992); and testing for competitive hierarchies in plant competition matrices (Shipley, 1993).

4.5　CHAPTER SUMMARY

- A Monte Carlo test is defined as one in which an observed test statistic calculated from a set of data is assessed by comparing it with a distribution obtained by generating alternative sets of data from some assumed model. With this definition, randomization tests and bootstrap tests are special cases of Monte Carlo tests.
- Generalized Monte Carlo tests are those in which the alternative sets of data are obtained by some form of stepwise changes to the initial observed set. Two algorithms for testing are described for this situation. With the serial algorithm one sequence of test statistics is generated by making forward and backwards changes from the initial set of data. With the parallel algorithm r backwards steps are made to produce a base set of data. Repetitions of r forwards steps are then made to produce alternative sets of data for comparison with the initial one.
- An implicit statistical model is one that is only defined in terms of a simulation model for producing sets of data. Methods for approximate maximum likelihood estimation exist for this situation.
- A number of applications of Monte Carlo methods in biology are briefly described.

5

Some general considerations

5.1 QUESTIONS ABOUT COMPUTER-INTENSIVE METHODS

The purpose of the present chapter is to discuss a number of questions related to randomization, bootstrap and Monte Carlo methods in order to set the stage for the chapters that follow. Specifically:

1. How efficient are these methods in comparison with alternative methods, assuming that alternatives exist?
2. If an observed test statistic is compared with a distribution generated by randomizing data (for a randomization test), bootstrap resampling (for a bootstrap test), or simulating data (for a Monte Carlo test), how many replications are required to ensure that the estimated significance level for the data is close to what would be found from determining the sampling distribution exactly?
3. Likewise, what number of replications are needed for determining confidence limits by randomization or bootstrapping?
4. If a randomization distribution is determined by a complete enumeration, how can this be done systematically?
5. Still considering the complete enumeration of a randomization distribution, are there ways to reduce the number of calculations by, for example, recognizing symmetries in the patterns of arrangements that are possible?
6. What are good computer algorithms for generating pseudo-random numbers and producing random permutations for data rearrangement?

The answers to questions 1, 4 and 5 depend upon the particular situation being considered. As such, it is only possible to make a few general statements. The answers to questions 2, 3 and 6 apply to all tests.

5.2　POWER

The power of a statistical test is the probability of obtaining a significant result when the null hypothesis being tested is not true. This probability depends upon the amount of difference that exists between the true situation and the null hypothesis, and the level of significance that is being used. An efficient test is a test that has high power in comparison with a test that is known to be as powerful as possible. An efficiency of 1 is the best obtainable.

As was mentioned in section 1.3.3, it is a general rule that a randomization test and a classical parametric test will give approximately the same significance level when the data appear to justify the assumptions of the parametric test. Therefore it is reasonable to expect that the power of randomization and classical tests should be about the same when these assumptions are true. This was confirmed by Hoeffding (1952), who found that in many cases the asymptotic (large sample) efficiency of randomization tests is 1 when these tests are applied with random samples from populations, and Robinson (1973) who studied the particular case of the randomized block design. More recently, Romano (1989) has shown that under fairly general conditions randomization and bootstrap tests become equivalent for large sample sizes in situations where either type of test can be applied.

When data are from non-standard distributions, there is some evidence to suggest that randomization tests have more power than classical tests. Edgington (1995, p. 88) discusses some cases where randomization tests consistently give more significant results than t-tests with highly skewed data. Presumably the t-test is losing power in these examples because the assumptions behind the test are not valid. Also, Kempthorne and Doerfler (1969) concluded from a study involving sampling eight distributions (normal, uniform, triangular etc.) that Fisher's randomization test, which was described in section 1.3.4 is generally superior to the Wilcoxon signed ranks test, the sign test and the F test for testing for a treatment effect with a paired comparison design.

It is difficult to say anything general about the power of Monte Carlo tests, since each one of these tends to be a special case.

5.3　NUMBER OF RANDOM SETS OF DATA NEEDED FOR A TEST

A test that involves sampling a randomization distribution is exact in the sense that using a $100\alpha\%$ level of significance has a probability of α or less of giving a significant result when the null hypothesis is true (section 1.3.1). From this point of view, a large number of randomizations is not necessary.

However, generally it will be considered important that the significance level estimated from randomizations is close to the level that would be obtained from considering all possible data rearrangements. This requires that a certain minimum number of randomizations should be made. It is the determination of this number that will now be considered.

One approach to this problem was proposed by Marriott (1979). He was discussing Monte Carlo tests, but the principle applies equally well with randomization tests. He argued as follows. Assume that the observed data give a test statistic u_0 and a one-sided test is required to see if this is significantly large. Another N values of u are generated by randomization, and the null hypothesis (u_0 is a random value from this distribution) is rejected if u_0 is one of the largest $m = \alpha(N + 1)$ when the $N + 1$ values are ordered from smallest to largest.

The probability of rejecting the null hypothesis in this situation is given by

$$P = \sum_{r=0}^{m-1} {}^N C_r p^r (1 - p)^{N-r}, \tag{5.1}$$

where p is the probability that u_0 is less than a randomly chosen value from the randomization distribution. The right-hand side of equation (5.1) is the sum of the probability of being less than none ($r = 0$) of the random values, plus the probability of being less than exactly one ($r = 1$) of them, and so on, up to the probability of being less than exactly $m - 1$ of them, each of these probabilities being given by the binomial distribution. For large values of N the normal approximation to the binomial can be used to approximate P.

If p is less than or equal to α then u_0 is significantly high at the $100\alpha\%$ level when compared with the full randomization distribution. It is therefore desirable that the test using N randomizations is also significant. In other words, P should be close to 1. On the other hand, if p is greater than α then u_0 is not significant at the $100\alpha\%$ level when compared with the full distribution, and P should be close to 0. One way of assessing the use of N randomizations therefore involves calculating P for different values of N and p, and seeing what the result is.

For example, suppose that α is taken to be 0.05, so that a test at the 5% level of significance is used. Marriott's Table 1, supplemented by some additional values, then gives the P values shown in part (a) of Table 5.1. It seems that 1000 randomizations are almost certain to give the same result as the full distribution except in rather borderline cases with p very close to 0.05. The results of similar calculations, but for significance at the 1% level, are given in part (b) of the table. These are again largely based on Marriott's Table 1. Here 5000 randomizations are almost certain to give

Table 5.1 Probabilities (P) of a significant result for tests as they depend on the number of randomizations (N) and the significance level (p) that would be given by the full randomization distribution

(a) Test at the 5% level of significance ($\alpha = 0.05$)

m^a	$N+1$	p 0.1	0.075	0.05	0.025	0.01
1	20	0.135	0.227	0.377	0.618	0.826
2	40	0.088	0.199	0.413	0.745	0.942
5	100	0.025	0.128	0.445	0.897	0.997
10	200	0.004	0.065	0.461	0.971	1.000
50	1000	0.000	0.001	0.474	1.000	1.000

(b) Test at the 1% level of significance ($\alpha = 0.01$)

m^a	$N+1$	p 0.02	0.015	0.01	0.005	0.001
1	100	0.135	0.224	0.370	0.609	0.906
2	200	0.091	0.199	0.407	0.738	0.983
5	500	0.028	0.131	0.441	0.892	1.000
10	1000	0.005	0.069	0.459	0.969	1.000
50	1000	0.000	0.001	0.472	1.000	1.000

$^a m = \alpha(N+1)$

the same significance as the full distribution, except in borderline cases.

Equation (5.1) applies for a two-sided test with p defined as the significance level of the data when it is compared with the full randomization distribution. Hence Table 5.1 applies equally well to this case.

An alternative approach to the problem of fixing the number of randomizations involves calculating the limits within which the significance value estimated by sampling the randomization distribution will lie 99% of the time, for a given significance level p from the full distribution (Edgington, 1995, p. 50). The estimated significance level will be approximately normally distributed with mean p and variance $p(1-p)/N$ when the number of randomizations, N, is large. Hence 99% of estimated significance levels will be within the limits $p \pm 2.58\sqrt{\{p(1-p)/N\}}$. These limits are shown in Table 5.2 for a range of values on N for $p = 0.05$ and $p = 0.01$.

Table 5.2 Sampling limits within which estimated significance levels will fall 99% of the time when the significance level of the observed data compared with the full randomization distribution (p) is 0.05 and 0.01

	$p = 0.05$			$p = 0.01$	
N	Lower limit	Upper limit	N	Lower limit	Upper limit
200	0.010	0.090	600	0.000	0.020
500	0.025	0.075	1000	0.002	0.018
1000	0.032	0.068	5000	0.006	0.014
5000	0.042	0.058	10000	0.007	0.013

From this table it can be seen that if the observed result is just significant at the 5% level compared with the full distribution then 1000 randomizations will almost certainly give a significant or near-significant result. On the other hand, when the result is just significant at the 1% level 5000 might be considered a realistic number of randomizations. These numbers are consistent with what was found with Marriott's approach. It seems therefore that 1000 randomizations is a reasonable minimum for a test at the 5% level of significance, while 5000 is a reasonable minimum for a test at the 1% level. See also Edwards (1985) and Jöckel (1986) for further discussions concerning sample sizes and power.

Lock (1986) and Besag and Clifford (1991) have pointed out that, if a test involving the sampling of the randomization distribution is carried out to determine whether a sample statistic is significant at a certain level, it will almost always be possible to decide the outcome before the sampling is completed. For example, with 999 randomizations an observed statistic is significant at the 5% level on a one-sided test if it is one of the top 50 values (or possibly one of the bottom 50 values). Hence it is known that a significant result will not occur as soon as 49 randomized values greater than the observed value are obtained. This led Lock to propose making the randomization test a sequential probability ratio test of the type proposed by Wald (1947).

A sequential version of a randomization test certainly offers the possibility of a substantial saving in the number of randomizations. It seems likely, however, that most tests will continue to use a fixed number of randomizations, with the idea being to estimate the significance level rather than just see whether it exceeds some pre-specified level.

The discussion so far in this section has been in terms of carrying out a randomization test by sampling the distribution of a test statistic that is found by random data permutations. However, the discussion applies equally well to bootstrap and Monte Carlo tests where the reference distribution is estimated by resampling data or by simulation. Indeed, as noted earlier, Marriott (1979) was discussing Monte Carlo tests rather than randomization tests when he developed equation (5.1). Therefore, for a bootstrap or Monte Carlo test at the 5% level of significance a minimum of 1000 simulated sets of data are desirable, while for a test at the 1% level 5000 sets of simulated data are desirable.

Jackson and Somer (1989) have suggested that many biological studies have used too few randomizations. They suggest 10 000 randomizations as a minimum, with up to 100 000 in critical cases. However, accepting this recommendation depends on how seriously the need to determine exact p-values is viewed. They quote one example where 1000, 1500 and 2000 randomizations give estimates of 4.7%, 6.0% and 4.8%, respectively, of a true significance level of 5.1%. The differences between these three estimates are minimal if the significance level is merely regarded as a measure of the strength of evidence against the null hypothesis.

A rather special situation occurs when randomization or other computer-intensive tests are run as part of a study to evaluate the properties of these tests, rather than for the analysis of one particular set of real data. Often what is required is to estimate the probability of a significant result (the power of the test) for data arising under certain specified conditions. In that case there may be some interest in deciding what the best balance is between using computer time to generate sets of data or using the time to do more randomizations for each set of data. Oden (1991) has discussed this problem and provides a solution based on the cost of generating one set of data and the cost of doing one randomization.

5.4 DETERMINING A RANDOMIZATION DISTRIBUTION EXACTLY

The testing of hypotheses by systematically determining all possible permutations of data is not a procedure that is emphasized in this book because sampling the randomization distribution of the test statistic is usually much simpler that enumerating all possibilities and allows the significance level to be determined to whatever accuracy is considered reasonable. However, it is worth making a few comments on how this type of complete enumeration can be made, and noting that it is sometimes possible to reduce the number of cases to be considered by taking an appropriate subset of the full set of permutations.

Suppose a set of data contains n values, and that it is necessary to calculate a test statistic for each of the $n!$ possible ways to rearrange these. This can be done fairly easily by associating each permutation with an n digit 'number', and working through from the smallest to the largest number.

An example should make this idea clearer. Consider therefore a set of five data values labelled 1 to 5, so that the initial order is 12345. This is the smallest possible number from these digits. The next largest is 12354, which stands for the permutation with values 4 and 5 interchanged. Listing the numbers like this gives the following orders:

$$12345 \quad 12354 \quad 12435 \quad 12453 \quad 12534 \quad 12543 \ldots 54321.$$

There are $5! = 120$ numbers in the list, each one standing for a different permutation of the original order.

In some cases, working through all the possible permutations like this will mean a good deal of redundancy, because many of the permutations must necessarily give the same test statistic. For example, suppose that the five values just considered consist of two from one sample and three from another, and the problem is to find the full randomization distribution of the difference between the two-sample means. Then the original data order can be written as (12)(345). The test statistic for this must be the same as that for (12)(354) because the sample means are not affected by changing the order of observations within samples. There are, in fact, only 10 different allocations to samples. These are found by considering all possible pairs in the first sample, namely (12), (13), (14), (15), (23), (24), (25), (34), (35) and (45). Each one of these produces 12 of the 120 permutations of all five numbers. Thus the randomization distribution can be determined by enumerating 10 cases rather than 120. These 10 cases can be worked through systematically by starting with the five-digit 'number' (12)(345) and then finding successively larger 'numbers', keeping digits in order of magnitude within the two samples. The sequence is then (12)(345), (13)(245), (14)(235), ..., (45)(123). The same procedure generalizes in a fairly obvious way with more than two samples.

A further economy is possible when there are equal sample sizes and the test statistic being considered is unaffected by the numbering of the samples. For example, consider the situation where there are three samples of size three and the test statistic is the F-value from an analysis of variance. The original ordering of observations can then be represented as (123)(456)(789), with the values labelled from 1 to 9. Thinking of this as a nine digit 'number', the next largest 'number' with digits in order of magnitude within samples is (123)(457)(689). But this order gives the same F-values as any of the alternative orders (123)(689)(457), (457)(123)(689), (457)(689)(123), (689)(123)(457) and (689)(457)(123). Generally, any allocation of observations to the three samples will be equivalent to six other allocations. Hence the full randomization distribution can be found by evaluating the F statistic for only one sixth of the total number of possible allocations.

The same principle can be used with two equal size samples if exchanging the observations in the two samples just reverses the sign of the test statistic being considered. Thus if the statistic is the difference between the sample means and the two samples are of size four then the mean difference for the allocation (1234)(5678) must be the same as the difference for (5678)(1234), but with the opposite sign. The randomization distribution of the difference can therefore be found with only half of the total possible allocations.

A rather different approach to reducing the number of rearrangements of the data that need to be considered was proposed by Chung and Fraser (1958). They suggested that there is no reason why all possible rearrangements of the data must be considered when determining the significance of a test statistic. Instead, any rules for rearranging the data can be adopted. All the permutations that these rules produce can be determined with their corresponding values for the test statistic. The distribution thus obtained can then be used as the reference distribution for determining significance. For example, suppose there are two samples with sizes three and two, as indicated with observation numbers (1,2,3) and (4,5). The random allocations considered could then involve rotating the observations to obtain the sample allocations indicated below:

$$(5, 1, 2), (3, 4)$$
$$(4, 5, 1), (2, 3)$$
$$(3, 4, 5), (1, 2)$$
$$(2, 3, 4), (5, 1)$$

and

$$(1, 2, 3), (4, 5).$$

Further rotation just leads to repeats, so that there are five possible allocations to be considered. This contrasts with the $^5C_2 = 10$ possible ways of allocating five observations to a sample of three and a sample of two. This concept can drastically cut down the number of possibilities that have to be considered, although it does not appear to have been used much in practice.

Other possibilities for cutting down on the computations involved in the complete enumeration of randomization distributions are discussed by Gabriel and Hall (1983), John and Robinson (1983), Pagano and Tritchler (1983), Tritchler (1984), Hall (1985), Welch and Gutierrez (1988) and Spino and Pagano (1991). See also the review by Good (1994, Chapter 13). Algorithms for finding full bootstrap distributions for small samples are provided by Fisher and Hall (1991).

5.5 THE NUMBER OF REPLICATIONS FOR CONFIDENCE INTERVALS

The standard bootstrap method for constructing confidence limits that is discussed in section 3.2 takes these limits to be

$$\text{Estimate} \pm z_{\alpha/2}(\text{bootstrap standard deviation}),$$

where $z_{\alpha/2}$ is the value that is exceeded with probability $\alpha/2$ for the standard normal distribution, and the bootstrap standard deviation is estimated by resampling the original sample a large number of times.

What exactly constitutes a 'large number of times' depends to some extent on the distribution being sampled. It is a standard result (Weatherburn, 1962, p. 137) that the standard deviation of the estimated standard deviation s_B from repeated samples of size B taken from a population with mean μ, variance σ^2, and kurtosis $\Delta = E[(X - \mu)^4]/\sigma^4 - 3$ is approximately given by

$$SD(s_B) \approx \sigma\sqrt{\{(\Delta + 2)/(4B)\}}. \tag{5.2}$$

In the context of bootstrapping, the 'population' is an observed sample of values, and the mean, variance and kurtosis in equation (5.2) are what are obtained from resampling this observed sample. Furthermore, $\sigma = s_\infty$ because resampling the sample an infinite number of times would give an estimate of σ without any error. What is therefore required is that

$$SD(s_B) \approx s_\infty\sqrt{\{(\Delta + 2)/(4B)\}}$$

should be small in comparison with s_∞, which will be achieved providing that $\sqrt{\{(\Delta + 2)/(4B)\}}$ is close to zero.

The kurtosis Δ has a minimum value of -2 for distributions with very short tails, but can be arbitrarily large for distributions with long tails. It is zero for a normal distribution, and in practice values above about 9 are unusual. Taking $\Delta = 9$ as a worst-case scenario gives $SD(s_{50}) = 0.23s_\infty$, $SD(s_{100}) = 0.17s_\infty$, $SD(s_{200}) = 0.12s_\infty$, and $SD(s_{400}) = 0.08s_\infty$. On the other hand, taking $\Delta = 0$ as a best-case scenario gives $SD(s_{50}) = 0.10s_\infty$, $SD(s_{100}) = 0.07s_\infty$, $SD(s_{200}) = 0.05s_\infty$, and $SD(s_{400}) = 0.04s_\infty$. Overall it seems that taking $B = 200$ bootstrap samples will usually result in errors in the estimation of the bootstrap distribution that are relatively small. This is the conclusion reached by Efron and Tibshirani (1993, p. 52) using a modification of the argument given here. Efron (1987) suggested that in practice there will often be little point in taking more than 100 bootstrap samples.

It must be stressed that recommendations of using only 100 or 200 bootstrap samples only apply with the standard bootstrap method for calculating confidence limits. Much larger sample sizes are required for the methods that involve the estimation of percentage points of bootstrap distributions. At least, this is true if the end-points of a confidence interval are required to be close to the end-points that would be obtained from an infinite number of bootstrap samples. Under certain circumstances a small number of bootstrap samples will give nearly the correct probability of covering the true value of the parameter of interest, although the end-points may differ to a considerable extent from what an infinite number of samples would give (Hall, 1986).

Usually it will be desirable for the end-points of confidence intervals to be close to what would be obtained from an infinite number of bootstrap samples. The sample size B that is required to achieve this depends on the nature of the bootstrap distribution, but some calculations presented by Efron (1987) and Efron and Tibshirani (1993) suggest that B should be at least 1000 in order to get good estimates of the limits for a 90% confidence interval with a distribution that is close to normal. Larger values of B are required for intervals with a higher level of confidence, so that taking $B = 2000$ for a 95% confidence interval may be more realistic.

It is, of course, always possible to repeat the bootstrap process several times to see how variable are the results obtained. Alternatively, the accuracy of estimation for any bootstrap statistic can be assessed by a type of jackknife-after-bootstrap procedure (Efron, 1992; Efron and Tibshirani, 1993, p. 275).

For confidence limits obtained by the randomization argument as discussed in section 1.4 it is necessary to have enough randomizations to get a good representation of the full randomization distribution. Because the limits are based on finding what changes have to be made to one-sample in order to just obtain significant result on randomization tests, it seems that the conclusions reached in section 5.3 are pertinent in this respect: at least 1000 randomizations should be made for a test at the 5% level, and at least 5000 randomizations should be made for a test at the 1% level. This suggests that 1000 randomizations is a minimum for 95% confidence limits and 5000 randomizations is a minimum for 99% confidence limits.

5.6 MORE EFFICIENT BOOTSTRAP SAMPLING METHODS

A number of authors have discussed methods for improving the efficiency of bootstrap sampling. One of these methods, balanced bootstrap sampling, has already been described in section 3.11, where this involves ensuring that each of the original data values occurs the same number of times in the bootstrap samples. Other possibilities include antithetic sampling (Hall, 1989), importance sampling (Johns, 1988; Hinkley and Shi, 1989; Do and Hall, 1991a), quasi-random sampling (Do and Hall, 1991b), and the replacement of sampling by the use of exact or approximate equations for properties of bootstrap distributions (DiCiccio and Tibshirani, 1987; Huang, 1991). For details see the reviews of Hall (1992a, Appendix II) and Efron and Tibshirani (1993, Chapter 23)

5.7 THE GENERATION OF PSEUDO-RANDOM NUMBERS

Many computer languages have an instruction for generating pseudo-random numbers uniformly distributed between 0 and 1. If this is not the case, then subroutines are readily available to produce these numbers. Care must be taken, however, to ensure that the generator used with randomization tests is satisfactory. It is a good idea to do at least some simple checking before taking one to be reliable, and preferable only to accept one that has undergone quite stringent testing along the lines discussed, for example, by Morgan (1984, Chapter 6).

Many of the randomization and Monte Carlo tests described in this book have used a generator developed by Wichmann and Hill (1982). Some aspects of the algorithm were overlooked by Wichmann and Hill in the original publication (Wichmann and Hill, 1984; McLeod, 1985; Zeisel, 1986), but the subroutine is still satisfactory for general applications. According to Wichmann and Hill (1982), the serial test, the poker test, the coupon collector's test and the runs test have been used on their generator with satisfactory results. Wichman and Hill's generator will be too slow for some applications needing the generation of extremely large numbers of random numbers, in which case faster alternatives are available (Ripley, 1990).

Unfortunately, even with the best of pseudo-random number generators there is the possibility of small systematic errors due to subtle correlations in the sequences. This was demonstrated by Ferrenberg et al. (1992), for example, when they applied what they considered to be the best available simulation algorithms to a situation where an analytical solution is available. They recommend that tests should be carried out to assess a generator for use with all new algorithms, irrespective of how many tests the generator has already received. In truth, this recommendation is seldom followed.

5.8 THE GENERATION OF RANDOM PERMUTATIONS

A fast algorithm for generating random permutations of n numbers $X_1, X_2, \ldots, X_n$ is important for sampling a randomization distribution. Algorithm P of Knuth (1981, p. 139) will service this need. It operates as follows:

1. Set $j = n$.
2. Generate a random number U uniformly distributed between 0 and 1.
3. Set $k = \text{INT}(jU) + 1$, so that k is a random integer between 1 and j.
4. Exchange X_j and X_k.
5. Set $j = j - 1$. If $j > 1$ return to step (b); otherwise stop.

This algorithm can be used repeatedly to randomly permute n numbers. There is no need to return them to the original order before each entry.

For randomization tests done in a spreadsheet a simple way to generate a random permutation for a column of numbers involves sorting the column using a second column of random numbers as the sort key.

5.9 CHAPTER SUMMARY

- Randomization tests and classical parametric tests tend to have similar power when the conditions for the parametric test are justified. With data from non-standard distributions there is some evidence to suggest that randomization tests are more powerful than classical alternatives.
- For a test of significance at the 5% level 1000 randomizations or bootstrap resamples is a realistic minimum. For a test at the 1% level 5000 randomizations or bootstrap resamples is a realistic minimum.
- Methods for determining a randomization distribution exactly are described and referenced.
- For the standard bootstrap confidence interval 100 to 200 bootstrap resamples may be quite adequate. However, for bootstrap percentile types of limits more resamples are desirable, with perhaps 2000 for a 95% confidence interval.
- Some methods for improving the efficiency of bootstrap resampling (balanced bootstrap sampling, antithetic sampling, quasi-random sampling and mathematical approximations) are referenced.
- The generation of random numbers and random permutations is discussed.

6
One- and two-sample tests

6.1 THE PAIRED COMPARISONS DESIGN

Fisher (1935) introduced the idea of the randomization test with paired comparison data in his book *The Design of Experiments*, using as an example some data collected by Charles Darwin on the plant *Zea mays*. Darwin took 15 pairs of plants, where within each pair the two plants 'were of exactly the same age, were subjected from the first to last to the same conditions, were descended from the same parents'. One individual in each pair was cross-fertilized and the other was self-fertilized. The heights of offspring were then measured to the nearest eighth of an inch. Table 6.1 gives the results in eighths of an inch over 12 inches. The question of interest to Darwin was whether these results confirm the general belief that the offspring from crossed plants are superior to those from either parent.

Because of the pairing used by Darwin, it is natural to take the differences shown in Table 6.1 as indicating the superiority of crossing over self-fertilization under similar conditions. Fisher (1935) argued that if the cross- and self-fertilized seeds are random samples from identical populations, and that their sites (in growing pots) were assigned to members of pairs independently at random, then the 15 differences were equally likely to have been positive or negative. He therefore determined the significance of the

Table 6.1 The heights of offspring of *Zea mays* as reported by Charles Darwin in eighths of an inch over 12 inches

Pair:	1	2	3	4	5	6	7	8	9	10	11	12	13	14	15
Cross-fertilized	92	0	72	80	57	76	81	67	50	77	90	72	81	88	0
Self-fertilized	43	67	64	64	51	53	53	26	36	48	34	48	6	28	48
Difference	49	−67	8	16	6	23	28	41	14	29	56	24	75	60	−48

observed sum of differences (314) with reference to the distribution that is obtained by taking both signs for each difference. There are $2^{15} = 32\,768$ sums in this distribution, and Fisher concluded that in 863 cases (2.6%) the sum is as great as or greater than that observed. Hence the mean difference between cross- and self-fertilized plants is significantly high at the 2.6% level on a one-sided test.

With this example the assessment of the data depends to an appreciable extent on the statistic used to evaluate the group difference. For example, if the test statistic used is the number of positive differences then the sample comparison by a randomization test reduces to the sign test. The distribution of the test statistic under randomization is the binomial distribution with 'success' probability 0.5 and $n = 15$ trials. There are 13 positive differences and the probability of 13, 14 or 15 positive differences is 0.0037. A one-sided test therefore gives a result that is significant at the 0.37% level. By way of comparison, it can be noted that the paired t-test gives a test statistic of $t = 2.15$, with a corresponding upper tail probability of 0.025, while the Wilcoxon signed ranks test gives an upper tail probability of 0.020. The result from a sign test is therefore distinctly more significant than the results from the other tests.

Basu (1980) argued that the dependence of the significance level on the test statistic used is one reason why a randomization test does not provide a satisfactory method for sample comparisons. However, as Hinkley (1980) pointed out in the discussion of Basu's paper, a significance level is associated with a test statistic plus the data, and not the data only. It is up to the data analyst to decide on the appropriate measure of sample difference. In the example being considered it is clear that if the observations for cross- and self-fertilized plants come from the same distribution then the mean of the paired differences is expected to be zero and the probability of a paired difference being positive is 0.5. On the other hand, if the observations come from different distributions then it is possible that the distribution of differences has a mean of zero, but the probability of a positive difference is higher than the probability of a negative one. If this is the case then it is desirable for a randomization test on the sum of sample differences to give a non-significant result and the sign test to give a significant result.

Many years ago Pearson (1937) stressed that the appropriate test statistic for a randomization test depends upon the alternative hypothesis that is considered to be plausible if the null hypothesis is not true. Indeed, the ability to be able to choose an appropriate test statistic is one of the advantages of such tests. The *Zea mays* example merely highlights the importance of this consideration more than for some other data sets.

In deciding on a test statistic, it may be important to ensure that it does not depend too crucially on extreme sample values. In the context of the paired comparison design obvious possibilities are to use the median as the test statistic (Welch, 1987) or a trimmed total with the m largest and m smallest differences not being included (Welch and Gutierrez, 1988).

A problem with trimming is that it is necessary to decide on the value of m. As Welch and Gutierrez note, it is not appropriate to try several values and choose the one that gives the best results! They show that taking $m = 2$ works well with six example sets of data, but in all except one of these examples $m = 1$ gave more or less the same result. Because $m = 1$ involves particularly straightforward calculations (just omitting the maximum and minimum differences from the sum of differences), this may be a reasonable choice.

With Darwin's data there is no reason to doubt the accuracy of any of the measurements, and it is an overall comparison of crossed and self-fertilized plants that is of interest, not a comparison with extreme cases omitted. Hence it is difficult to justify using a trimmed total rather than the overall total of differences as a test statistic. Furthermore, it is a fair guess that trimming will have the effect of making the observed sum of differences more significant, because even taking $m = 1$ will trim off the largest of the two negative differences. However, a trimmed test was run just to see how the outcome compares with the untrimmed result. Taking $m = 1$, the trimmed sum of observed differences is 306. There are 2.1% of randomized trimmed sums this large. Hence trimming has changed the significance level of the data from 2.6% to 2.1%. This change is quite minimal and both the trimmed and untrimmed sums of sample differences are providing about as much evidence against the hypothesis that the mean difference is zero.

With n paired comparisons, there are 2^n elements in the randomization distribution of a test statistic. For small values of n these are easy enough to enumerate on a computer. For example, with Darwin's data, $n = 15$ and the randomization distribution has 32 768 elements, including the one observed. These can be evaluated by the following algorithm.

Algorithm for an exact test

1. Input paired differences $d(1), d(2), \ldots, d(n)$. Calculate the test statistic $S(0)$ for these differences. Set $j = 1$ and Count $= 0$.
2. Calculate the test statistic $S(j)$ as a function of the differences. If $S(j) \geq S(0)$ then set Count $=$ Count $+ 1$.
3. Set $i = 1$.
4. If the fractional part of $j/2^{i-1}$ is 0, change the sign of $d(i)$. Set $i = i + 1$.

---

5. If $i > n$ then go to step 6; otherwise return to step 4.
6. Set $j = j + 1$. If $j > 2^n$ then go to step 7; otherwise return to step 2.
7. Calculate $P =$ Count$/2^n$, the proportion of randomization statistics that equal or exceed the observed statistic and stop.

In the above algorithm, step 4 has the effect of changing the sign of $d(1)$ for every calculation of a test statistic, changing the sign of $d(2)$ for every other calculation, changing the sign of $d(3)$ for every fourth calculation, ..., changing the sign of $d(n)$ for every 2^{n-1}th calculation. In this way all possible signs of all differences have occurred by the time that the 2^nth calculation is made. The assumption made is that the significance level of the observed statistic $S(0)$ is the proportion of randomized values greater than or equal to $S(0)$. For a lower tail test, step 2 has to be changed so that Count is incremented if $S(j) \leq S(0)$.

For n larger than about 20 this method of completely enumerating the randomization distribution becomes impractical. Either the randomization distribution will have to be sampled, or a more complicated algorithm for determining the significance level will have to be used. See Robinson (1982), Gabriel and Hall (1983), Gabriel and Hsu (1983), John and Robinson (1983), Pagano and Tritchler (1983), Tritchler (1984), Hall (1985), Welch and Gutierrez (1988), and Spino and Pagano (1991) for descriptions of possible approaches.

Confidence intervals for the mean difference between paired observations can be made using an obvious extension of the method described in section 1.4 for the comparison of two unpaired samples (Kempthorne and Doefler, 1969). In the context of Darwin's example it is necessary to assume that the difference between cross- and self-fertilized plants is a constant μ_D in the sense that for any plant where self-fertilization will give a height X for offspring, cross-fertilization will change this value to $X + \mu_D$. In that case, subtracting μ_D from each of the cross-fertilized observations, which has the effect of reducing all the paired differences by the same amount, will make the null hypothesis of no difference between the samples hold. A 95% confidence interval for μ_D therefore consists of the range of values for this parameter for which a randomization test based on the sum of paired differences is not significant at the 5% level on a two-sided test. Or, in other words, since there is a probability of 0.95 that the test does not give a significant result, the 95% confidence interval comprises of those values of μ_D that meet this condition. Intervals with other levels of confidence can be derived by the same principle.

The limits of the 95% confidence interval are the values which, when

Table 6.2 Significance levels obtained when different effects are subtracted from each of the cross-fertilized heights for Darwin's *Zea mays* data

Effect	−18.0	−15.6	−13.2	−10.7	−8.3	−5.9	−3.4	−1.0	1.4	3.9
Upper %	0.2	0.2	0.3	0.4	0.6	0.9	1.5	2.4	3.4	5.4
Effect	35.6	38.0	40.4	42.9	45.3	47.7	50.2	52.6		
Lower %	8.8	5.0	3.0	1.5	0.8	0.4	0.2	0.1		

subtracted from the paired differences, result in sample sums falling at the lower 2.5% and upper 2.5% points in the distribution obtained by randomization. Determining these points by enumerating the full randomization distribution for different values of μ_D may be quite computer-intensive even for small sample sizes. However, sampling the randomization distribution does provide a realistic approach. The results shown in Table 6.2 were determined by generating 4999 random allocations of signs to the 15 observations in Darwin's data, and seeing where the sample sum of paired differences falls within the distribution obtained for these randomizations plus the observed sum. For example, subtracting the effect −18.0 from each of the paired differences resulted in a sample sum that was equalled or exceeded by 10 (0.2%) of the randomized sums. By linear interpolation, the upper 2.5% point in the randomization distribution corresponds to −0.8 and the lower 2.5% point corresponds to 41.2. An approximate 95% confidence interval is therefore −0.8 < μ_D < 41.2. John and Robinson (1983) state the exact limits to be −0.2 to 41.0, and the limits given here compare well with those calculated by John and Robinson by three other approximate methods.

As mentioned above, in his discussion of Darwin's data Fisher assumed that the cross- and self-fertilized plants were random samples from populations of possible plants. The null hypothesis was that these two populations are the same, and it is this that justifies comparing the test statistic with the distribution obtained by giving each observed difference a positive on negative sign. With these assumptions it is perfectly valid to generalize the test result and the confidence interval to the populations of plants. If Fisher had not been prepared to make the assumption of random sampling, but was prepared to assume that the plants within each pair were effectively allocated at random to crossing and self-fertilization, then the randomization test and confidence interval would still be valid because of this randomization in the design. However, the extent to which the results could be generalized to plants of this species in general would be debatable.

It is appropriate to conclude this section with a few remarks on the power of randomization tests. As was mentioned in section 5.2, there are some theoretical results that suggest that for large samples randomization tests have the same power as their parametric counterparts. A useful study was carried out by Kempthorne and Doerfler (1969). They generated small samples of paired differences from eight different distributions and compared the randomization test based on the sum of differences with the Wilcoxon test, the sign test and the F-test (which is equivalent to the paired t-test). They concluded that the only justification for using the Wilcoxon test is computational convenience, and the sign test should never be used. The F-test, which requires stronger assumptions, gives almost no increase in power.

Of course, in assessing studies like this the earlier discussions in this chapter should be kept in mind. With real data it is up to the data analyst to decide on an appropriate test statistic, taking into account what the plausible alternative hypotheses are.

6.2 THE ONE-SAMPLE RANDOMIZATION TEST

The significance test discussed in the previous section is sometimes suggested for use in situations where only a single sample is concerned rather than paired comparisons. The question of interest is then whether this sample could reasonably have come from a population with a mean equal to a particular value μ. This is Fisher's one-sample randomization test, which was mentioned in section 1.3.

To take a specific example, suppose that in many previous experiments on Z. *mays* grown under similar conditions to those used with his experiment to compare cross- and self-fertilized plants, Darwin had found that the average height of offspring was 19 inches. He might then have been interested in testing the hypothesis that the mean height of the self-fertilized plants in the fertilization experiment was consistent with these previous results.

A reasonable approach seems to be to calculate the differences between the observed heights and the hypothetical mean height, and compare the sum of these differences with the distribution obtained by randomizing their signs. The data differences, in units of one eighth of an inch, are -13, 11, 8, 8, -5, -3, -3, -30, -20, -8, -22, -8, -50, -28 and -8, with a sum of -171. In comparison with the full randomization distribution the probability of getting a sum this far or further from zero is 0.0156. Hence the significance level obtained with this test statistic is 1.6%, and there is clear evidence that the mean height of the progeny of the self-fertilized plants is not 19 inches when grown under these conditions.

In this example there is no way that the comparison of the test statistic with the randomization distribution can be justified just from a randomization that is part of the experimental design. Even if plants were randomly allocated to cross- and self-fertilization this is irrelevant to testing the mean height of the self-fertilized plants. The only way that the test can be justified is by assuming that (a) the self-fertilized plants are effectively a random sample from the population of possible plants, and (b) the distribution of progeny heights is symmetric in this population. In other words, the assumptions are the same as for a t-test except that the distribution does not have to be normal. Therefore, this is not a randomization test that is valid with non-random samples. However, given assumptions (a) and (b), a trimmed sum can be used just as well as a complete sum. Also, a $100(1 - \alpha)\%$ confidence interval for the true population mean can be calculated by finding the smallest and largest values that just give a non-significant result on a two-sided test at the $100\alpha\%$ level.

The calculations for Fisher's randomization test are exactly the same as required for the paired comparisons design. The RT program (Manly, 1996a) was used to do the calculations for the *Zea mays* examples.

6.3   THE TWO-SAMPLE RANDOMIZATION TEST

The two-sample randomization test is the one that was described in Example 1.1. Recall that this example involved the comparison of mandible lengths for samples of 10 male and 10 female golden jackals. Significance was determined by comparing the observed mean difference with the distribution of differences obtained by randomly allocating the 20 observed mandible lengths to two samples of size 10.

This test can be justified in general for comparing the means of two groups of items if either the experimental design involves a random allocation of the items to the two groups before they are treated differently (Pitman, 1937a), or if the two groups are independent random samples from two populations (Fisher, 1936). In the first case, if the treatment has no effect then clearly the experiment is just choosing one of the possible sample allocations of items at random from the randomization distribution. This immediately justifies the randomization test. In the second case, justification comes from the fact that if the two populations sampled are the same then all the possible allocations of observations to samples are again equally likely. With random samples inferences can be made about the populations.

The calculations for the two-sample randomization test can be reduced considerably by recognizing that the sum of the observations in one-sample is an equivalent test statistic to the difference in sample means. To see this, let $\bar{x}_1$ and $\bar{x}_2$ denote the sample means, let S_1 and S_2 denote the sample sums, and let n_1 and n_2 denote the sample sizes. The mean difference can then be written:

$$D = \bar{x}_1 - \bar{x}_2 = S_1/n_1 - S_2/n_2,$$
$$= S_1(1/n_1 + 1/n_2) - S/n_2,$$

where $S = S_1 + S_2$, the sum of all the observations. With D written this way, everything is constant for all randomizations of the observations except S_1. Therefore, when sample randomizations are ordered by mean differences they will be in the same order as for the first sample sum S_1. As S_1 is simpler to calculate than the mean difference, this is a preferable statistic. It is also possible to show that S_1 is equivalent to the conventional t-statistic

$$t = (\bar{x}_1 - \bar{x}_2)/\{s\sqrt{(1/n_1 + 1/n_2)}\},$$

where s is the pooled, within-sample estimated standard deviation.

It is sometimes assumed that because the randomization test requires fewer assumptions than parametric alternatives it is a relatively robust test. For example, it might be supposed that it is a useful alternative to a conventional t-test for comparing means in situations where the two samples being compared are random samples from populations with different variances. However, this may not be the case, because a test statistic designed to detect mean differences may also have some power to detect variance differences. In fact, results reported by Boik (1987) suggest that a randomization test may or may not perform better than a t-test when variances are unequal, depending on the ratio of the sample sizes. This matter is discussed further in section 7.5.

Confidence limits for the difference between two groups are discussed in section 1.4 for the two-sample case. In calculating these it is necessary to assume that the group difference is such that moving an item from one group to the other has the effect of shifting the observation on that item an amount that is constant for all items. Limits based on a model for which moving from one group to another results in scores being multiplied by a constant amount are discussed by Hall (1985).

Example 6.1: prey consumption by two sizes of lizard   Randomization tests are particularly suitable for analysing data that are from clearly non-normal distributions with an appreciable number of tied values. In such cases the non-normal distributions suggest that the use of a t-test is suspect, and the tied values suggest that tables for critical values for non-parametric tests are only approximately correct.

An example of this type arises with some data collected as part of a study by Powell and Russell (1984, 1985) of the diet of two size morphs of the eastern horned lizard *Phrynosoma douglassi brevirostre*, and published by Linton *et al.* (1989). Stomach contents of 45 lizards collected over a period of time were determined. Here the collection time will not be considered and attention will be restricted to the question of whether there is any evidence of a mean difference between lizards in the two size classes in terms of milligrams of dry biomass of Coleoptera in the stomachs. For the first size class (adult males and yearling females), the values obtained from 24 lizards were as follows: 256, 209, 0, 0, 0, 44, 49, 117, 6, 0, 0, 75, 34, 13, 0, 90, 0, 32, 0, 205, 332, 0, 31 and 0. For the second size class (adult females) the values obtained from 21 lizards were: 0, 89, 0, 0, 0, 163, 286, 3, 843, 0, 158, 443, 311, 232, 179, 179, 19, 142, 100, 0 and 432.

The mean difference between the two samples is -108.2. A t-test for the significance of this difference, using a pooled estimate of the within-group variance, gives a test statistic of -2.29 with 43 degrees of freedom. From the t-distribution the probability of a value this far from zero is 0.027. A Mann–Whitney non-parametric test, which is just a randomization test carried out on ranked data, gives a probability of 0.080 for a sample difference as large as that observed. A randomization test, with significance estimated from 5000 randomizations, gives an estimated probability of 0.018 for a sample mean difference as large as that observed.

In this example the significance level from the t-table is not much different from the level obtained by randomization. However, using the t-table would certainly be questionable if the result of the randomization test was not known. The level of significance from the Mann–Whitney test was calculated using a standard statistical computer package. The large discrepancy between this level and the randomization level can be accounted for by the high proportion of tied zero observations affecting the Mann–Whitney probability calculation, although the fact that the test statistic is of a different type may also contribute to the discrepancy.

6.4   BOOTSTRAP TESTS

A bootstrap-t test for a significant difference between two-sample means is a computer-intensive alternative to a randomization test. The idea with this approach is to use bootstrapping to approximate the distribution of a suitable test statistic when the null hypothesis is true. The significance of the observed test statistic is then assessed in comparison with the bootstrap distribution.

The obvious approach is to compute the usual t-statistic

$$t = (\bar{x}_1 - \bar{x}_2)/\{s\sqrt{(1/n_1 + 1/n_2)}\},$$

where $\bar{x}_i$ is the mean of sample i, of size n_i, and s is the usual pooled estimate of standard deviation from the two samples. This is then compared with a bootstrap distribution for which the null hypothesis is made to be true.

A simple way to make the null hypothesis true involves adjusting the data values in sample j to

$$x'_{ij} = x_{ij} - \bar{x}_j + \bar{x},$$

where x_{ij} is the ith original value in the sample, $\bar{x}_j$ is the original sample mean, and $\bar{x}$ is the mean of the original values in both samples. With this adjustment both samples have a mean of $\bar{x}$, although in fact any other common mean will serve just as well because this cancels out in the calculation of the test statistic. Indeed, one way to adjust the sample values involves replacing them by their residuals $r_i = x_{ij} - \bar{x}_j$.

Having adjusted the data values, these can be combined into a single bootstrap population of size $n_1 + n_2$. A bootstrap sample 1 is then obtained by selecting n_1 of these values at random, with replacement. Similarly, a bootstrap sample 2 is obtained by selecting n_2 of them at random, with replacement. The two samples can then be used to produce t_B, a bootstrap value for t. By repeating the bootstrap sampling many times the bootstrap distribution of t is generated.

It is implicit in the test just described that the two samples being compared come from distributions that are the same, apart from a possible difference in their means. It is this assumption that allows the combining of the adjusted values to produce a single set of $n_1 + n_2$ values for the bootstrap population. However, it is possible to construct a test without making this assumption (Efron and Tibshirani, 1993, p. 222) by modifying the test statistic and the bootstrap sampling method. Bootstrapping then provides a possible solution to what is called the Behrens–Fisher problem, which is the comparison of two-sample means without assuming that the samples are from populations with a common variance.

A suitable test statistic for this situation is

$$t = (\bar{x}_1 - \bar{x}_2)/\sqrt{(s_1^2/n_1 + s_2^2/n_2)},$$

where $s_j^2 = \sum(x_{ij} - \bar{x}_j)^2/(n_j - 1)$ is the usual unbiased estimator of variance for sample j. This statistic does not have a t-distribution, but it is often used to compare sample means without assuming equal population variances because for large samples it approximately has a standard normal distribution when the population means are equal.

To produce bootstrap samples the original data values can be adjusted as before to remove the sample mean difference. A bootstrap sample 1 is then obtained by resampling the adjusted values only from the original sample 1, with replacement. Similarly, a bootstrap sample 2 is obtained by resampling the adjusted values only from the original sample 2, with replacement. A bootstrap t-statistic is then computed as for the original data.

The reason for using t-statistics here is the same as it was for Example 3.5. The hope is that a statistic that has the form of an estimate of a parameter divided by an estimate of its standard error will have a distribution that is nearly the same whatever the true value of the parameter may be. If this is the case then the distribution is said to be approximately pivotal. The point is, of course, that if a statistic has a pivotal distribution then it should be possible to approximate this well by bootstrap resampling providing the sample being bootstrapped is large enough to give a good representation of the population that it is from.

For a solution to the Behrens–Fisher problem based on a randomization argument see Manly (1995b, 1996c).

6.5 RANDOMIZING RESIDUALS

Ter Braak (1992) has proposed a general approach for hypothesis testing that can be thought of as a hybrid between randomizing and bootstrap testing in the sense that the justification can come from either point of view. He described the procedure in terms of a general regression model. However, it is useful to introduce it here in terms of the simplest type of situation where it might be used, which is the comparison of two-sample means.

Ter Braak's test involves randomizing the sample residuals in order to generate a distribution to which a sample statistic can be compared. In the two-sample comparison the procedure has the following steps.

A two-sample test to compare means by randomizing residuals

1. Calculate a suitable test statistic such as $t_1 = (\bar{x}_1 - \bar{x}_2)/\{s\sqrt{(1/n_1 + 1/n_2)}\}$ for the observed data.
2. Calculate the residuals for samples 1 and 2, where these are the deviations of the original observations from the sample means. Thus the ith residual in sample j is $r_{ij} = x_{ij} - \bar{x}_j$, for $i = 1, 2, \ldots, n_j$.
3. Randomly allocate n_1 of the $n_1 + n_2$ residuals to sample 1 and the remainder to sample 2.
4. Calculate the test statistic t_2 for the two samples of residuals.
5. Repeat steps 3 and 4 a large number N of times, in order to generate values $t_2, t_3, \ldots, t_N$ from the randomization distribution of t.

On a one-sided test, t_1 is significantly large at the $100\alpha\%$ level if it exceeds $100(1 - \alpha)\%$ of the values $t_1, t_2, \ldots, t_N$, or it is significantly small if it is less than $100(1 - \alpha)\%$ of these values. For a two-sided test it is a question of whether $|t_1|$ is significantly large in comparison with the distribution of $|t_1|, |t_2|, \ldots, |t_N|$.

Although randomization is involved with this procedure, it is not a randomization test in the traditional sense because the data are modified before the randomization takes place. One of the effects of this modification is that there is no longer a guarantee that a test carried out at the $100\alpha\%$ level will have a probability of α or less of giving a significant result when the null hypothesis is true (section 1.3), although in practice this may still be the case.

There are two arguments that can be put forward to justify randomizing residuals. First, for samples of a reasonable size the residuals $r_{ij} = x_{ij} - \bar{x}_j$ should be close to the deviations $x_{ij} - \mu_j$ of the observations from their population means. A test based on randomizing these deviations is valid on the assumption that the deviations have the same distribution for both samples. Therefore randomizing residuals can be thought of as an approximation to a valid randomization test.

The second argument is that randomizing residuals is just a type of balanced bootstrap sampling, as discussed in section 3.11. Instead of randomly sampling residuals to produce bootstrap sets of data, each residual is used once for every set of data. Indeed, this is the only important difference between the bootstrap test that was described in the last section and the ter Braak procedure.

Randomizing residuals is not necessarily a sensible procedure. In particular, there may be problems with count data, data covering a restricted range or data where certain particular values often occur. This is because the residual corresponding to a particular data value will be different in the two samples, unless the sample means happen to be identical. Randomizing the residuals then immediately produces sets of data that are implausible.

To make this clearer, consider the data from Example 6.1 on Coleoptera consumption of the eastern horned lizard. The mean for sample 1 (adult males and yearling females) is 62.2 mg. The residuals for the 10 lizards that did not consume Coleoptera are therefore -62.2. For sample 2 (adult females) the mean is 170.4 mg. The residuals for the six lizards that did not consume Coleoptera are therefore -170.4. Now, suppose a randomization is carried out and sample 1 ends up with five each of the residuals of -62.2 and -170.4. This could only have happened with the real data if there were two values of prey consumption repeated five times each. This seems extremely unlikely, because the only prey consumption that can be expected to be repeated is zero.

This argument makes a test based on either randomizing or bootstrap resampling of residuals seem unrealistic for these data, although randomizing or bootstrap resampling of the original values can be justified. It suggests that the use of tests based on residuals should be restricted to those situations where the residuals in the two samples appear as if they could come from the same distribution.

6.6 COMPARING THE VARIATION IN TWO SAMPLES

In some cases the important comparison between two samples is in terms of the amounts of variation that they possess. It is then fairly obvious to consider using the ratio of the two-sample variances as a statistic and assessing this for significance by randomization or bootstrapping. One particular reason for adopting a computer-intensive approach is the abundant evidence that the use of the F-distribution for assessing the significance of the variance ratio gives poor results with non-normal data (Miller, 1968).

Bailer (1989) found that assessing the significance of the variance ratio using the randomization distribution has comparable power to other possible procedures. However, Boos and Brownie (1989) and Boos et al. (1989) have noted that such a test may lose power when there is a large difference between the two-sample means because this difference contributes to variation in the randomization distribution of the variance ratio. They suggested that this problem may be largely overcome by carrying out either a randomization or a bootstrap test on the sample residuals, and that these two tests have similar properties.

An alternative to using the variance ratio as a test statistic involves using the method that was proposed by Levene (1960). With this approach x_{ij}, the jth data value for the ith sample, is transformed to either $|x_{ij} - \bar{x}_i|$, the absolute deviation from the sample mean, or $|x_{ik} - M_i|$, the absolute deviation from the sample median. The transformed values are then subjected to an analysis of variance to produce the usual F-statistic for a test of whether the means vary significantly between the samples, as discussed more fully in section 7.1. Usually, the F-value is tested against the F-distribution. A randomization version of the test can compare the observed value with the distribution that is obtained by randomly reassigning observations or residuals (deviations from sample means) to the samples. The basis of the test is the idea that a large difference in the variances for two samples will be converted to a large difference in the means of the transformed data.

The non-randomization version of Levene's test was found by Conover et al. (1981) to be relatively powerful and robust to non-normality in comparison with many alternatives. The test can also be used with more than two samples. Furthermore, simulation results that are discussed later indicate that this test performs well in the randomization setting. It has been found that the good properties seem most apparent when absolute deviations from medians are used. Therefore it is these deviations that are used for all the examples that are later. For this purpose, the median for a sample is defined as the middle value when an odd number of observations are ordered from smallest to largest, and midway between the two middle values when there are an even number.

It should be noted that work completed recently indicates that the Levene test based on randomizing absolute deviations from sample medians does not always have satisfactory performance with very non-normal data, and that a modified version is slightly better (Manly and Gonzalez, 1997). Furthermore, another randomization test that appears to have good properties has been proposed recently by Baker (1995).

Example 6.2: nest selection by fernbirds   As an example of comparing the amount of variation in two samples, consider the selection of nest sites in an area in Otago, New Zealand, by fernbirds (*Bowdleria puncta*). Harris (1986) compared several measurements on the clumps of vegetation selected for nests by the birds with randomly chosen clumps in the same region. Here only the perimeter of the clumps is considered and, in particular, whether there is any evidence that the variation in this variable differs for random clumps and nest sites.

For 24 nest sites the perimeters of the clumps of vegetation in metres were found to be 8.90, 4.34, 2.30, 5.16, 2.92, 3.30, 3.17, 4.81, 2.40, 3.74, 4.86, 2.88, 4.90, 4.65, 4.02, 4.54, 3.22, 3.08, 4.43, 3.48, 4.50, 2.96, 5.25 and 3.07, with the variance $s_1^2 = \sum(x_i - \bar{x})^2/(n - 1) = 1.877$. For 25 random clumps the perimeters were 3.17, 3.23, 2.44, 1.56, 2.28, 3.16, 2.78, 3.07, 3.84, 3.33, 2.80, 2.92, 4.40, 3.86, 3.48, 2.36, 3.08, 5.07, 2.02, 1.81, 2.05, 1.74, 2.85, 3.64 and 2.40, with variance $s_2^2 = 0.712$. The ratio of the largest variance to the smallest variance is therefore $R = 1.877/0.712 = 2.63$. The median for the nest sample is 3.88, the median for the random sample is 2.92, and the F-statistic from Levene's test is $L = 2.55$.

If the variance ratio of 2.63 is compared with the F-distribution with $n_1 - 1 = 23$ and $n_2 - 1 = 24$ degrees of freedom then it is found to be significantly large at almost the 1% level ($p = 0.011$). There is clear evidence that the variance is different for the two types of clumps of vegetation. On the other hand, comparing Levene's statistic $L = 2.55$ with the F-distribution with 1 and $n_1 + n_2 - 2 = 47$ degrees of freedom does not give a result that is significant at the 5% level ($p = 0.12$), so that there is no real evidence of a variance difference.

Three computer-intensive tests were carried out using the variance ratio statistic: (a) by randomly allocating the original 49 data values to the two samples without any adjustments for the initial mean difference between the two samples; (b) by randomly allocating the 49 residuals (i.e. the differences between the original data values and their sample means) to the two samples; and (c) by bootstrap resampling the 49 residuals, with replacement. For each of the tests 9999 bootstrap or randomized samples were generated and corresponding variance ratios were obtained. The p-value for each test was then calculated as the proportion of the times that the 10 000 F-values consisting of the observed one plus the 9999 randomized ones were as large or larger than the observed one. These p-values were found to be $p_a = 0.262$, $p_b = 0.297$ and $p_c = 0.217$, all of which suggest that the observed F-value is quite likely to occur if there is no variance difference between the two types of clumps of vegetation. These tests therefore provide no evidence at all of a difference in the variance.

For Levene's test only the randomization of observations and the randomization of residuals were used, again with 9999 randomized sets of data. Randomizing observations gave $p = 0.11$, and randomizing residuals gave $p = 0.086$. In this case there is reasonable agreement with what is obtained using the F-distribution ($p = 0.12$).

One explanation for the difference between the test on the variance ratio using the F-distribution and the computer-intensive tests with the same statistic is that it is due to the effect of the very large value of 8.90 in the sample of nest sites. If this value is omitted from the sample then the F-statistic changes from 2.63 to 1.18 and the two samples show about the same level of variation. The effect of this observation is automatically allowed for in the computer-intensive tests but not in the test using the F-table, which assumes that the data values are normally distributed within samples.

The variance ratio calculations for this example were carried out using the program RESAMPLING STATS (Simon and Bruce, 1991), which provides a convenient medium for setting up special-purpose bootstrap and randomization tests. The program RT (Manly, 1996a) was used for Levene's test.

6.7 A SIMULATION STUDY

An observed set of data such as the one considered in the last example can be used as a basis for comparing the properties of different computer-intensive procedures. Thus suppose that there really is no difference in the distribution of the perimeter length of clumps of vegetation used by fernbirds and the distribution for clumps that are not used. The 49 sample values given above then all come from the same distribution, with the 24 values for the nest sites being a random selection from the total. Viewing the data in this light makes it possible to generate samples based on alternative scenarios for how the birds might have chosen clumps of vegetation.

It is interesting, for example, to consider the data that might have been obtained if the distribution of the perimeter length for clumps chosen by fernbirds was the same as for clumps in general except that the mean was Δ metres higher. This can be simulated by randomly allocating 24 of the observed clump perimeters to nest sites and adding Δ on to each of these distances, with the remaining 25 observations, unmodified, being considered as the perimeters for non-nest sites. The data generated in this way can be analysed using different methods and the results compared. By generating and analysing a large number of sets of data it is then possible to assess the relative performance of different tests on data of the type being considered.

This approach for comparing tests was applied to the fernbird data. Sample mean differences of $\Delta = 0.0, 0.5, 1.0$ and 2.0 metres were used, with 1000 sets of data generated for each of these differences. The opportunity was taken to compare tests on both mean differences and variance differences, as follows:

(a) For each generated set of data, the data values were randomly reallocated to a sample of 24 'nest sites' and 25 'non-nest sites' 99 times. Three test statistics were calculated for both the initial set of data and its randomizations. The first was $|t|$, the absolute value of the usual t-statistic with a pooled estimate of variance. The second was R, the ratio of the largest to the smallest sample variance. The third was L, the F-statistic from Levene's test. A statistic for the initial set of data was then considered to be significantly large if it was among the largest 5 of the 100 values comprising itself and the 99 values from randomized data.

(b) The second method for testing involved randomizing residuals. Thus for each generated set of data the differences between observations and their sample means were calculated. These residuals were then randomly allocated to the sample of 24 'nest sites' and the sample of 25 'non-nest sites' 99 times to give randomized sets of data. Otherwise, the tests were the same as for (a).

(c) The third method for testing involved bootstrap resampling of residuals. Tests were carried out as for (b) except randomized sets of data were obtained from an initial set by resampling the 49 residuals with replacement. This method was not used with Levene's test statistic.

(d) The fourth method for testing involved assessing the significance of $|t|$, R and L using the t-distribution and the F-distribution in the conventional way.

Table 6.3 shows the results obtained from the procedure just described. It is seen that all four of the methods for testing for a mean difference have similar performance, with about the correct 5% of significant results when the null hypothesis is true ($\Delta = 0$), and 100% significant results when there is a large sample mean difference ($\Delta = 2$).

There are, however, substantial differences between the tests for equal variance based on the variance ratio R. For these tests the null hypothesis

Table 6.3 Results of a study of the performance of different types of computer-intensive test to compare means and variances. The table gives the percentages of significant results for tests at a nominal 5% level for a *t*-test (*t*), a test based on the ratio of the largest to smallest sample variance (*R*), and Levene's test (*L*). The significance of test statistics was based on (a) randomizing observations, (b) randomizing residuals, (c) bootstrap resampling of residuals and (d) using the *t*-distribution and the *F*-distribution in the conventional way. Cases are shown in italic where the desired result is 5% but the actual result is significantly different from this at the 5% level (outside the range 3.6% to 6.4%)

	Percentage significant results from										
Mean difference	Randomizing observations			Randomizing residuals			Bootstrapping residuals		Using *t*- and *F*-distributions		
Δ	*t*	*R*	*L*	*t*	*R*	*L*	*t*	*R*	*t*	*R*	*L*
0.0	5.8	3.9	5.5	5.3	4.3	5.3	5.1	*1.6*	5.4	*0.8*	4.5
0.5	25.3	5.7	4.7	25.3	4.6	4.5	26.0	*1.8*	24.7	*2.0*	*2.9*
1.0	78.7	*10.3*	3.6	78.9	*6.6*	3.6	77.6	*1.9*	78.8	*1.5*	*2.4*
2.0	100.0	*32.6*	5.3	100.0	4.9	5.0	100.0	*2.4*	100.0	*2.3*	*3.4*

was always true because of the procedure used to generate data. Nevertheless, test (a), based on randomizing observations, has given an increasing number of significant results as the mean difference has increased. This demonstrates that although the test is supposed to be for variance differences, it is also sensitive to mean differences. Test (b) based on randomizing residuals has given a fairly satisfactory performance, with just one case with rather more than 5% of tests that are significant. Test (c) based on bootstrap resampling of residuals and test (d) based on using the *t*-distribution and the *F*-distribution have given too few significant results. It seems, therefore, that randomizing residuals should be the method of choice for testing variance ratios with data of the type considered in this example.

Unfortunately, the reasonably good performance of the test on the variance ratio based on randomizing residuals does not always occur. Simulation studies that are described in Chapter 7 give a different result, indicating that this test is also unreliable.

Levene's test has given more consistent results for the different methods of assessing significance. The agreement is very good for randomizing observations or residuals, with close to the desired 5% of significant results. However, the use of the *F*-distribution has given too few significant results. As noted earlier, the use of Levene's test is discussed more fully by Manly and Gonzalez (1997).

6.8 THE COMPARISON OF TWO SAMPLES ON MULTIPLE MEASUREMENTS

A common situation is that two groups are compared on several measurements at the same time and it is important to determine which variables display important differences between the groups. Two approaches are then possible: a multivariate method can be used to test for overall differences between the two groups, or a test can be carried out on each variable, with some adjustment made to reduce the number of significant results that are otherwise expected from multiple testing. Multivariate approaches are reviewed in Chapter 12. In this section the univariate approach will be considered. Some alternatives to the randomization approach that will be discussed are considered by O'Brien (1984) and Pocock *et al.* (1987) in the context of clinical trials and more generally by Westfall and Young (1993).

Manly *et al.* (1986) describe a randomization method for comparing two groups on multiple measurements that allows for missing values and also any correlation that is present between the measurements. The particular application that they were interested in involved a group of 21 patients with multiple sclerosis and 22 normal individuals. With some exceptions, the individuals in both groups were given 38 semi-animated computer tests. It was important to decide which, if any, of these tests give different results for the two groups. Test results seemed to have non-normal distributions, with many tied values.

One approach that is often used in a case like this involves testing each variable independently for a mean difference between the two groups, possibly using a randomization test. To control the overall probability of declaring differences to be significant by mistake, the significance level used is set at $5\%/m$, where m is the number of variables tested. The Bonferroni inequality then ensures that the probability of declaring any difference significant in error is 0.05 or less. In general, using a significance level of $100(\alpha/m)\%$ with the tests on individual variables will ensure that when there are no differences between groups the probability of declaring anything significant is α or less.

The Bonferroni adjustment to significance levels is appropriate when the variables being tested have low correlations. In the case of perfect correlations either all the tests will yield a significant result or none will. The appropriate significance level for all tests is then $100\alpha\%$. It is therefore clear that in order to obtain a probability of exactly α of obtaining any significant results by chance the level of significance that should be used with the individual tests should be somewhere within the range $100(\alpha/m)\%$ and $100\alpha\%$.

The testing procedure proposed by Manly *et al.* (1986) was designed to take into account the correlation between variables. The test statistic used to compare the group means for the ith variable was

$$t_i = (\bar{x}_{1i} - \bar{x}_{2i})/\sqrt{(s_{1i}^2/n_{1i} + s_{2i}^2/n_{2i})},$$

where $\bar{x}_{ji}$, s_{ji} and n_{ji} are the mean, standard deviation and size, respectively, for the sample of values of this variable from the jth group. They proposed determining the significance level of the mean difference for each variable by comparing t_i with the randomization distribution obtained by randomly allocating individuals, with all their measurements, to the two groups. At the same time, it is possible to determine the minimum significance level found over all the variables for each random allocation. An appropriate level of significance for declaring the results of individual variables either significant or not significant is then the minimum level of significance that is exceeded by 95% of randomized allocations of individuals. Using this level ensures that the probability of declaring anything significant by chance alone is only 0.05. Similarly, if the level of significance used with individual tests is the minimum level that is exceeded by 99% of randomized allocations of individuals then the probability of declaring anything significant by chance alone will be 1%.

It will usually be convenient to determine significance levels by sampling the randomization distribution rather than enumerating it exactly. For reasons discussed by Manly *et al.* (1986) concerning repeated values in the distribution of minimum significance levels from randomization, a minimum of 100m randomizations are desirable when there are m variables. However, in practice many more should probably be used. Following the recommendations discussed in section 5.3, a minimum of 1000 randomizations is suggested for tests at the 5% level of significance and 5000 randomizations for tests at the 1% level of significance.

Example 6.3: shell type distributions of Cepaea hortensis   As an example of the use of the method for comparing several variables that has just been described, consider the data shown in Table 6.4. These data are percentage frequencies of different shell types of the snail *Cepaea hortensis* in 27 colonies where *Cepaea nemoralis* was also present, and 33 colonies where *C. nemoralis* was absent. The data were first published by Clarke (1960, 1962) with a different classification for shell types. An interesting question here is whether the mean shell type percentages differ in the two types of colony, which may indicate some interaction between the two species. It is assumed that the sampled colonies can be considered to be randomly chosen from those in southern England.

If the sample means for the ith shell type percentage in the two groups are compared using the t-statistic

Table 6.4 Percentage frequencies for shell types of the snail *Cepaea hortensis* in 27 colonies in areas where *Cepaea nemoralis* was also present and 33 colonies where *C. nemoralis* was absent (YFB = yellow fully banded; YOB = yellow other banded; YUB = yellow unbanded; P = pink; B = brown)

	Mixed colonies					*C. nemoralis* missing				
	YFB	YOB	YUB	P	B	YFB	YOB	YUB	P	B
	32.9	0.0	50.6	16.5	0.0	16.7	0.0	31.5	14.8	37.0
	53.2	0.0	45.2	0.0	1.6	84.6	1.9	13.5	0.0	0.0
	44.4	0.0	38.6	17.0	0.0	77.3	0.0	0.0	18.2	4.5
	77.1	7.9	15.0	0.0	0.0	100.0	0.0	0.0	0.0	0.0
	100.0	0.0	0.0	0.0	0.0	63.4	7.3	19.5	4.9	4.9
	75.9	0.0	24.1	0.0	0.0	62.7	30.7	6.6	0.0	0.0
	58.8	0.0	41.2	0.0	0.0	72.7	0.0	20.5	6.8	0.0
	54.1	0.0	45.9	0.0	0.0	51.5	9.1	27.3	12.1	0.0
	55.6	0.0	44.4	0.0	0.0	61.7	11.3	22.6	4.3	0.0
	94.3	5.7	0.0	0.0	0.0	78.0	1.7	20.3	0.0	0.0
	47.4	15.8	36.8	0.0	0.0	19.1	0.0	80.9	0.0	0.0
	90.6	7.1	2.4	0.0	0.0	76.0	0.0	17.0	7.0	0.0
	92.3	7.7	0.0	0.0	0.0	89.3	10.7	0.0	0.0	0.0
	87.9	12.1	0.0	0.0	0.0	24.0	0.0	28.0	48.0	0.0
	18.0	1.2	80.9	0.0	0.0	61.2	0.8	26.9	0.0	11.2
	50.1	0.0	49.9	0.0	0.0	68.0	0.0	26.2	0.0	5.7
	54.2	0.0	45.8	0.0	0.0	50.5	0.0	45.1	2.2	2.2
	26.3	0.0	73.7	0.0	0.0	67.9	0.0	29.5	0.0	2.6
	86.7	0.0	13.3	0.0	0.0	18.6	1.2	80.2	0.0	0.0
	46.8	11.7	41.6	0.0	0.0	23.3	0.0	73.3	0.0	3.3
	42.1	0.0	26.3	31.6	0.0	37.8	10.8	24.3	27.0	0.0
	38.9	0.0	36.1	25.0	0.0	24.1	0.0	69.6	0.0	6.3
	25.0	0.0	75.0	0.0	0.0	44.6	0.0	49.2	6.2	0.0
	93.8	6.3	0.0	0.0	0.0	50.8	1.8	43.7	2.2	1.5
	78.7	1.6	19.7	0.0	0.0	60.2	0.8	36.1	0.0	3.0
	93.9	6.1	0.0	0.0	0.0	70.0	0.0	0.0	16.7	13.3
	100.0	0.0	0.0	0.0	0.0	53.6	2.1	36.1	8.2	0.0
						60.6	0.0	39.4	0.0	0.0
						58.0	5.7	36.4	0.0	0.0
						93.2	4.5	2.3	0.0	0.0
						37.5	0.0	62.5	0.0	0.0
						45.8	4.2	45.8	4.2	0.0
						38.1	0.0	61.9	0.0	0.0
Mean	63.7	3.1	29.9	3.3	0.1	55.8	3.2	32.6	5.5	2.9
SD	25.6	4.6	25.1	8.5	0.3	22.6	6.1	23.3	10.1	6.9

Table 6.5 Values for t-statistics and significance levels for the comparison of *Cepaea hortensis* shell type frequencies in colonies with and without *C. nemoralis* being present

	Shell type				
	YFB	YOB	YUB	P	B
t	−1.27	0.06	0.44	0.90	2.11
$p(\%)$	21.6	95.7	66.1	38.2	0.5

$$t_i = (\bar{x}_{1i} - \bar{x}_{2i})/\sqrt{\{s_i^2(1/n_{1i} + 1/n_{2i})\}},$$

where s_i is the pooled-sample standard deviation, then the values obtained are as shown in Table 6.5. From 4999 randomizations of the 60 colonies to one group of 27 and another of 33, plus the observed allocation, the significance levels that are also shown in Table 6.5 were found, where these are the percentages of randomized data sets giving statistics as far as or further from zero than the observed values.

The only observed statistic that indicates any difference between the two groups is for the percentage of brown snails. Bonferroni's inequality suggests that to have an overall chance of 0.05 of declaring any result significant by chance the individual shell types should be tested for significance at the $5\%/5 = 1\%$ level. On that basis the result for brown ($p = 0.5\%$) is still significant.

When the 5000 sets of data (4999 randomized, 1 real) are ranked in order according to the minimum significance level obtained from the five variables it is found that 95% (4750) of these minimum significance levels exceed 1.1%. In other words, if the variables are tested individually at the 1.1% level then the probability of obtaining any of them significant by chance is 0.05. On this basis the percentage of brown snails is again clearly significantly different for the two groups.

Another way of looking at this procedure involves regarding the observed minimum significance level for the five variables as an indication of the overall difference, if any, between the groups. It can then be noted that 236 (4.7%) of the 5000 randomizations gave a minimum significance level as low or lower than the one observed. The difference between the two groups is therefore significant at this level.

The use of the Bonferroni inequality with this example gives almost the same result as the randomization method as far as determining the appropriate significance level to use with individual tests because the difference between the randomization value of 1.1% and the Bonferroni value of 1.0% can be considered to be of no practical importance. This is in spite of the fact that the variables being tested add up to 100%. Some calculations carried out by Manly and McAlevey (1987) indicate that this is a general result, and that variables must have high correlations before the Bonferroni inequality becomes seriously conservative.

6.9 FURTHER READING

Maritz (1981) reviews a range of one- and two-sample tests from a more theoretical point of view than the material that is presented in this chapter, and more recently (Maritz, 1995) discusses exact distribution theory for a randomization version of the paired comparison test with some missing values. Lambert (1985) discusses the construction of two-sample randomization tests for comparing the null model of randomness with specific alternatives. The idea is to find tests that are the best compromise between being robust to departures from the alternative models and having high power. Rosenbaum (1988) discusses randomization tests for paired comparison tests allowing for covariate differences because the matching within pairs is not exact.

Green (1977) provides a computer program for the one- and two-sample tests that have been discussed in this chapter. By using suitable heuristics, this program is able to count many of the possible enumerations from a full randomization distribution implicitly and hence avoid the computation of up to 90% of the test statistics. See also the algorithms in the C language that have been produced by Baker and Tilbury (1993). Efficient computational methods for obtaining the full randomization distributions for a range of two-sample tests are also discussed by Mehta et al. (1988a,b).

The text by Westfall and Young (1993) covers multiple testing using randomization and bootstrap methods in great detail for continuous and discrete data in a great variety of different situations. Anyone with a particular interest in this area should consult their book.

In this area of multiple testing the method proposed by Gates (1991) may be found to be useful. This can be applied in any situation where a set of p test statistics $S = (s_1, s_2, \ldots, s_p)$ for a real set of data are to be compared with alternative sets of statistics generated by randomization, bootstrapping or Monte Carlo simulation. For instance, s_1 to s_p might be t-statistics for mean differences between two samples on p different variables, as was the case with Example 6.3. The question of interest is whether the set of statistics $S_1 = (s_{11}, s_{12}, \ldots, s_{1p})$ for an observed set of data is unusual in comparison with the alternative sets $S_2, S_3, \ldots, S_n$.

A problem here is to define what 'unusual' means when there is no simple and obvious way to put the sets of statistics S_1 to S_n in order. Assuming that large values are evidence against the null hypothesis of interest for all of the p individual test statistics, Gates suggests ordering S_1 to S_n such that $S_i > S_j$ when all of the components in S_i are larger than those of S_j. Thus the observed S_1 is considered to be significant at the $100\alpha\%$ level if it is among the largest $100\alpha\%$ values in this respect. If necessary, supplementary statistics are simulated in order to obtain a sufficient number of statistics that can be ordered in this way.

There are times when there is interest in seeing whether two samples seem to have come from the same distribution without specifying any particular alternative hypothesis of interest, such as a difference in means. In this situation the two-sample version of Orlowski *et al.*'s (1991) empirical coverage test may be used, as discussed further in section 7.7.

6.10   CHAPTER SUMMARY

- Fisher (1935) introduced the idea of a paired comparison test based on randomizing the sign of the differences of paired scores using some data collected by Charles Darwin on cross-fertilized and self-fertilized plants.
- With Fisher's example the outcome of the test depends on the test statistic used. This is as it should be, because in general the outcome of a statistical test should depend on the data and the test statistic. It is an advantage of randomization tests that the user can choose the appropriate statistic.
- Fisher's paired comparison randomization test can be used with trimming of extreme values if necessary.
- For small samples the complete randomization distribution for Fisher's paired comparison test can be generated. An algorithm is provided.
- Confidence intervals for a population mean difference can be derived from Fisher's paired comparison test.
- A simulation study has demonstrated the good performance of Fisher's paired comparison test in comparison with alternatives.
- Fisher's one-sample randomization test is similar to the paired comparison randomization test, except that the differences between the observations in one-sample and a hypothetical population mean are calculated and their mean is compared with the distribution for this statistic that is obtained by randomly reallocating signs to the differences.
- The two-sample randomization test involves comparing the mean difference between two samples, or some other suitable statistic, with the distribution of the statistic that is obtained by randomly reallocating observations to samples.

- The bootstrap-t test is an alternative to the two-sample randomization test. With the bootstrap test the t-statistic for comparing two-sample means is compared with a distribution obtained by resampling the original data after making an adjustment to ensure that the null hypothesis is true for bootstrap samples. An allowance for unequal variances is possible by modifying this test.
- A test is sometimes based on the randomization of residuals instead of the randomization of observations. This can be justified as an approximation to an exact test based on randomizing true residuals, or as being equivalent to a balanced bootstrap resampling test. When using residuals care should be taken to ensure that the data sets based on randomized residuals could reasonably have occurred in practice.
- A number of computer-intensive tests for comparing two sample variances have been proposed because it is known that the standard test based on comparing the ratio of the variances with the F-distribution gives poor performance with non-normal data. Tests of the type suggested by Levene (1960) give better performance. These are discussed more fully by Manly and Gonzalez (1997).
- With the Bonferroni method for making an allowance for carrying out several tests at the same time the significance level of each test is adjusted according to how many tests are being made. A randomization alternative is described whereby the significance level for each test is chosen based on the randomization distribution of the significance level for the most significant test.
- Some further extensions and methods are briefly reviewed.

Table 6.6 Consumption (mg dry weight) of Orthoptera by 24 adult male and yearling female lizards

Ants	Orthoptera	Difference	Ants	Orthoptera	Difference
488	0	488	245	0	245
1889	142	1747	50	340	−290
13	0	13	8	0	8
88	52	36	515	0	515
5	94	−89	44	0	44
21	0	21	600	0	600
0	0	0	2	0	2
40	376	−336	242	190	52
18	50	−32	82	0	82
52	429	−377	59	0	59
20	0	20	6	60	−54
233	0	233	105	0	105

6.11 EXERCISES

6.1 Example 6.1 was concerned with data arising from a study by Powell and Russell (1984, 1985) on prey consumption by eastern horned lizards *Phrynosoma douglassi brevirostre*. Table 6.6 shows the prey consumptions of the 24 adult male and yearling females of ants and Orthoptera in milligrams dry biomass, and the difference between these. Test whether the mean difference is significantly different from zero using (a) a paired *t*-test, (b) the Wilcoxon signed ranks test and (c) the paired comparison randomization test described in section 6.1.

6.2 From the same study that is referred to in Exercise 6.1 the prey consumptions in milligrams dry weight for two size classes of lizard

Table 6.7 Consumption (mg dry weight) of Coleoptera and Orthoptera by two size classes of lizards

Size class 1		Size class 2	
Coleoptera	Orthoptera	Coleoptera	Orthoptera
256	0	0	0
209	142	89	0
0	0	0	0
0	52	0	0
0	94	0	0
44	0	163	10
49	0	286	0
117	376	3	0
6	50	843	8
0	429	0	1042
0	0	158	0
75	0	443	137
34	0	311	7
13	340	232	110
0	0	179	0
90	0	179	965
0	0	19	0
32	0	142	0
0	0	100	110
205	190	0	1006
332	0	432	1524
0	0		
31	60		
0	0		

(adult males and yearling females, and adult females) are shown in Table 6.7 for Coleoptera and Orthoptera. The mean difference between the size classes for Coleoptera was tested in Exercise 6.1.

(a) Use a randomization Levene's test to see if there is any evidence that the two size classes differ in their levels of variation for the consumption of Coleoptera. Compare the significance level with what is obtained using the usual F-distribution tables. Note that randomizing residuals suffers from the difficulties discussed at the end of section 6.5.
(b) Test for a significant difference between the consumption level of Orthoptera for the two size classes using a t-test, a Mann–Whitney two-sample test and a randomization test.
(c) Use a randomization Levene's test to compare the levels of variation in the consumption of Orthoptera for the two size classes. Compare the significance level with what is obtained from using the usual F-distribution tables.

7
Analysis of variance

7.1 ONE-FACTOR ANALYSIS OF VARIANCE

The two-sample randomization test described in Chapter 6 generalizes in an obvious way to situations involving the comparison of three or more samples, and to more complicated situations of a type that are usually handled by analysis of variance (Welch, 1937; Pitman, 1937c, Kempthorne, 1952, 1955).

An example of a four-sample situation is provided by considering more data from the study of stomach contents of eastern horned lizards *Phrynosoma douglassi brevirostre* (Powell and Russell, 1984, 1985; Linton *et al.*, 1989) that has been referred to in Example 6.1 and the exercises in Chapter 6. The values shown below are milligrams of dry biomass of ants for 24 adult males and yearling females, for samples taken in four months in 1980:

> June: 13, 242, 105
> July: 8, 59, 20, 2, 245
> August: 515, 488, 88, 233, 50, 600, 82, 40, 52, 1889
> September: 18, 44, 21, 5, 6, 0

Here a randomization test can be used to see whether there is any evidence that the consumption changed with time. This involves choosing a test statistic that is sensitive to the sample means varying with time, and comparing the observed value with the distribution of values that is obtained by randomly allocating the 24 data values to months with the same sample sizes as observed (three in June, five in July etc.).

In a general situation of this kind there will be g groups of items to compare, with sizes of n_1 to n_g, and values for a total of $\sum n_i = n$ items. A randomization test will then involve seeing how an observed test statistic compares with the distribution of values obtained when the n items are randomly allocated, with n_1 going to group 1, n_2 going to group 2, ..., n_g going to group g.

As was the case for the two-sample test, comparing the test statistic with the randomization distribution can be justified in two different situations:

1. In some cases it can be argued that the groups being compared are random samples from distributions of possible items, and the null hypothesis is that these distributions are the same. In that case the group labels are arbitrary and all allocations of items to groups are equally likely to have arisen. This is what has to be assumed with the lizard data just considered.
2. If items are randomly allocated to groups before they receive different treatments then randomization is justified on the null hypothesis that the different treatments have no effect.

A one-factor analysis of variance is summarized in the usual analysis of variance table, which takes the form shown in Table 7.1. Here x_{ij} denotes the jth observation in the ith group, T_i is the sum of the n_i observations in the ith group, and $T = \sum T_i$ is the sum of all the observations.

Usually the F-ratio of the between-group mean square to the within-group mean square is compared with an F-distribution table to determine the level of significance, but this F-ratio can be used just as well as the test statistic for a randomization test. There are a number of equivalent test statistics for the randomization test, including $\sum T_i^2/n_i$, the between-group sum of squares, and the between-group mean square. The randomization version of the one-factor analysis of variance therefore involves calculating either F or an equivalent test statistic and seeing what percentage of times this is exceeded by values from randomized data.

The number of possible randomizations with g groups is the number of choices of observations for group 1, $n!/\{n_1!(n - n_1)!\}$, times the number of choices for observations in group 2 when those for group 1 have been determined, $(n - n_1)!/\{n_2!(n - n_1 - n_2)!\}$, times the number of choices for group 3 when groups 1 and 2 have been determined, and so on, which gives a total of $n!/\{n_1!n_2!\ldots n_g!\}$. For example, with the lizard data above there are $24!/(3!5!10!6) \approx 3.3 \times 10^{11}$ possibilities and complete enumeration is obviously not practical. If complete enumeration is required for a small

Table 7.1 A one-factor analysis of variance table

Source of variation	Sum of squares	Degrees of freedom	Mean square	F
Between groups	$SSB = \sum_i T_i^2/n_i - T^2/n$	$g - 1$	$MSB = SSB/(g - 1)$	MSB/MSW
Within groups	$SSW = \sum_i \sum_j x_{ij}^2 - \sum_i T_i^2/n_i$	$n - g$	$MSW = SSW/(n - g)$	
Total	$SST = \sum_i \sum_j x_{ij}^2 - T^2/n$	$n - 1$		

data set then the Fortran subroutine provided by Berry (1982) may prove useful.

It must be remembered that a randomization test to compare group means may be affected by differences in the amount of variation within the groups (Boik, 1987), as discussed for two groups in section 6.3. This problem arises with some of the examples in this chapter, and is considered further in section 7.5.

7.2   TESTS FOR CONSTANT VARIANCE

One of the assumptions of analysis of variance is that the amount of variation is constant within groups so that the differences in the variance from group to group are only due to sampling errors. The conventional F-test for comparing group means is known to be robust to some departure from this assumption, but large apparent differences in variation between groups must lead to questions about the validity of this test. For this reason, it is common to test for significant differences between the variances in different groups using Bartlett's test.

The Bartlett test statistic is

$$B = C \sum_{i=1}^{g} (n_i - 1) \log_e(s^2/s_i^2), \qquad (7.1)$$

where s_i^2 is the variance calculated from the n_i observations in the ith group, $s^2 = \{\sum(n_i - 1)s_i^2\}/(n - g)$ is the pooled sample variance (MSW in Table 7.1), and

$$C = 1 + [1/\{3(g - 1)\}]\left[\sum\{1/(n_i - 1)\} - 1/\sum(n_i - 1)\right]$$

is a constant that is close to 1 (Bartlett, 1937; Steel and Torrie, 1980, p. 471).

For random samples from a normal distribution the significance of B can be determined by comparison with the chi-squared distribution with $g - 1$ degrees of freedom. However, Bartlett's test using chi-squared tables is known to be very sensitive to non-normality in the distributions from which groups are sampled (Box, 1953). For this reason it may seem reasonable to determine the significance level by comparison with the randomization distribution. In doing this, it should, however, be appreciated that the test will be affected by changes in mean values with factor levels, in the same way as the F-test is for two samples as illustrated in section 6.7. In particular it should be recognized that if observed mean values vary widely with factor levels then the variation within these levels may be substantially lower than what is expected from a random allocation of observations to factor levels, and that this may affect the test for different variances to an undesirable extent. To attempt to overcome this problem, observations can be replaced by deviations from sample means before being randomized (Boos and Brownie, 1989; Boos *et al.*, 1989), although this will result in a test that has only the approximately correct properties when the null hypothesis is true and (as discussed in section 6.5) may result in many randomized data sets being implausible. In fact, as will be shown below, replacing observations with residuals may produce rather unfortunate results.

An alternative to using the Bartlett statistic involves using Levene's (1960) test, as discussed in section 6.6 for the case of two samples. Evidence is provided below which indicates that this sometimes gives a more robust procedure. It involves replacing the observations with absolute deviations from group medians and carrying out a one-factor analysis of variance. The *F*-value from this analysis of variance is then the test statistic. The idea is that large differences in the amount of variation displayed within groups will be converted by the transformation to large differences between group means.

7.3 TESTING FOR MEAN DIFFERENCES USING RESIDUALS

The randomization of residuals that has been proposed for testing for variance differences can also be used for testing for mean differences, as suggested by ter Braak (1992) and discussed in section 6.5 for the case of two groups. The idea is then that the residuals will be approximately equal to the deviations of the observations from the mean values for the populations from which those observations are drawn. Randomization is then justified by the assumption that all the deviations from population means have the same distribution. This justification is not as simple as the one for the usual randomization test, which is essentially that the observed data come in a random order. Nevertheless, it is plausible that the randomization of residuals will produce a test with good properties. An algorithm for a test based on residuals has the following steps.

Algorithm for testing for a mean difference using residuals

1. Calculate the *F*-statistic for the observed data, F_1, in the usual way.
2. Find the residual for each observation as the difference between that observation and the mean for all observations in the same group.
3. Randomly allocate the residuals to groups of the same size as for

---

the original data and calculate F_2, the F-statistic from an analysis of variance on the data thus obtained.
4. Repeat step 3 a large number N of times to generate values $F_2, F_3, \ldots, F_N$.
5. Declare F_1 to be significantly large at the $100\alpha\%$ level if it exceeds $100(1 - \alpha)\%$ of the values $F_1, F_2, \ldots, F_N$.

A modification to this algorithm involves replacing step 3 with the resampling of residuals with replacement to produce a new set of data. This then gives a bootstrap test of significance which seems likely to have rather similar properties to the test that uses each residual once. There is, however, a difference in principle between the randomization test and the bootstrap test, because the randomization test is based on the idea that residuals come in a random order, while bootstrapping is thought of as approximating the distribution of the F-statistic that would be obtained by resampling the populations that the data came from originally.

Example 7.1: ant consumption by lizards in different months   For the data shown above on the milligrams dry biomass of ants in the stomachs of lizards, the analysis of variance table takes the form shown in Table 7.2. The significance level of the F-value by comparison with the F-distribution with three and 20 degrees of freedom is 21.1%. The Bartlett test statistic is $B = 44.69$. This is very highly significant, corresponding to a probability of zero to four decimal places in comparison with tables of the chi-squared distribution with three degrees of freedom. On the other hand, Levene's test gives a statistic of $L = 1.51$, corresponding to a significance level of 24.3% in comparison with the F-distribution with three and 20 degrees of freedom. This therefore gives no evidence of a variance difference.

When the observed F-value of 1.64 is compared with the distribution found from this value itself and 4999 alternative F-values obtained by randomly reallocating the data values to months, it is found that the observed value is significantly large at the 18.9% level. If residuals are

Table 7.2 Analysis of variance for the consumption of ants by lizards

Source of variation	Sum of squares	Degrees of freedom	Mean square	F
Between months	726 694	3	242 231	1.64
Within months	2 947 024	20	147 351	

randomized instead, then the significance level is 20.3%. Therefore, either method of randomization gives a result that is quite close to what is obtained using the F-distribution to assess significance. There seems to be little evidence of changes in the consumption level from month to month.

Using the same 4999 random allocations of observations to months as were used for the F-test it was found that this produced 3.3% of Bartlett test statistics as large or larger than the observed 44.69, while randomizing residuals (i.e. replacing each observation by the deviation that it has from the mean for the observations in that month), again with the same 4999 randomizations, gave the most extreme significance level possible of 0.02% (1 in 5000). Thus the level of significance from the randomization of observations is much less extreme than the use of the chi-squared distribution suggests, while the level of significance from the randomization of residuals agrees with the assessment using the F-distribution.

With Levene's test there is much better agreement between the results for different methods of testing. The significance level is 25.9% by randomizing observations and 20.9% by randomizing residuals. Neither of the methods gives evidence of variance differences between months.

In order to understand better the properties of the different methods for testing means and variances, a simulation study along the same lines as the one discussed in section 6.7 was carried out. What was done was to begin with the assumption that the ant consumption figures, which are given at the start of this chapter, were all from the same source, with no real differences between months. Based on this assumption, the allocation to months has no real meaning, and any other allocation was just as likely to have occurred. Therefore, sets of data with various scenarios for the mean differences between months that might have occurred can be obtained by randomly allocating the 24 consumption figures to months and adding a constant to the values for each month corresponding to a month effect. If many sets of data are generated in this way with the same month effects, and each set is analysed using several different tests, then this provides a means of comparing the performance of the tests under conditions similar to what might have occurred in reality.

The results obtained from carrying out the experiment are quite informative. The monthly effects on consumption chosen for use were (0,0,0,0), (0,50,100,200), (0,100,200,400), (0,200,400,800) and (0,400, 800,1600) for the months June to September, in order. Thus the month of June never had any effect added, the month of July had effects from 50 mg to 400 mg added, and so on. For each of these five sets of effects, 1000 sets of data were generated, and each of these sets was analysed as follows:

(a) An F-test for a mean difference between months was carried out with the significance of the observed statistic assessed as being significantly large at the 5% level if it was among the largest five of the set of 100 values comprising itself and 99 alternative values obtained by randomly reallocating the data values to months. Bartlett's test and Levene's tests were carried out as well.

(b) The same statistics as for (a) were assessed for significance by randomizing residuals between months.

(c) The F-statistic and Bartlett's statistic as for (a) were assessed for significance in comparison to sets of 100 values comprising themselves and 99 values obtained from bootstrap resampling of residuals to produce alternative sets of data.

(d) Finally, the F, Bartlett and Levene statistics were assessed for significance in comparison with the F-distribution with three and 20 degrees of freedom, and the chi-squared distribution with three degrees of freedom.

The surprising results from this simulation experiment are shown in Table 7.3. It seems that all four methods for testing for mean differences give fairly similar results, and all give about the desired 5% significant results when there are no month effects. This is reassuring, and suggests that the method chosen to test for mean differences is not crucial.

The null hypothesis is always true for the tests to compare the level of variation in the observations within months. Therefore, it was desirable that about 5% of sets of data gave significant results for these tests. Instead, the Bartlett's test with significance based on randomizing observations has proved to be more powerful than the F-test for detecting changes in the mean! The randomizing of residuals and the bootstrapping of residuals have also given far too many significant results for this statistic, and the use of the chi-squared distribution has given a significant result for every set of data that was generated.

Levene's statistic for testing for variance differences has given results that are much more consistent, although the use of the F-distribution has tended to give too few significant results and randomizing observations has tended to give too many significant results. Overall, Levene's test with randomized residuals gave the best performance.

It is clear from a little thought that assessing the significance of Bartlett's statistic by randomizing residuals should give about the same results no matter what month effects are added to the data because these month effects are removed when the residuals are calculated. The same is true for bootstrapping residuals. It is rather surprising that using residuals gives such poor results. This is presumably due to the unusual nature of the data, with one very large value, although it must also be remembered that there

Table 7.3 Results of a study of the performance of different tests to compare means and variances. The table gives the percentages of significant results for tests at a nominal 5% level for an F-test (F), Bartlett's test (B), and Levene's test (L), based on (a) randomizing observations, (b) randomizing residuals, (c) bootstrap resampling of residuals and (d) the use of the F- and chi-squared distributions. Cases shown in italic are where the desired result is 5% but the actual result is significantly different from this at the 5% level (outside the range 3.6% to 6.4%). Bold type shows where there was a month effect and the test gave the lowest observed power for detecting this

Percentage significant results from

Month effect				Randomizing observations			Randomizing residuals			Bootstrapping residuals		Use of the F- and chi-squared distributions		
Jun	Jul	Aug	Sep	F	B	L	F	B	L	F	B	F	B	L
0	0	0	0	*5.7*	*5.1*	*4.6*	*5.4*	47.0	*5.2*	*5.1*	34.6	*4.8*	100.0	*3.9*
0	50	100	200	6.5	21.4	*5.2*	*5.6*	46.2	*4.9*	*5.3*	35.0	**4.5**	100.0	*3.7*
0	100	200	400	18.3	49.3	*6.4*	17.8	45.1	*5.2*	16.0	33.6	**15.0**	100.0	*3.7*
0	200	400	800	**70.8**	80.0	7.9	74.1	46.8	*4.2*	72.0	34.6	71.8	100.0	3.0
0	400	800	1600	100.0	96.2	8.7	100.0	47.1	*4.8*	100.0	34.6	100.0	100.0	3.4

are no guarantees that methods based on randomizing residuals will have the desired behaviour when the null hypothesis is true.

Boos and Brownie (1989) suggest that using residuals from 20% trimmed means may improve the properties of tests in situations with extreme distributions and small sample sizes. However, they also caution against the use of tests for distributions that are more non-normal than the exponential. The advantages of using trimmed means were not investigated with the lizard data on the grounds that the main potential advantage of computer-intensive methods (applicability to all data) disappears when it becomes necessary to make changes to procedures in order to cope with unusual distributions.

In terms of the analysis of the actual data on ant consumption by lizards, the simulation study suggests that the reality may be that there are no differences in variation from month to month because of the results using Levene's statistic, and some differences in means because of the F-tests.

The randomization analyses described in this example for the observed set of data on ant consumption by lizards were carried out using the package RT (Manly, 1996a). However, the simulation study required a special computer program to be written.

The results from this simulation study should be treated with some caution. More extensive simulations indicate that Levene's test based on residuals is not completely reliable with highly skewed data, and that a modification using a type of partial residuals has slightly better properties (Manly and Gonzalez, 1997).

7.4   EXAMPLES OF MORE COMPLICATED TYPES OF ANALYSIS OF VARIANCE

A complete coverage of randomization tests within the area of statistics that is usually referred to as experimental design and analysis of variance is beyond the scope of the present book. A fuller treatment is provided by Edgington (1987), with the emphasis being on the types of design that are useful with experiments in psychology. The approach here is to use some examples to discuss the general principles involved. To begin, with a two-factor situation is considered.

Example 7.2: ant consumption by two sizes of lizard in four months (two-factor analysis of variance with replication)   Consider some more data from the study by Powell and Russell (1984, 1985) on the diet of the eastern horned lizard *Phrynosoma douglassi brevirostre*, which has already been used as the basis for Example 7.1. Table 7.4 shows the stomach contents of ants for 24 lizards classified according to the size of the lizard (adult males and yearling females, or adult females), and the month of collection

Table 7.4 Ant consumption (milligrams dry biomass) by two sizes of lizard in four months

	Adult males and yearling females			Adult females		
June	13	242	105	182	21	7
July	8	59	20	24	312	68
August	515	488	88	460	1223	990
September	18	44	21	140	40	27

Table 7.5 Analysis of variance from the data on the consumption of ants by lizards

Source of variation	Sum of squares	Degrees of freedom	Mean square	F
Months	1 379 495	3	459 831	14.06
Size classes	146 172	1	146 172	4.47
Interaction	294 010	3	98 003	3.00
Residual	523 222	16	32 701	

(June, July, August or September), with three lizards in each size–month category.

There are three types of effect that there might be interest in detecting: differences between months, differences between size classes, and an interaction between these two effects. Here, the interaction can be interpreted as a difference between the two size classes changing in magnitude from month to month or, alternatively, as a difference between months that is not the same for each size class. A conventional two-factor analysis of variance provides the results shown in Table 7.5. Both of the factors 'months' and 'size classes' are regarded as having fixed effects. That is to say, no other months or size classes are being considered. If one or either of the factors was regarded as having levels that were selected at random from a population of levels then some of the mean square ratios for the F-values would be different (Montgomery, 1984, p. 215).

If the F-statistics shown in Table 7.5 are compared with the F-distribution then it is found that the significance level for months is 0.01%, the significance level for size classes is 5.1%, and the significance level for interaction is 6.2%. On this basis there is very strong evidence of mean differences between months, while the difference between size classes and the interaction between months and size classes are almost significant at the 5% level.

In the context of this example there are several different approaches that have been suggested for randomization or bootstrap tests. These do not give the same levels of significance for the three different effects that are of interest, and it is therefore necessary to consider which of these methods is best. Because the different methods can be applied to other sets of data with two or more factors this matter is of crucial importance and deserves careful consideration.

The methods are:

(a) If the null hypothesis is considered to be that the observations for each month–size combination are random samples from the same distribution, then the appropriate randomization involves freely allocating the 24 observations to the month–size combinations, with the only constraint being that three observations are needed for each combination. There are $24!/(3!)^8 \approx 3.7 \times 10^{17}$ ways of making this allocation. Obvious choices for the test statistics to be used for testing the effects of interest are F_M, the ratio of the between month mean square to the error mean square; F_S, the ratio of the between size class mean square to the error mean square; and $F_{M \times S}$, the ratio of the interaction mean square to the error mean square. The significance levels obtained using these statistics when 4999 random allocations plus the observed allocation were used to approximate the null hypothesis distributions of the test statistics were 0.04% for F_M, 4.4% for F_S, and 4.7% for $F_{M \times S}$. This analysis therefore gives very strong evidence of a mean difference between months, with some evidence of a difference between size classes and some evidence of an interaction.

(b) A second analysis uses the same randomization method as for (a), but the test statistics are the mean squares MS_M, MS_S, and $MS_{M \times S}$ for months, size classes and interaction, respectively. With the same randomizations as used for (a) above, this gives significance levels of 0.04% for MS_M, 27.0% for MS_S, and 47.1% for $MS_{M \times S}$. This then gives very strong evidence of differences between months, but no evidence of differences between size classes or of an interaction.

(c) Edgington (1995, Chapter 6) suggests a different method of randomization. His approach for the lizard data involves first considering the effect of months. To test this, observations are interchanged between the four months, keeping the size class constant. For example, the 12 observations from size class 1 are interchanged between months, but are never moved to size class 2. By this restricted randomization, the effect of size is controlled while testing the effect of months. Either of the test statistics F_M and MS_S can be used. To test for a difference between the two size classes, observations are randomized between sizes, keeping the month constant, in a similar way as described for testing for a difference between months. Either of the statistics F_S and MS_S can be used. Edgington argues that true randomization testing of interactions is not possible in the two-factor situation. He does, however, suggest that if observations are randomized freely over factor combinations then $F_{M \times S}$ can be used as a test statistic that is sensitive to interactions, which is the same test for interactions as suggested in (a). Using the observed data plus 4999 restricted randomizations (for tests on F_M and F_S), or 4999 unrestricted randomization (for the test on $F_{M \times S}$), the significance level for F_M was found to be 0.04%, the significance level for F_S was found to be 8.8%, and the significance level for $F_{M \times S}$ was found to be 4.7%. This analysis gives very strong evidence of a difference between months, no real evidence of an overall difference between size classes, and some evidence of an interaction.

(d) Still and White (1981) followed Edgington (1980) in suggesting that the overall effects of factors should be tested by restricted randomizations. However, they proposed that in order to test for the interaction between the factors in a two-factor analysis of variance situation it is appropriate to randomize the observations after adjusting them to remove the overall effects of the factors. That is to say, x_{ijk}, the observation for the kth lizard for month i and size class j, should be replaced by $r_{ijk} = x_{ijk} - \bar{x}_{i..} - \bar{x}_{.j.} + \bar{x}_{...}$, where $\bar{x}_{i..}$ is the mean of the observations in the ith month, $\bar{x}_{.j.}$ is the mean of the observation in the jth size class and $\bar{x}_{...}$ is the mean of all observations. The significance of the observed statistic $F_{M \times S}$ for interaction is then compared with the distribution consisting of itself plus alternative values found by randomly reallocating the r_{ijk} values to factor combinations. The idea behind this is that the r_{ijk} values will be unaffected by any overall effects of months or the size of the lizards, and that therefore these effects will not influence the test for interaction. From (c) 4999 restricted randomizations gave a significance level of 0.04% for F_M and 8.8% for F_S, while from 4999 unrestricted randomizations of the residuals r_{ijk} it was found that the significance level for $F_{M \times S}$ is 5.0%. The Still and White analysis therefore gives strong evidence of a mean difference between months, no real evidence of an overall mean difference between size classes, and some evidence of an interaction.

(e) Ter Braak's (1992) method (which has already been discussed in section 7.3) involves replacing each of the observations in the eight different combinations of month and size class by its corresponding residual (the deviation that the observation is from the mean for the factor combination). The observed statistics F_M, F_S and $F_{M \times S}$ are then compared with the distributions of these statistics that are obtained by constructing new sets of data by randomly reallocating the residuals to the combinations. Using the observed statistics from the lizard data plus 4999 values from randomized data gave significance levels of 0.02%, 4.6% and 5.0% for F_M, F_S, and $F_{M \times S}$, respectively. There is very strong evidence for a difference between months and some evidence of differences between size classes and an interaction.

Variations of these methods involving the bootstrapping of observations or residuals are also possible, but will not be considered here.

The differences between the outcomes from the five analyses (a) to (e) are considerable, and emphasize the importance of using an appropriate method of randomization. The only thing that all methods agree about is the existence of a difference between months.

To get some idea of the relative merits of the alternative methods a small simulation experiment was run along the same lines as one described in Example 7.1. The procedure was to assume that the 24 observations in Table 7.4 represent typical values for the stomach contents of ants for eastern horned lizards. These values were allocated in a random order to four months and two size classes of lizards, with three observations for each month–size combination. Month effects, size effects and interaction effects were then added to produce plausible sets of data with the specified levels of these effects.

Nine situations were considered, with 1000 sets of randomized data generated for each of these situation, and each of the sets of data analysed using the various randomization procedures described in (a) to (e) above. The nine situations were (1) no month or size effects; (2) a low difference between months, with the data values increased by 100 in June, 200 in July and 300 in August; (3) a high difference between months, with data values increased by twice the values used in (2); (4) a low difference between size classes, with data values increased by 300 for the second size class; (5) a high difference between months, with data values increased by 600 for the second size class; (6) low differences between months and size classes introduced by adding both of the effects described for (2) and (4); (7) high differences between months introduced by adding both of the effects described for (3) and (5); (8) low differences between months and size classes as described for (6), plus an interaction obtained by increasing all observations in size class 2 in August and September by 300; and (9) high effects for months and size classes plus a high interaction, obtained by doubling all of the effects described for (6).

The results of this simulation experiment are shown in Table 7.6. It can be seen that freely randomizing observations, randomizing residuals as suggested by ter Braak, using Edgington's restricted randomization, and using critical values from the F-distribution all gave very similar results when F-statistics are used for testing. Using sums of squares as test

Table 7.6 Results of a simulation experiment to see how different methods of randomization compare. The table shows the percentage of significant results at the 5% level for 1000 sets of data generated using the effects 1 to 9. Percentages shown in italic are where it is desired that the percentage should be 5% but the observed percentage is significantly different from 5% (outside the range 3.6% to 6.4%). Bold type indicates the test with the lowest observed power when the null hypothesis was not true. Test statistics used are the F-values F1 for the effect of months, F2 for the effect of sizes and F12 for interaction, and the sums of squares S1 for months, S2 for sizes and S12 for interaction. Edgington's method involves restricted randomization, Still and White's method involves randomizing observations after an adjustment to remove the overall effects of months and sizes, ter Braak's method involves randomizing residuals for month–size combinations, and the F-distribution method involves assessing significance using critical values from the F-distribution

Effects added to data	Randomizing observations						Edgington method		Still and White method	ter Braak method			*F*-distribution		
	F1	F2	F12	S1	S2	S12	F1	F2	F12	F1	F2	F12	F1	F2	F12
1 None	4.7	5.1	5.9	5.5	4.4	6.1	5.4	4.7	4.7	5.0	5.7	5.8	4.2	4.7	5.0
2 Low for months	20.7	5.6	5.6	22.5	3.7	2.8	23.4	5.5	4.4	**19.4**	5.3	5.6	**19.4**	5.1	5.1
3 High for months	72.1	4.5	*4.3*	73.9	*1.3*	*0.3*	74.2	4.3	5.3	72.1	4.3	5.5	**71.1**	3.7	4.3
4 Low for size	4.8	**56.1**	4.7	3.1	57.8	*1.8*	6.1	58.1	4.2	5.3	**55.5**	4.9	4.4	56.4	3.7
5 High for size	4.2	**99.8**	4.2	0.1	99.9	0.3	4.5	**99.8**	4.2	5.2	100.0	5.1	4.5	100.0	4.0
6 Low for both	21.5	54.4	3.8	**14.5**	**51.3**	0.9	24.5	54.7	4.2	22.2	53.1	4.4	20.7	53.3	3.7
7 High for both	70.7	99.7	3.3	**29.2**	**97.8**	0.0	74.2	100.0	3.8	72.2	99.9	4.7	73.0	100.0	3.2
8 Interaction with 6	49.5	89.8	11.2	**23.0**	**81.4**	1.3	49.9	88.6	10.6	52.7	90.1	12.9	51.1	90.2	10.6
9 Interaction with 7	99.6	100.0	30.9	**42.6**	100.0	**0.2**	98.5	100.0	33.3	99.9	100.0	33.1	100.0	100.0	31.7

statistics with observations being freely randomized has not given good results because the resulting tests have relatively low power and the presence of one effect has led to the probability of a significant effect falling below the nominal 0.05 level for the tests on other effects. There is also a suggestion that a similar effect occurs with the randomization of observations and scenario (7).

In practical terms, the simulation experiment suggests that effects can be tested by freely randomizing observations, by freely randomizing residuals, or by using Edgington's restricted randomization method, providing that F-statistics are used. For the data in Table 7.5 these approaches all give fairly consistent results. Differences between months are always very highly significant. Differences between size classes vary from being significant at the 4.4% level (randomizing observations) to being significant at the 8.8% level (for restricted randomization). The significance level for interaction is either 4.7% (randomizing observations) or 5.0% (randomizing residuals).

It is hard to say from the simulation experiment which method is best. For the testing of main effects the restricted randomization has the desirable property of being exact for testing one effect in the presence of the other effect. In practice, however, randomizing observations or residuals seems to work just as well, while these alternative approaches can be used for testing interactions as well as main effects. Furthermore, this conclusion has also been confirmed by a more extensive simulation study (Gonzalez and Manly, 1997). Randomization of observations and residuals will therefore be stressed for the remainder of this chapter.

Because there are three lizards available with each of the 12 month–size class combinations for the data in Table 7.4, it is reasonable to expect to be able to test for the level of variation being constant for each of these combinations, although the results of the simulation experiment that are discussed in Example 7.1 suggest that caution is required. What is found is that Bartlett's test statistic for the observed data is 27.09. This is significantly large at the 0.03% level in comparison with the chi-squared distribution with seven degrees of freedom, but this method for assessing significance is, of course, suspect with non-normal data of the type being considered. The significance level in comparison with the distribution consisting of the observed value plus 4999 values obtained by randomly reallocating observations to the eight factor combinations is 69.2%. This is in stark contrast with the significance level of 4.7% found in comparison with the distribution consisting of the observed value plus 4999 values obtained by randomly reallocating residuals to the 8 factor combinations.

Levene's test did not give such extreme differences. The value of $L = 1.02$ has a significance level of 46% in comparison with the F-distribution with 7 and 16 degrees of freedom, 22% from randomizing observations and 19% from randomizing residuals.

In order to assess the properties of tests for differences in variation based on randomizing observations and tests based on randomizing residuals, these were included in the simulation experiment described above. Basically the results were similar to those for Example 7.1. Randomizing observations gave about the correct 5% significant results for Bartlett's test statistic when there were no mean differences between the month–size combinations, but gave far too many significant results when mean differences were present. That is to say, Bartlett's test with observations randomized is sensitive to mean differences as well as differences in variation. On the other hand, testing Bartlett's statistic by the randomization of residuals gave a significant result for about 75% of data sets although the null hypothesis was always true.

Levene's test using the F-distribution to assess significance did not give a single significant test result for all the simulated data. Randomizing observations gave 4.3% significant results when there were no mean differences, but up to 44% when mean differences existed between factor combinations. Both of these procedures therefore failed to work well. By contrast, randomizing residuals gave between 3.5% and 4.7% significant results. These percentages are all less than the desired 5%, but it is clear that this test is much better than the alternatives in terms of the performance when there are no variance differences between factor levels. As noted earlier, these results should be interpreted taking into account the further studies described by Manly and Gonzalez (1997).

Because the simulation results suggest that the only test for variance differences that can be trusted is Levene's test with randomized residuals, and this test gives a non-significant result for the lizard data, it must be concluded that there is no real evidence that the level of variation changed with the different factor combinations.

Example 7.3: Orthoptera consumption by lizards (two-factor analysis of variance without replication)   The randomization methods that were used in the last example can be applied equally well when there is no replication, so that each factor combination has only a single observation. The null hypothesis of no factor effects suggests that test statistics should be compared with the distributions of the same statistics that are obtained when observations or residuals are exchanged freely between factor combinations. It is possible to test for the existence of the main effects of factors, but not for the existence of an interaction.

As an example, consider the data shown in Table 7.7, where each observation is the stomach content of Orthoptera for a lizard in a month and size class. These data are again from Powell and Russell's (1984, 1985) study, as published by Linton *et al.* (1989). Here a conventional analysis of variance produces the results shown in Table 7.8. By comparison with the

Table 7.7 Total consumption of Orthoptera (milligrams of dry biomass) for groups of three lizards in each of four months and two size classes (1, adult males and yearling females; 2, adult females)

Month	Size 1	Size 2
June	190	10
July	0	110
August	52	8
September	50	1212

Table 7.8 Analysis of variance for the data in Table 7.7

Source of variation	Sum of squares	Degrees of freedom	Mean squares	F
Months	491 244	3	163 748	0.88
Size class	137 288	1	137 288	0.73
Residual	561 052	3	187 017	

F-distribution, the significance level for the F-value of 0.88 for differences between months is 54.06% and the significance level for the F-value of 0.73 for differences between size classes is 45.57%. There is no evidence of the existence of either of these effects.

There are 8! ≈ 40 320 possible allocations of the eight observations to the month and size classes. However because the F-values are unchanged by relabelling the months in any of the 4! = 24 possible orders, or relabelling the size classes in any of the two possible orders, it is apparent that there are really only 40 320/(24 × 2) = 840 different configurations that may give different F-values. For a test based on randomizing observations it is therefore quite conceivable to determine all the possible F-values, and hence determine the exact significance levels for the observed F-values. However, this was not done for this example. Instead, the randomization distribution was sampled because this is much easier to do, and the result obtained was quite clear: 4999 randomizations of the data values gave a significance level of 53.78% for the F-value for differences between months, and a significance level of 64.82% for the F-value for differences between size classes. There is no evidence at all that either of these differences exist.

For Edgington's restricted randomization method the observed F-value for differences between months is compared with the distribution obtained by randomizing the values within size classes only. For this distribution there are only 4! = 24 possible F-values, and these can be obtained by keeping the values in size class 1 fixed and reordering the values in size class 2. This is because the F-value for a data configuration depends only on the four pairs of values that are assigned to the four months. It is apparent, therefore, that a test based on restricted randomizations can only give a significant result if the F-value between months for the observed data is the largest of the 24 possible values, in which case the significance level is $(100/24)\% = 4.17\%$. This is not what occurs with the real data.

There are even fewer possible F-values for the restricted randomization test for a difference between the two size classes. There are $2^4 = 16$ possible orders for the pairs of observations within months, but these give only eight possible F-values because these only depend on which four data values are assigned to each size class. It is therefore not even possible to obtain an F-value that is significantly large at the 5% level. The largest possible F-value has a significance level of $(100/8)\% = 12.5\%$.

In this situation, the residuals for ter Braak's (1992) type of randomization are the differences between the original observations and the expected values from a model that allows for effects of months and size classes. These residuals are calculated by subtracting from each observation the estimated effect for the month and the estimated effect for the size class, where these estimated effects are the deviations of the month means and the size classes means from the overall mean. Algebraically, the residual for month i and size class j becomes $r_{ij} = x_{ij} - \bar{x}_{i.} - \bar{x}_{j} + \bar{x}_{..}$, where x_{ij} is the original data value, $\bar{x}_{i.}$ is the mean of the observations in the ith month, $\bar{x}_{j}$ is the mean of the observations in the jth size class, and $\bar{x}_{..}$ is the overall mean. Based on 4999 randomizations of the residuals calculated in this way, the significance level for the observed F-value for differences between months is 55.74%, and the significance level for the observed F-value for differences between size classes is 44.26%. Again, there is no evidence at all of any effects.

To see what happens, a simulation experiment similar to the ones described with Examples 7.1 and 7.2 was conducted. Thus sets of data for four months and two size classes were constructed with the eight observations in Table 7.7 allocated in a random order. Month effects and size class effects were then added at various levels and randomization tests carried out with 100 randomizations and a significance level of 5%. A total of 1000 sets of data were generated for each level of the effects and the percentages of significant results were determined. Without going into more details, it was found that:

(a) Tests based on randomizing observations gave about 5% of significant results when there were no month or size effects, and gave relatively good power for detecting effects when they were present. There was, however, some interference between the tests on the F-values for months and size classes so that, for example, when a month effect was present but a size effect was absent there were no significant results for size differences, although it is desirable that 5% of the data sets should give a significant size effect.

(b) The restricted randomization test for differences between months performed reasonably well, but, as expected, the restricted randomization test for differences between size classes was incapable of giving a significant result.

(c) Ter Braak's method of randomizing residuals gave no significant results when there were no month or size class effects and gave a poor performance in comparison with the results obtained by randomizing observations. For example, when month and size effects were both present at low levels, randomizing observations gave a significant result for differences between months for 7.7% of data sets and a significant result for differences between size classes for 21.9% of data sets. The corresponding percentages from randomizing residuals were 0% for months and 3.4% for size classes.

(d) Tests based on the use of F-distribution tables to assess significance gave very few significant results. For example, there were no significant results when no effects were present instead of the desired 5% significant results.

Overall, it seems fair to say that randomizing observations gave the best results, although the interference between the test for month differences and the test for size class differences is unfortunate.

Example 7.4: proportions of species numbers in different streams, seasons and positions in streams (three-factor analysis of variance)   The data for the next example are shown in Table 7.9. They come from a study of pollution in New Zealand streams, as indicated by the relative numbers of ephemeroptera (E) to numbers of oligochaetes (O) in samples taken from the streams. The question to be considered is whether there is any evidence that the $E/(E + O)$ proportion varied with the stream (A,B,C), the position within the stream (bottom, middle, top) or the season of the year (summer, autumn, winter, spring).

An analysis of variance on the ratios gives the results shown in Table 7.10. Here the F-ratios all have the 'error' mean square as the denominator, because (as discussed further in the next section) fixed effects are being assumed. In comparison with the F-distribution, all the ratios are very highly significant, corresponding to percentage levels of less than 0.1%.

For a randomization analysis the significance of the F-ratios was determined with reference to the distribution given by 4999 random allocations of the observations to the factor combinations and the observed

Table 7.9 The proportion of ephemeroptera out of counts of Ephemeroptera and Oligochaetes from samples of New Zealand streams. The three proportions shown for each stream–position–season combination come from three independent samples taken in the same area at the same time (source: Dr Donald Scott, University of Otago)

Stream	Position	Summer	Autumn	Winter	Spring
A	Bottom	0.17 0.17 0.00	0.00 0.00 0.00	0.00 0.00 0.00	0.00 0.00 0.00
	Middle	0.72 0.41 0.41	0.23 0.38 0.29	0.00 0.00 0.00	0.00 0.00 0.00
	Top	0.41 0.89 0.88	0.33 0.52 0.58	0.09 0.09 0.17	0.17 0.09 0.29
B	Bottom	0.00 0.00 0.00	0.00 0.00 0.00	0.00 0.00 0.00	0.00 0.00 0.00
	Middle	0.23 0.17 0.23	0.00 0.00 0.00	0.00 0.00 0.00	0.00 0.00 0.09
	Top	0.55 0.41 0.44	0.91 0.95 0.90	0.67 0.86 0.83	0.67 0.62 0.66
C	Bottom	0.09 0.17 0.09	0.00 0.00 0.09	0.29 0.00 0.23	0.00 0.00 0.00
	Middle	0.50 0.72 0.66	0.00 0.00 0.00	0.09 0.17 0.00	0.00 0.00 0.00
	Top	0.88 0.91 0.90	0.98 0.95 0.96	0.55 0.44 0.86	0.17 0.09 0.09

allocation. On this basis, all of the F-values in Table 7.10 are again all significantly large at the 0.1% level. The same result was also found when residuals (the original proportions minus the mean of the three proportions for the same stream, position and season) were randomized instead of the original observations. There is therefore clear evidence that the proportion of ephemeroptera varied for all of the 36 factor combinations as well as varying systematically with the levels of the three factors.

The replicated sampling in this study permits an assessment of the assumption of a constant level of variation for different factor combinations

Table 7.10 Analysis of variance on proportions of Ephemeroptera in New Zealand streams

Source of variation	Sum of squares	Degrees of freedom	Mean square	F
Stream	0.1792	2	0.0896	11.88
Position	5.8676	2	2.9338	388.97
Season	0.7468	3	0.1867	24.75
Stream × Position	1.3467	4	0.4489	59.52
Stream × Season	0.8688	6	0.1448	19.20
Position × Season	1.0098	6	0.1683	22.31
Stream × Position × Season	0.7200	12	0.0600	7.95
Error	0.5431	72	0.0075	

that is implicit in the analysis of variance. Bartlett's test statistic cannot be used because there are estimated variances of zero for some factor combinations and, in any case, the results obtained from simulations with earlier examples have shown that this test is not reliable. However, Levene's test as described in section 6.6 can still be used.

For the data being considered there are 36 different factor combinations, each with three observations, and the statistic for Levene's test is $L = 0.94$. By comparison with the F-distribution with 35 and 72 degrees of freedom this has a significance level of 56.7%, by comparison with the distribution generated from 4999 randomizations of observations it has a significance level is 0.2% level, and by comparison with the distribution generated by 4999 randomizations of residuals it has a significance level of 13.5%.

To examine the properties of the different tests on means and variances, another simulation study of the type used in the previous examples in this chapter was carried out. Briefly, artificial sets of data were constructed by taking the observations in Table 7.9 in a random order and then adding to these observations effects for the three factors of stream, position and season. For each set of effects, 1000 sets of data were generated and analysed, using the 5% level of significance for tests. The results of this study indicate that:

(a) Tests based on randomizing observations, randomizing residuals and using the F-distribution all gave very similar performance for assessing the significance of the three factors and their interactions. All three methods had similar power for detecting effects, and gave about the correct 5% of significant results when effects were not present.

(b) Randomizing observations gave 5.4% of significant results for Levene's test when there were no mean differences between the 36 factor combinations, and up to 7.8% significant results when differences between mean values were introduced. The performance was therefore fairly good for this method of testing. Use of the F-distribution was not satisfactory, with no significant results at all, and randomizing residuals was only slightly better, with between 0.3% and 0.8% significant results.

It seems from this simulation study that randomizing observations or residuals gives a reliable method for testing for the effects of factors and their interactions with data of the type being considered. However, Levene's test for changes in the level of variation for different factor combinations only works reasonably well with the randomization of observations. The poor performance of Levene's test with randomization of residuals is unfortunate, because with the previous examples this was the best method. As noted before, the use of Levene's test has been studied in more detail by Manly and Gonzalez (1997). This has shown that the randomization of a type of partial residual has slightly better properties than the randomization of full residuals, but that these properties are still not altogether satisfactory.

Because Levene's test with the randomization of observations gives a highly significant result with the data in Table 7.9 it seems that the amount of variation may well have changed with the different factor combinations. This then casts some doubt on the validity of the tests for factor effects on the mean levels. It is hard to say how important this is in the present example, but just looking at the data suggests that three factor interactions were involved with changes in both the mean level and the amount of variation.

Example 7.5: plasma fluoride concentrations (repeated measures analysis of variance)   Table 7.11 shows some data that resulted from an experiment to compare plasma fluoride concentrations (PFC) for litters of rats given different treatments (Koch *et al.*, 1988). There were 18 litters that were

Table 7.11 Average logarithms of plasma fluoride concentrations from pairs of baby rats in a repeated measures experiment

Age (days)	Dose (μg)	Litter	Minutes after injection		
			15	30	60
6	0.50	1	4.1	3.9	3.3
		2	5.1	4.0	3.2
		3	5.8	5.8	4.4
	0.25	4	4.8	3.4	2.3
		5	3.9	3.5	2.6
		6	5.2	4.8	3.7
	0.10	7	3.3	2.2	1.6
		8	3.4	2.9	1.8
		9	3.7	3.8	2.2
11	0.50	10	5.1	3.5	1.9
		11	5.6	4.6	3.4
		12	5.9	5.0	3.2
	0.25	13	3.9	2.3	1.6
		14	6.5	4.0	2.6
		15	5.2	4.6	2.7
	0.10	16	2.8	2.0	1.8
		17	4.3	3.3	1.9
		18	3.8	3.6	2.6

Table 7.12 Analysis of variance for the repeated measures data in Table 7.11

Source of variation	Sum of squares	Degrees of freedom	Mean squares	F
Between groups				
Age	0.02	1	0.02	0.01
Dose	20.33	2	10.17	6.65
Age × dose	0.21	2	0.10	0.07
Litter within group	18.32	12	1.53	
Within groups				
Time	35.45	2	17.72	109.40
Time × age	1.53	2	0.77	4.74
Time × dose	0.99	4	0.25	1.53
Time × age × dose	0.88	4	0.22	1.36
Within litters	3.88	24	0.16	
Total	79.80	53		

assigned to three dose levels within each of two age levels. Levels of PFC were determined for each litter at three different times after injection. The table shows logarithms of these levels.

Repeated measures experiments like this are a little different from the cases that have been considered before because randomization is involved at two levels. The observations for litters can be randomized to the different combinations of age and dose on the null hypothesis that neither factor has an effect. However, in addition, the observations after different amounts of time can be randomized for each litter on the null hypothesis that time has no effect.

A conventional analysis of variance on the data provides the results given in Table 7.12. Comparison of the F-values with critical values from the appropriate F-distributions shows that there is strong evidence of a time effect, and some evidence of an effect of dose and a time by age interaction.

As in the previous examples, the total sum of squares is unchanged by reallocating the data to the factor levels. However, with the two-stage randomization mentioned above (litters to age–dose combinations, and then randomization of the data to the three observation times) the total of the between-group sums of squares will also remain constant because litter totals will not change. This means that the total of the within-group sums of squares must remain constant as well, so as to maintain the overall total sum of squares constant. It follows that the first level of randomization provides distributions against which to test the three between-group F-ratios, while the second level of randomization provides distributions against which to test the four within-group F-ratios.

As an alternative to the two-stage randomization of observations, ter Braak's (1992) method of randomizing residuals can be used. The residuals in this case are the observations after removing any estimated effects of age, dose, time and the litter involved. Thus for the observation x_{ijkl} on litter l, for level k of the time after injection, level j of the dose, and level i of age, the residual is

$$r_{ijkl} = x_{ijkl} - \bar{x}_{ij.l} - \bar{x}_{ijk.} + \bar{x}_{ij..}$$

where $\bar{x}_{ij.l}$ is the mean of the three observation for litter l, $\bar{x}_{ijk.}$ is the mean of the observations for level i of age, level j of dose and level k of time, and $\bar{x}_{ij..}$ is the mean of the observations at level i of age and level j of dose. For a randomization test these residuals can be randomized freely to produce alternatives to the observed set of data.

The significance of the observed F-ratios was determined by comparing them with randomization distributions as approximated by 4999 randomizations plus the observed data order. This produced the results shown in Table 7.13. The results are very similar when the significance level is determined using the F-distribution, randomizing observations or randomizing residuals. In each case, the effects of dose and time and the time × dose interaction come out significant.

To examine the properties of the tests on repeated measures data of the type being considered, a simulation experiment was run in a similar way to what was done in the other examples in this chapter. That is to say, the

Table 7.13 Significance levels obtained for the F-values in Table 7.12 as determined using F-distributions, a two stage randomization of observations, and a free randomization of residuals

Source of variation	Percentage significance level using		
	F-distributions	Randomizing observations	Randomizing residuals
Age	91.4	90.8	91.8
Dose	1.1	1.4	1.4
Age × dose	93.5	93.9	93.4
Time	0.0	0.0	0.0
Time × age	1.8	2.0	2.0
Time × dose	22.4	22.2	22.9
Time × age × dose	27.8	27.9	27.6

observed values in Table 7.11 were randomly reallocated to the different factor levels and different litters and factor effects of various types were added. The data were then analysed to determine the significance levels of the F-values for the different effects using F-distributions, 99 randomization of observations and 99 randomization of residuals. For each type of effect (e.g. a mean difference between ages 6 days and 10 days), 1000 sets of data were generated. A count was then made of the number of sets of data for which an effect was significant at the 5% level.

This experiment produced very similar results for each of the three methods of determining the significance level for F-values, and all tests had appropriate behaviour for detecting each effect in the presence or absence of the other effects. There was also a suggestion that determining significance using F-distribution tables gives slightly higher power than the use of randomization.

Because none of the data values in Table 7.11 are very extreme, it seems possible that the good performance found using the F-distribution in the simulation experiment may not be a good guide to the situation with all sets of data. Therefore a second simulation was run using as initial data the PFC values without a logarithmic transformation. These values are much more extreme than the logarithms and reflect better a situation where there might be some concern about using F-distributions to assess whether effects are significant. In brief, the results of this simulation indicated a tendency for the percentage of significant results to be less than 5% when an effect was not present and testing used the appropriate F-distribution or the randomization of residuals. On the other hand, the randomization of observations generally gave about the correct behaviour under these conditions, and appears to be somewhat more robust. Results were mixed in terms of power to detect effects of the magnitude considered, but in any case did not differ much with the three testing methods.

The analysis of variance used in this example has not taken any account of the ordering of the three levels of the factors of dose and time. There are extensions to the analysis to take this ordering into account, but these will not be considered here.

7.5   PROCEDURES FOR HANDLING UNEQUAL VARIANCES

As has been mentioned in sections 6.3 and 7.1, the randomization test for a difference between two or more sample means may be upset if the samples come from sources that have the same means but different variances. This is apparent because the null hypothesis for the randomization test is that the samples come from exactly the same source, which is not true if variances are not constant.

There have been various suggestions for computer-intensive methods for comparing means from sources with different variances, where an attempt is made to allow for the changing variance. The bootstrap solution for two samples proposed by Efron and Tibshirani (1993) has already been discussed in section 6.4; see also Fisher and Hall (1990) and Good (1994, p. 47). A randomization solution is suggested by Manly (1995b, 1996c). However, all of these methods require further study to know when they can be relied upon, particularly with grossly non-normal data.

7.6   OTHER ASPECTS OF ANALYSIS OF VARIANCE

In Example 7.2, mention was made of the difference between fixed and random effects factors with analysis of variance. It may be recalled that a fixed effects factor is one where all the levels of interest are included in the experiment. An example would be a temperature factor that is varied over a predetermined range. On the other hand, a random effects factor is one where the levels included in the experiment are a random selection from a population of possible levels. Thus in an experiment involving human subjects, the subjects used in the experiment might be considered to be a random sample from the population of subjects of interest. The distinction between the two types of effect is important when it comes to deciding on what ratios of mean squares should be used with F-tests (Montgomery, 1984).

With randomization testing as carried out for the examples discussed in section 7.4, factors must necessarily be regarded as having fixed effects. The null hypothesis is that none of the factors considered has any effect on the observations, so that the assignment of the observations to the different factor combinations is effectively at random. Testing is conditional on the factor combinations used, irrespective of how these were chosen.

Analysis of variance calculations may be considerably complicated when there are either missing observations or unequal numbers of replicates with different factor combinations (Montgomery, 1984, p. 236). These complications will not be considered here, but it can be noted that in principle any of the standard methods for handling them can be used, with randomization being used to determine the significance level of sums of squares. Also, Edgington (1995, p. 159) suggests using a different statistic for testing main effects when there are unequal numbers of replicates for different factor combinations.

In the examples that have been considered in this chapter it has been found that there is generally a close agreement between the significance levels for F-values from analysis of variance as determined by reference to F-distribution tables and the levels determined by randomization of observations or residuals. This agrees with the findings of Baker and Collier

(1966) that standard tests and randomization tests give good agreement for reasonable ranges for skewness and kurtosis in the data being analysed. Furthermore, when discrepancies are found between these two approaches, these are reduced by increasing sample sizes.

The situation for tests for different amounts of variation for different factor combinations is less satisfactory. Using Bartlett's test statistic has not worked at all well, and using Levene's test statistic has produced mixed results in terms of the performance when the null hypothesis is true. As noted before, Levene's test is discussed further by Manly and Gonzalez (1997).

7.7 FURTHER READING

Analysis of variance is often supplemented by multiple comparisons, to see, for example, which factor levels have significantly different means. A problem here is the large number of comparisons that it may be possible to make, and the need to allow for this in the analysis. Essentially this is the same problem as has been discussed in section 6.8. Randomization and bootstrap methods in this area are discussed at length by Westfall and Young (1993).

Another area where randomization methods can be used is with the comparison of several samples when the data available are zeros and ones indicating 'successes' and 'failures'. In this situation, a one-factor analysis of variance as discussed in section 7.1 can be employed. However, it is better to carry out an analysis that takes account of the special nature of the data.

Soms and Torbeck (1982) discussed this situation and noted that when there are s samples the null hypothesis is usually that all are from populations where the probability of a 'success' is the same. That is, if P_i is the probability of a success in sample i then the null hypothesis is that $P_i = P$ for all i. There are three common alternative hypotheses:

1. $P_i \neq P_1$, for some i (comparisons with sample 1, a control);
2. $P_i \neq P_j$, for some i and j (pairwise comparisons); and
3. $P_1 \leq P_2 \leq \ldots \leq P_s$, with at least one inequality (ordered alternatives).

Soms and Torbeck suggest different test statistics for each of these alternatives, and provide a Fortran program to carry out the testing. See their paper for further details.

There are times when what is required is a comparison between several samples to decide whether these seem to come from the same distribution without specifying any particular form of alternative hypothesis or hypotheses (e.g. different means or different variances). In that case, one possibility is to use a g-sample empirical coverage test as described by

Orlowski *et al.* (1991). This test is able to detect any distributional differences that may exist between the g source populations without making any assumptions other than the randomness of the samples and that the data are continuous.

The test involves first combining and ranking the data. The coverage values for a sample then depend on how the sample is spread out over the full set of data, and the test statistic measures the difference between the observed and expected coverage if all samples are from the same distribution. Orlowski *et al.* suggest estimating the variance of the test statistic by randomly reallocating the observed data values to the samples and then using this for a test based on the normal distribution. It seem simpler, however, just to run a standard randomization test to see whether the observed test statistic is significantly large.

Finally, Legendre *et al.*'s (1990) method for carrying out a one-factor analysis of variance on spatial data can be mentioned. This compares the sum of squares for deviations of a variable from its mean calculated within contiguous areas with the distribution obtained by keeping the sampled locations fixed but varying the positions of the areas. For examples they considered the distribution of the number of species in an area of Quebec in relationship to the nature of the surface deposits in different subareas, and gene frequencies of different linguistic groups in Europe.

7.8 CHAPTER SUMMARY

- A randomization test with one-factor analysis of variance can be carried out by comparing a suitable test statistic, such as the usual F-value, with the distribution found when observations are randomly reallocated to the factor levels.
- A randomization test can be justified by either random sampling from populations or a random allocation of sample units to treatment groups.
- Randomization tests for variance differences between several groups have been proposed based on Bartlett's test and Levene's test.
- Tests for a mean difference between several groups can be based on the randomization of residuals instead of the randomization of observations. Also, resampling of residuals with replacement gives a bootstrap one-factor analysis of variance.
- An example shows that randomizing observations, randomizing residuals, bootstrap resampling of residuals and the use of F-distribution tables all give very similar results for comparing means by one factor analysis of variance.
- With the same example, Levene's test with randomized residuals was found to give better results than several alternatives. However, it is noted

that more extensive studies have indicated that this test does not necessarily have good performance with highly non-normal data.

- An example of two-factor analysis of variance with replication indicates that randomizing observations, randomizing residuals, or using a restricted randomization of observations gives similar results for testing for factor effects and interactions, providing that F-values are used for test statistics. In this case Levene's test with randomized residuals was found to give a fairly satisfactory performance for variance comparisons.

- An example of a two-factor analysis of variance without replication indicated some problems with the use of randomization of residuals, and low power with the restricted randomization of observations. The unrestricted randomization of observations gave the best performance, although the tests for the two-factor effects were not quite independent.

- An example of a three-factor analysis of variance with replication showed that randomizing observations, randomizing residuals and using the F-distribution tables gave similar results for testing factor effects and interactions. Testing for variance differences by randomizing observations with Levene's test worked fairly well in this application. However, randomizing residuals and the use of the F-distribution for assessing the significance of variance differences gave far fewer significant results than expected.

- An example of analysis of variance with repeated measures showed that the randomization of observations, the randomization of residuals and the use of F-distribution tables produced similar results for testing for the significance of factor effects.

- Overall, the several examples considered in the chapter suggest that different methods for carrying out randomization tests to compare factor effects and interactions tend to give very similar results. However, the comparison of sample variances is more complicated, with none of the methods considered being completely satisfactory. This problem is considered more thoroughly by Manly and Gonzalez (1997).

- It is noted that procedures are available for allowing for unequal variances with different groups when testing for mean differences, based on bootstrapping and randomization.

- A number of extensions to the methods discussed in the chapter are briefly mentioned.

7.9 EXERCISES

7.1 Table 7.14 shows Cain and Sheppard's (1950) data on the percentages of yellow *Cepaea nemoralis* snails found in 17 colonies from six different types of habitat in southern England. Use an F-test with

Table 7.14 Percentages of yellow *Cepaea nemoralis* in colonies from six habitat types in southern England

Habitat	Colony				
	1	2	3	4	5
Downland beech	25.0	26.9			
Oakwood	8.1	13.5	3.8		
Mixed deciduous wood	9.1	30.9	17.1	37.4	26.9
Hedgerows	76.2	40.9	58.1	18.4	
Downside long coarse grass	64.2	42.6			
Downside short turf	45.1				

randomization to see if there is any evidence that the mean percentage of yellow snails changed with the habitat. Use Levene's test to compare the variation within different habitats. See how the significance levels obtained compare with those found using tables of the *F*-distribution. Because the data are percentages, the arcsine transformation, $X' = \arcsin\{\sqrt{(X/100)}\}$, can be expected to make the distribution within habitats more normal. See how much results are changed by making this transformation before analysing the data.

7.2 Table 7.15 shows counts of total numbers of aquatic insects collected in two streams in North Carolina in each of four months. For each stream in each month there are six replicate values that were each obtained from a standard square-foot bottom sampler. These data were used as an example by Simpson *et al.* (1960, p. 284) and were collected by W. Hassler. Use an *F*-test with randomization to see if there is any evidence that the mean counts of insects varied significantly with the month or the stream. Use Levene's test to compare the variation within

Table 7.15 Number of aquatic insects taken by a square-foot bottom sampler in Shope Creek and Ball Creek, North Carolina, in the months of December 1952 and March, June and September 1953

Month	Shope Creek						Ball Creek					
December	7	9	19	1	18	15	25	9	16	28	10	14
March	29	37	114	49	24	64	35	45	22	29	18	27
June	124	51	63	81	83	106	20	44	26	127	38	52
September	72	87	100	68	67	9	40	45	263	100	129	115

Table 7.16 Number of quack-grass shoots per square foot 52 days after spraying with maleic hydrazide

Days delay in cultivation	Maleic hydrazide per acre (lb)	Block			
		1	2	3	4
3	0	246	213	272	216
	4	96	213	142	154
	8	62	106	94	92
10	0	324	303	228	207
	4	185	112	139	177
	8	77	67	128	125

the eight different samples. See how the significance levels obtained compare with those found using tables of the F-distribution. Because the data are counts, the square root transformation, $X' = \sqrt{X}$, can be expected to make the distribution within month–stream combinations more like normal distributions with the same variance. See how much the results are changed by making this transformation before analysing the data.

7.3 The data that are shown in Table 7.16 come from an experiment conducted by Zick (1956), and were used as an example by Steel and Torrie (1960, p. 225). The observations are the number of quack-grass shoots per square foot, counted 52 days after spraying with maleic hydrazide. The first experimental factor was the rate of application of the spray (0, 4 or 8 lb per acre). The second factor was the number of days delay in cultivation after spraying (3 or 10 days). The experiment was repeated on four blocks of land. Use an F-test, with randomization, to see if the mean counts vary significantly with the days of cultivation, amount of maleic hydrazide or the block of land. See how the significance levels obtained compare with those found using the F-distribution table, and see how results change if a square root transformation is applied before carrying out an analysis.

It should be noted that in an experiment carried out in blocks like this one it is conventional to combine the sums of squares for blocks × days of cultivation, blocks × rate of application and blocks × days of cultivation × rate of application to form an error sum of squares, on the assumption that any block effects do not interact with other effects.

8
Regression analysis

8.1 SIMPLE LINEAR REGRESSION

Pitman (1937b) discussed a randomization test for assessing the significance of a correlation coefficient in one of his three classic papers on 'significance tests which may be applied to samples from any populations'. Because (as will be shown below) the regression coefficient for a simple linear regression is an equivalent test statistic to the correlation coefficient, his test is in effect the same as the one that will now be considered.

The usual simple linear regression situation occurs when two variables X and Y are both measured on n individuals so that the data have the form $(x_1, y_1), (x_2, y_2), \ldots, (x_n, y_n)$. The model assumed is

$$Y = \alpha + \beta X + \varepsilon$$

where α and β are constants and the ε values are 'errors' that are independent, with the same distribution for each observation. The standard estimators of α and β are then

$$a = \bar{y} - b\bar{x}, \tag{8.1}$$

and

$$b = S_{xy}/S_{xx}, \tag{8.2}$$

respectively, where $\bar{x} = \sum x_i/n$ and $\bar{y} = \sum y_i/n$ are the mean values for X and Y, $S_{xy} = \sum(x_i - \bar{x})(y_i - \bar{y})$, and $S_{xx} = \sum(x_i - \bar{x})^2$, with all the summations being over the n data values. These estimators minimize the error sum of squares

$$SSE = \sum(y_i - a - bx_i)^2.$$

In most examples, the constant α is a nuisance parameter that just reflects the mean values of X and Y. The important parameter is β, which indicates the strength of the relationship between X and Y. Often there is interest in knowing whether b is significantly different from zero, because if it is not there is no real evidence of any linear relationship between X and Y. In other cases there is interest in knowing whether b is significantly different from some hypothetical value β_0 for β.

A randomization test for $\beta = 0$ can be conducted by comparing the observed value of b with the distribution of values that is obtained by pairing the X and Y values at random. This can be justified on one of three grounds:

1. The (X, Y) pairs are a random sample from a population where X and Y are independent so that all possible pairings are equally likely to occur; random sampling justifies the test.
2. The data are obtained from an experiment where the X values are randomly assigned to n units and an experimental response Y is obtained. All possible pairings of X and Y are then equally likely if the distribution of Y values is the same for all values of X; the random allocation of X values justifies the test.
3. The circumstances suggest that if X and Y are unrelated then all possible pairings of values are equally likely to occur as a result of the mechanism generating the data. This is a weaker justification than the previous two, but is nevertheless a possible justification in some cases.

From the form of equation (8.2) it can be seen that S_{xy} is an equivalent test statistic to b for a randomization test because S_{xx} does not change with randomization. The simple correlation between X and Y, $r = S_{xy}/\sqrt{(S_{xx}S_{yy})}$ where $S_{yy} = \sum(y_i - \bar{y})^2$, is an equivalent test statistic for the same reason. It is less obvious that the usual t-statistic for testing for whether the regression coefficient b is significantly different from zero is also an equivalent statistic for a randomization test. However, this statistic has the form $t = b/SE(b)$, where $SE(b) = \sqrt{\{(S_{yy} - b^2 S_{xx})/(n - 2)\}}$ is the estimated standard error of b. This is a monotonically increasing function of b, and therefore t and b are equivalent.

8.2 RANDOMIZING RESIDUALS

Instead of randomly reallocating the Y values to the X values it is possible to calculate the regression residuals $r_i = y_i - a - bx_i$ and compare the observed value of $t = b/SE(b)$ with the distribution of this statistic that is obtained when the residuals are randomly allocated to the X values and the residuals are used in place of the Y values for regressions. The benefits of this alternative type of test are not obvious in the simple regression situation, particularly as there is no longer any guarantee that the correct behaviour will be obtained when the null hypothesis $\beta = 0$ is true. Nevertheless, the possibility of randomizing residuals is mentioned here because there are potential benefits from this approach in the context of multiple regression, as will be discussed in section 8.6.

The justification for randomizing residuals is the idea that these residuals will be approximately equal to the errors ε in the assumed model $Y = \alpha + \beta X + \varepsilon$. Therefore, if $\beta = 0$, then randomizing the residuals and regressing them against the X values will generate a distribution for t that approximates what would be obtained if the true errors were equally likely to have occurred in any order. The constant α has no effect in this respect because the t-statistic is unchanged by adding a constant to all the values that are regressed against the X values.

Randomizing of residuals for regression and analysis of variance was proposed by ter Braak (1992) and has already been discussed in section 6.5 and Chapter 7.

Example 8.1: electrophoretic frequencies of Euphydryas editha *related to altitude* Table 8.1 shows the percentage frequencies obtained by McKechnie *et al.* (1975) for hexokinase (Hk) 1.00 mobility genes from electrophoresis of samples of the butterfly *Euphrydryas editha* in some colonies in California and Oregon, and the altitudes of these colonies in thousands of feet. (The precipitation and temperature variables will be ignored for the present.) Originally 21 colonies were studied, but here the

Table 8.1 Environmental variables and the frequencies of the Hk 1.00 gene for colonies of *Euphydryas editha* from California and Oregon. The colony labels are as used by McKechnie *et al.* (1975)

Colony	Altitude ('000 ft)	1/Altitude	Annual precipitation (inches)	Maximum temp. (°F)	Minimum temp (°F)	Hk 1.00 freq (%)
PD+SS	0.50	2.00	58	97	16	98
SB	0.80	1.25	20	92	32	36
WSB	0.57	1.75	28	98	26	72
JRC+JRH	0.55	1.82	28	98	26	67
SJ	0.38	2.63	15	99	28	82
CR	0.93	1.08	21	99	28	72
MI	0.48	2.08	24	101	27	65
UO+LO	0.63	1.59	10	101	27	1
DP	1.50	0.67	19	99	23	40
PZ	1.75	0.57	22	101	27	39
MC	2.00	0.50	58	100	18	9
HH	4.20	0.24	36	95	13	19
IF	2.50	0.40	34	102	16	42
AF	2.00	0.50	21	105	20	37
SL	6.50	0.15	40	83	0	16
GH	7.85	0.13	42	84	5	4
EP	8.95	0.11	57	79	−7	1
GL	10.50	0.10	50	81	−12	4

results have been combined for some close colonies with similar environments and gene frequencies in order to obtain more independent data. A plot of the Hk 1.00 frequency (Y) against altitude indicates a non-linear relationship, with the higher colonies tending to have lower gene frequencies (Figure 8.1(a)). However, a plot of the gene frequencies against $X = 1/\text{altitude}$ indicates an approximately linear relationship (Figure 8.1(b)). The present example is concerned with the significance and strength of the relationship between X and Y.

A linear regression of Y on X gives the fitted equation

$$Y = 10.65 + 29.15X,$$

where the estimated standard error of the coefficient of X calculated from standard regression theory is $SE(b) = 6.04$, with 16 degrees of freedom. A test for the significance of the regression therefore involves comparing $b/SE(b) = 29.15/6.04 = 4.83$ with the t-distribution with 16 degrees of freedom. On this basis the observed value is significantly different from zero at the 0.02% level. These calculations involve the assumption that the errors in the regression relationship are independently normally distributed with a mean of zero and a constant variance.

The randomization distribution of $b/SE(b)$ can be approximated by the observed value and 4999 values obtained by randomly pairing the observed X and Y values because it is not practical to determine the full distribution that consists of $18! \approx 6.4 \times 10^{15}$ possible values. Comparing the observed 4.83 with the randomization distribution indicates a significance level of 0.04% because 4.83 is further from zero than 4998 of the randomized values. This is virtually the same significance level as obtained using the t-distribution.

When residuals were randomized 4999 times instead of the Y values, as discussed in the previous section, the significance level of the t-statistic was found to be 0.02%. Essentially the same result as from randomizing observations was therefore obtained.

The randomization calculations for this example were carried out using the RT package (Manly, 1996a). This package can be used either for simple linear regression, as described above, or for multiple linear regression as described later in this chapter.

8.3 TESTING FOR A NON-ZERO β-VALUE

As mentioned before, there may be occasions when it is desirable to test whether the regression coefficient β is equal to a specified value β_0. This is usually done by seeing whether $(b - \beta_0)/SE(b)$ is significantly different from zero in comparison with the t-distribution with $n - 2$ degrees of freedom. A randomization test can be carried out by noting that if $\beta = \beta_0$

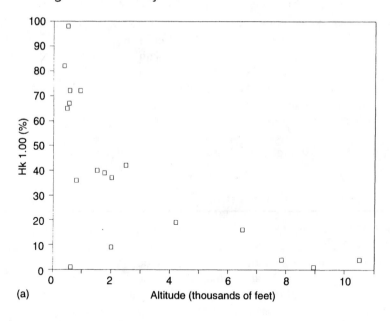

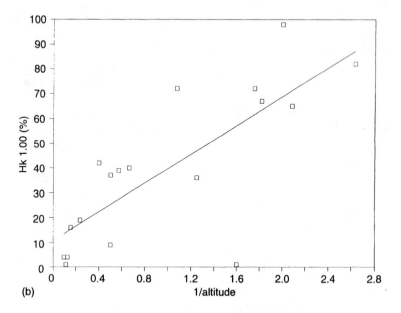

Figure 8.1 Plot of Hk 1.00 frequencies against (a) altitude in thousands of feet, and (b) 1/altitude. The regression line $Y = 10.74 + 29.05X$ is also plotted on (b).

then $Z = Y - \beta_0 X = \alpha + \varepsilon$ will be independent of X. Hence a test for a significant regression of $Y - \beta_0 X$ against X is equivalent to testing $\beta = \beta_0$.

For example, suppose that in the previous example there is some reason for believing that the true regression equation is $Y = \alpha + 20X + \varepsilon$. The values $Z = Y - 20X$ must then be calculated for the 18 cases in Table 8.1 and a test made to see whether there is a significant regression of Z on X. When this was done and the randomization distribution of b was approximated by the observed value and 4999 randomized values it was found that the observed value is further from zero than 85% of randomized values. The significance level is therefore about 15%, which provides little evidence of any linear relationship. Hence the observed data are consistent with the hypothesis that the regression coefficient of Y on X is 20.

8.4   CONFIDENCE LIMITS FOR β

The idea behind testing for $\beta = \beta_0$ when β_0 is not zero can be used to determine confidence limits β for based on randomization. The $(100 - \alpha)\%$ confidence interval consists of the range of values for β_0 for which testing $\beta = \beta_0$ does not give a significant result on a two-sided test at the $\alpha\%$ level. For example, a 95% confidence interval consists of all values of β_0 that do not give a significant result for a test at the 5% level. The limits consist of the value β_L such that the coefficient of $Y - \beta_L X$ regressed on X is exceeded by 2.5% of randomized values, and the value β_U such that the coefficient of $Y - \beta_U X$ regressed on X is greater than 2.5% of randomized values. Any value of β_0 within the range β_L to β_U will not give a significant result for a two-sided test of $\beta = \beta_0$ at the 5% level and can therefore be considered a plausible value for the regression coefficient.

Example 8.2: confidence limits from the Euphydryas editha *data*   Consider the determination of confidence limits for the regression coefficient of the percentage of Hk 1.00 genes on 1/altitude for the data of Example 8.1. To find these, a range of trial values of β_0 was tried for the regression of $Z = Y - \beta_0 X$ against X. As a result it was found that if β_0 is 16.0 or less then the regression coefficient of Z on X is exceeded by 2.5% or less of randomizations, and if β_0 is 42.6 or more then the regression coefficient of Z on X is greater than 2.5% or less of randomizations. The 95% confidence limits for the true regression coefficient are therefore approximately 16.0 to 42.6. These limits were determined by approximating the randomization distributions of regression coefficients by 4999 randomized pairings of X and Y values, plus the observed pairing.

A standard method for determining 95% confidence limits, assuming that the regression errors are independently and normally distributed with a mean of zero and a constant variance, involves determining these as

$b \pm tSE(b)$, where t is the upper 2.5% point for the t-distribution with $n - 2$ degrees of freedom. With the data being considered, this gives the limits as $29.05 \pm 2.12(6.04)$, or 16.2 to 41.9, which are slightly narrower than the ones that were obtained by the randomization.

8.5 MULTIPLE LINEAR REGRESSION

Multiple linear regression is the generalization of simple linear regression for cases where a Y variable is related to p variables $X_1, X_2, \ldots, X_p$. The model usually assumed is

$$Y = \beta_0 + \beta_1 X_1 + \beta_2 X_2 + \ldots + \beta_p X_p + \varepsilon,$$

where ε is a random error, with expected value zero, that has the same distribution for all values of the X variables. Here it will be supposed that n observations are available for estimating the model, with the ith of these comprising a value y_i for Y, with the corresponding values $x_{i1}, x_{i2}, \ldots, x_{ip}$ for the X variables.

Before discussing the use of randomization for making inferences in this situation, it is useful to give a brief review of pertinent aspects of the theory of multiple regression. First, recall that the usual estimators of the regression coefficients $\beta_0, \beta_1, \ldots, \beta_p$ are given by the matrix equation

$$\mathbf{b} = (\mathbf{X}'\mathbf{X})^{-1}\mathbf{X}'\mathbf{Y},$$

where

$$\mathbf{b} = \begin{bmatrix} b_0 \\ b_1 \\ \vdots \\ b_p \end{bmatrix}, \quad \mathbf{Y} = \begin{bmatrix} y_1 \\ y_2 \\ \vdots \\ y_n \end{bmatrix}, \quad \text{and} \quad \mathbf{x} = \begin{bmatrix} 1 & x_{11} & x_{12} & \cdots & x_{1p} \\ 1 & x_{21} & x_{22} & & x_{2p} \\ \vdots & & & \ddots & \vdots \\ 1 & x_{n1} & x_{n2} & \cdots & x_{np} \end{bmatrix}.$$

Here b_i is the estimator of β_i which will exist if the inverse of the matrix $\mathbf{X}'\mathbf{X}$ can be calculated. A necessary condition is that the number of data points is at least equal to the number of β values, so that $n \geq p + 1$.

The total variation in the Y values can be measured by the total sum of squares

$$SST = \sum_{i=1}^{n}(y_i - \bar{y})^2,$$

where $\bar{y}$ is the mean of Y. This sum of squares can be partitioned into an error or residual sum of squares

$$SSE = \sum_{i=1}^{n}\{y_i - (b_0 + b_1 x_{i1} + b_2 x_{i2} + \ldots + b_p x_{ip})\}^2,$$

Table 8.2 Analysis of variance table for a multiple regression analysis

Source of variation	Sum of squares	Degrees of freedom	Mean square	F
Regression	SSR	p	$MSR = SSR/p$	MSR/MSE
Error	SSE	$n - p - 1$	$MSE = SSE/(n - p - 1)$	
Total	SST	$n - 1$		

and a sum of squares accounted for by the regression $SSR = SST - SSE$. The proportion of variation accounted for is the coefficient of multiple determination, $R^2 = 1 - SSE/SST$.

There are a variety of inference procedures that can be applied in the multiple regression situation when the regression errors ε are assumed to be independent random variables from a normal distribution with a mean of zero and constant variance σ^2. A test for whether the fitted equation accounts for a significant proportion of the total variation in Y can be based on the analysis of variance shown in Table 8.2. Here the F-value can be compared with critical values from the F-distribution with p and $n - p - 1$ degrees of freedom to see if it is significantly large. If this is the case then there is evidence that Y is related to one or more of the X variables. Also, to test the hypothesis that β_j equals the particular value β_{j0}, the statistic

$$t = (b_j - \beta_{j0})/SE(b_j)$$

can be computed, where $SE(b_j)$ is the estimated standard error of b_j. This statistic is compared with the percentage points of the t-distribution with $n - p - 1$ degrees of freedom. If the statistic is significantly different from zero then there is evidence that β_j does not equal β_{j0}.

One of the difficulties that is often encountered with multiple regression involves the assessment of the relationship between Y and one of the X variables when the X variables themselves are highly correlated. In such cases there is often value in considering the variation in Y that is accounted for by a variable X_j when this is included in the regression after some of the other variables are already in. Thus if the variables X_1 to X_p are in the order of their importance it is interesting to successively fit regressions relating Y to X_1, Y to X_1 and X_2, and so on up to Y related to all the X variables. The variation in Y accounted for by X_j after allowing for the effects of the variables X_1 to X_{j-1} is given by the extra sum of squares accounted for by adding X_j to the model. Hence, letting $SSR(X_1, X_2, \ldots, X_j)$ indicate the regression sum of squares with variables X_1 to X_j in the equation, the extra

Table 8.3 Analysis of variance table showing extra sums of squares

Source of variation	Sum of squares	Degrees of freedom	Mean square	F
X_1	$SSR(X_1)$	1	$MSR(X_1)$	$F(X_1)$
$X_2|X_1$	$SSR(X_2|X_1)$	1	$MSR(X_2|X_1)$	$F(X_2|X_1)$
.				
.				
.				
$X_p|X_1, ..., X_{p-1}$	$SSR(X_p|X_1, ..., X_{p-1})$	1	$MSR(X_p|X_1, ..., X_{p-1})$	$F(X_p|X_1, ..., X_{p-1})$
Error	SSE	$n - p - 1$	$MSE = SSE/(n - p - 1)$	
Total	SST	$n - 1$		

sum of squares accounted for by X_j on top of X_1 to X_{j-1} is

$$SSR(X_j|X_1, X_2, \ldots, X_{j-1}) = SSR(X_1, X_2, \ldots, X_j) - SSR(X_1, X_2, \ldots, X_{j-1}).$$

This allows an analysis of variance table to be constructed as shown in Table 8.3. Here the mean squares are the sums of squares divided by their degrees of freedom, and the F-ratios are the mean squares divided by the error mean square. A test for the variable X_j being significantly related to Y after allowing for the effects of the variables X_1 to X_{j-1} involves seeing whether $F(X_j|X_1, \ldots, X_{j-1})$ is significantly large in comparison to the F distribution with 1 and $n - p - 1$ degrees of freedom.

If the X variables are uncorrelated then the F-ratios in this analysis of variance table will be the same irrespective of what order the variables are entered into the regression. However, usually the X variables are correlated and the order may be of crucial importance. This merely reflects the fact that with correlated X variables it is generally only possible to talk about the relationship between Y and X_j in terms of which of the other X variables are controlled for. The problems involved in interpreting regression relationships involving correlated X variables are well known (Neter *et al.*, 1983, Chapters 8 and 9; Younger, 1985, p. 405).

8.6 ALTERNATIVE RANDOMIZATION METHODS WITH MULTIPLE REGRESSION

With a randomization test the simplest method for assessing the significance of the t- and F-statistics just discussed is by comparison with the distributions that are obtained when the Y observations are randomly assigned to the sets of X observations. As with simple linear regression, there are then three possible arguments for justifying this approach:

1. from the assumption that the n observations are a random sample from a population of possible observations, where Y may be independent of the X variables;
2. from the random allocation of the X values to n experimental units after which the Y values are observed as responses that may not be affected by the X variables; or
3. from general considerations which suggest that if Y is independent of the X values then the mechanism generating the data makes any of the Y values equally likely to occur with any of the sets of X values.

There are, however, cases where some other form of randomization may seem to be more appropriate. For example, Oja (1987) considered the situation where an experimental design involves n subjects being randomly assigned a treatment variable X and a response Y being observed. In addition, each subject has measured values for k covariates $Z_1, Z_2, \ldots, Z_k$ and the question of interest is whether Y is related to X after allowing for the effects of the covariates. In this case, the appropriate randomization distribution for considering evidence of a relationship between X and Y is the one obtained by randomizing the X values to subjects, rather than the one obtained by randomizing the Y values, because it is the X randomization that is inherent in the experimental design. Oja's work was extended by Collins (1987) who suggested the use of simpler test statistics.

Both Oja and Collins were concerned with approximating randomization distributions with other distributions. A more computer-intensive approach, in the spirit of the present book, would involve carrying out a few thousand randomizations of X values to subjects. A suitable test statistic for a significant relationship between X and Y after allowing for the effect of the covariates would then be the F-value for the extra sum of squares accounted for by X after the covariates are included in the regression equation.

An objection to randomizing observations is that this is not appropriate for testing for the effects of one or more X variables, given that the other X variables may or may not have an effect. This has led various authors to propose the randomization of residuals to produce approximate tests that it is hoped will have better properties in this respect than the tests based on randomizing observations.

One such approach is ter Braak's (1992) method, which has been discussed several times already. This involves fitting the full multiple regression equation, calculating the residuals

$$r_i = y_i - (b_0 + b_1 x_{i1} + b_2 x_{i2} + \ldots, + b_p x_{ip}),$$

and then comparing the usual t- and F-statistics with the distributions that are generated by carrying out multiple regressions of the residuals in a random order against the X values in the order in the original data set. The justification for this approach is the idea that the residuals r_i will be approximately equal to the true errors ε_i in the regression model and that the generated randomization distributions will therefore approximate the distributions that would result from randomizing the true errors. An important point here is that it is the nature of the t- and F-statistics that they are unaffected by the presence of effects that they are not designed to test. For example, the t-statistic for testing whether β_1 is zero is not changed by adding a linear combination of the X variables other than X_1 to the Y values.

In the context of analysis of variance it was shown in the previous chapter that randomizing observations and randomizing residuals as proposed by ter Braak usually give more or less the same results. As will be seen in the examples that follow, this is also the case with multiple regression. Therefore, it seems to matter very little in practice which of these two methods is used for most sets of data, although it may be recalled from section 6.5 that an argument against randomizing residuals with some sets of data is that many randomized sets of data are implausible.

An alternative to using the residuals from the full fitted multiple regression model involves testing for the effects of one or more of the X variables by randomizing the residuals from the fitted regression equation that only contains the other X variables that are not being tested (Levin and Robbins, 1983; Gail *et al.*, 1988). Thus to test for an effect of X_1 and X_2, for example, the residuals from a regression of Y on X_3 to X_p are calculated. A randomization test is then carried out to see whether these residuals are related to X_1 and X_2.

Kennedy (1995) and Kennedy and Cade (1996) have criticized this method for randomizing residuals on the grounds that it is only appropriate when the X variables being tested are uncorrelated with the other X variables that are used to calculate the residuals. They suggested that a better test is provided by regressing Y on the X variables that are not being tested, calculating the residuals and then assessing the significance of test statistics for the variables of interest by comparison with the distributions found from regressions of the residuals against all of the X values (Beaton, 1978; Freedman and Lane, 1983). This method is relatively complicated because the residuals being randomized change according to the variables being tested. Furthermore, according to Kennedy and Cade (1996), it is no better than ter Braak's (1992) method. Consequently, it will not be considered further here.

Another computer-intensive method for multiple regression was proposed recently by Hu and Zidek (1995). This involves the bootstrap resampling of the combinations or residuals that arise in the equations that are solved to obtain the estimated regression coefficients. Initial results suggest that this method is promising, particularly in the situation where the variance of the errors in the regression model is not constant. Randomization could be used with this method instead of bootstrapping, but this does not seem to have been investigated yet.

Kennedy (1995) and Kennedy and Cade (1996) also criticized the randomization of observations for testing for a relationship between Y and some of the X variables when it is accepted that the remaining X variables may be related to Y. They argued that the distribution of t- or F- statistics that is obtained from this type of randomization does not properly take the relationships that may exist into account. Their criticism is quite correct, and in this sense the randomization of observations gives only an approximate test of the type desired. However, all randomization tests are approximate under these conditions and, indeed, tests based on residuals are approximate even when Y is not related to any of the X variables. The question of interest is therefore which tests are generally better approximations.

In fact, numerical examples indicate that tests based on randomizing observations generally have good properties. To provide examples where this is not the case, Kennedy and Cade (1996) had to simulate very extreme situations. Furthermore, they chose to assume normally distributed errors in regression models, which is a situation where use of the t- and F- distributions gives exact results and randomizing residuals can be expected to work particularly well. In reality, randomization tests are most appropriate for use when the regression errors have grossly non-normal distributions, and therefore a key requirement for a testing method is that it performs well for such situations. Simulations results presented below suggest that under these conditions the randomization of observations provides a somewhat more robust procedure for assessing significance than the use of the t- and F-distributions or the randomization of residuals.

In truth, simulations of a few particular situations do not prove that one method of assessing significance is generally better or worse than another method, and a systematic comparisons of methods over a wide range of conditions is badly needed. Luckily, for most real data sets randomization of either residuals or observations gives similar results.

Example 8.3: electrophoretic frequencies of Euphydryas editha *related to altitude, precipitation and temperature*   Consider again the data from the colonies of the butterfly *Euphydryas editha* that are presented in Table 8.1. The linear regression of the Hk 1.00 frequency against the reciprocal of altitude was discussed in Example 8.1. The present example is concerned with the relationship between the Hk 1.00 frequency and all the environmental variables shown in the table, which are $X_1 = 1/(\text{altitude})$, $X_2 = $ annual precipitation, $X_3 = $ maximum annual temperature, and

X_4 = minimum annual temperature. To begin with, a standard analysis of the data will be presented, using tests based on assuming a normal distribution for the regression errors. The same tests will then be considered using randomization to determine significance levels.

The multiple regression equation estimated in the usual way is found to be

$$Y = -88.6 + 26.12X_1 + 0.47X_2 + 0.87X_3 + 0.25X_4.$$
$$\quad\quad\quad (8.65)\quad\quad (0.50)\quad\quad (1.17)\quad\quad (1.02)$$

The values shown beneath the coefficients are the estimated standard errors associated with them. The statistics $t = b_i/SE(b_i)$ with $n - p - 1 = 18 - 4 - 1 = 13$ degrees of freedom are as follows, where the significance level for a two-sided test of the hypothesis that $\beta_i = 0$ based on the t-distribution is indicated in parentheses after each value:

$$b_1, t = 26.12/8.65 = 3.02\ (0.98\%)$$
$$b_2, t = 0.47/0.50\ = 0.94\ (35.8\%)$$
$$b_3, t = 0.87/1.17\ = 0.74\ (47.3\%)$$
$$b_4, t = 0.25/1.02\ = 0.25\ (81.0\%).$$

The coefficient of X_1 is quite significant, but the coefficients of the other variables are not. It seems, therefore, that X_1 alone is sufficient to account for the variation in Y. This is true, even though simple regressions of Y on X_3 and Y on X_4 (which are not presented here) are significant at the 5% level.

An analysis of variance partitioning the total variation of Y into regression and error is shown in Table 8.4. The F-ratio is significantly large at the 0.6% level, showing strong evidence of a relationship between Y and at least one of the X variables. The more detailed analysis of variance based on adding the variables into the equation one at a time is given in Table 8.5. Compared with tables of the F-distributions, the F ratio for X_1 is significantly large at the 0.04% level, but none of the other F-ratios are at all significant. As did the t-tests, this indicates that X_1 alone accounts for most of the variation that the four variables account for together.

Table 8.4 Analysis of variance for a multiple regression of electrophoretic frequencies on four variables

Source of variation	Sum of squares	Degrees of freedom	Mean square	F
Regression	10 454	4	2614	5.96
Error	5 704	13	439	

Table 8.5 Analysis of variance showing the extra sums of squares for the regression of electrophoretic frequencies on four variables

Source of variation	Sum of squares	Degrees of freedom	Mean square	F
X_1	9585	1	9585	21.85
$X_2\|X_1$	88	1	88	0.20
$X_3\|X_1, X_2$	754	1	754	1.72
$X_4\|X_1, X_2, X_3$	26	1	26	0.06
Error	5704	13	439	

Consider now the same analyses, but with significance levels determined by randomly allocating the Y values to the 18 colonies. As noted with Example 8.1, there are 6.4×10^{18} possible allocations, so that it is not realistic to attempt to determine the full distribution. For the purpose of this example the distribution has therefore been approximated by the result for the actual data plus 4999 random allocations. The justification for determining significance by randomization is that if Hk 1.00 frequencies are independent of the environmental variables then the allocation of these frequencies to the environmental variables will appear to be random.

It was found that the absolute t-statistics for assessing the significance of b_1 to b_4 are exceeded by 1.0%, 35.2%, 47.6% and 81.2%, respectively, of their corresponding randomized values. Hence the coefficient of X_1 is significantly different from zero at the 1.0% level, but there is no real evidence that the coefficients of the other X variables are not zero. The F-value of 5.96 for assessing the overall relationship between Y and the X values is significantly large at the 0.5% level, which gives clear evidence that Y is related to at least one of the X variables. Finally, when the F-values for the extra sums of squares shown in Table 8.5 are compared with the distributions for these statistics obtained by randomization, it is found that the observed value for X_1 (21.85) is only exceeded by 0.1% of randomized values, but none of the other F-values is at all significantly large.

If significance is assessed by randomizing residuals as proposed by ter Braak (1992) then the results obtained are almost identical with those that have just been described from randomizing observations. Hence for this example randomization of observations or residuals gives about the same significance levels as are obtained from standard t- and F-tests based on the assumption of a normal distribution for regression errors. Both randomization and standard analyses indicate that 1/altitude accounts for a significant amount of the variation in the Hk 1.00 frequency, with no important extra variation being accounted for by the other environmental variables that have been considered.

A simulation study was carried out based on this example. What was done was to construct artificial data sets with the same X values as for the 18 colonies of *E. editha*, and Y values constructed by assuming various values for the regression coefficients of the four variables. The errors in the Y values were the residuals for the real data, allocated in a random order. The significance of the t-values for the effects of the individual variables, the F-values for the extra sums of squares and the F-value for the overall regression were then assessed by comparison with t-distributions and F-distributions, the randomization of observations, and ter Braak's method for randomizing residuals. One thousand sets of data were generated for each of nine sets of regression coefficients, and 99 randomizations were used for the tests, which were all at the 5% level.

In summary, all three testing methods gave close to 5% significant results when the null hypotheses were true, with overall rates from 34 000 tests of 5.18% for randomizing observations, 5.10% from randomizing residuals and 5.35% from the use of t- and F-distributions. Actually, the figure of 5.35% is significantly higher than 5% on a test at the 5% level. However, this is hardly of practical importance.

In terms of power there was very little difference between the three testing methods overall. There were 48 000 tests where the null hypothesis was not true. The percentages of these yielding significant results were 56.69% from randomizing observations, 56.43% from randomizing residuals and 57.70% from using t- and F-distributions. Thus the use of distributions gave the highest power. However, this advantage can be expected to be partly due to the fact that only 99 randomizations were used for observations and residuals. It is apparent from these simulation results that all three methods of testing are about equally good for data of the type being considered, although the use of the t- and F-distributions has given very slightly too many significant results when null hypotheses were true.

Example 8.4: a case where randomization and classical methods give different results   To show that it is not always the case that randomization gives the same results as using t- and F-distributions to assess significance, consider the artificial set of data with 20 observations on two X variables and a Y variable that are provided in Table 8.6. Here the fitted regression equation of Y on X_1 and X_2 is

$$Y = -5.67 + 2.67X_1 + 9.63X_2,$$
$$\quad\quad\quad (0.70) \quad\; (5.64)$$

with the estimated standard errors of the regression coefficients shown in parentheses below the values. The statistic for determining whether the coefficient of X_1 is significantly related to Y is $t_1 = b_1/SE(b_1) = 2.67/0.70 = 3.81$, with 17 degrees of freedom. From the t-distribution this

Table 8.6 An artificial data set with 20 observations

Case	Y	X_1	X_2	Case	Y	X_1	X_2
1	99.00	33.00	2.09	11	38.09	1.28	1.79
2	5.94	1.97	0.87	12	25.62	1.50	1.53
3	103.45	2.65	2.83	13	7.54	2.33	2.35
4	8.33	2.72	0.42	14	2.51	0.82	0.19
5	3.83	0.70	2.80	15	10.13	2.80	0.72
6	2.82	0.94	1.93	16	4.66	1.55	2.74
7	4.19	0.76	0.82	17	4.15	0.76	2.29
8	2.86	0.95	0.31	18	19.97	0.46	2.93
9	4.18	1.39	1.20	19	4.23	0.70	2.02
10	3.47	0.69	1.17	20	8.83	1.98	1.94

has a significance level of 0.15% on a two-sided test. From a similar calculation, the t-statistic for the coefficient of X_2 is $t_2 = 1.71$, with a significance level of 10.54%. There is therefore very strong evidence that the coefficient of X_1 is non-zero, but little evidence that the coefficient of X_2 is non-zero. Also, the F-value for the overall variation accounted for by the regression is $F = 9.42$ with 2 and 17 degrees of freedom, which is significantly large at the 0.17% level by comparison with the F-distribution, showing very strong evidence of a relationship between Y and at least one of the X variables.

When significance levels of t_1, t_2 and F were estimated from 4999 random allocations of the Y values to the X values, plus the actual allocation, rather different results were found: the estimated significance levels of t_1, t_2 and F were 2.74%, 11.52% and 0.74%, respectively. There is still some evidence that the coefficient of X_1 is non-zero and the equation still accounts for a significant amount of the variation in Y, but the levels of significance are much less extreme. When residuals were randomized 4999 times using ter Braak's method, the results were different again: the estimated significance levels for t_1, t_2 and F are 5.10%, 9.84% and 5.10%, respectively.

This example provides a useful basis for a comparison between the randomization of observations and residuals, as well as demonstrating that either of these methods may produce an outcome that is different from that obtained using t- and F-tables. It is modelled on an example used by Kennedy and Cade (1996) to argue against the validity of randomizing observations.

The process used to generate the data in Table 8.6 was as follows. First, 19 values for X_1 were chosen from the uniform distribution between 0 and 3. The other value of X_1 (for case 1) was an extreme 'outlier' set at 33. Next, 20 values for X_2 were chosen from the uniform distribution between 0 and 3. Finally, Y values were calculated from the equation $Y = 3X_1 + e$, where

Table 8.7 Results of a simulation experiment to examine the properties of three methods for analysing multiple regression data. The tabulated values are the percentage of results from 5000 sets of data where the test statistics t_1 (for β_1), t_2 (for β_2) and F (for the whole equation) were significant at the 5% level. Percentages that are shown in italic (outside the range 4.4% to 5.6%) are significantly different at the 5% level from what is desired (also 5%). Bold type indicates that the method concerned gave the lowest observed power when the null hypothesis was not true

		Results from								
		Randomizing observations			Randomizing residuals			Using t- and F-tables		
β_1	β_2	t_1	t_2	F	t_1	t_2	F	t_1	t_2	F
0	0	5.1	5.3	5.0	*7.1*	*4.1*	5.6	5.2	*2.4*	5.4
2	0	**25.3**	*3.8*	**18.5**	46.9	*4.2*	36.5	94.8	*2.1*	42.5
4	0	91.8	*3.8*	91.1	**49.6**	*4.3*	**49.8**	100.0	*2.5*	100.0
0	5	4.9	18.9	8.5	*6.2*	18.0	**8.0**	4.7	**14.3**	8.3
0	10	5.3	43.4	**16.5**	*6.7*	44.9	17.9	5.1	**39.3**	24.2
2	5	**21.8**	15.2	**24.0**	47.5	16.7	40.5	94.8	**13.0**	54.5
4	10	93.2	40.3	94.9	**48.0**	44.6	**50.2**	100.0	**38.6**	100.0

e was a random disturbance. The error terms were purposely chosen to be highly non-normally distributed. They were random values from the exponential distribution raised to the third power (0.00, 0.00, 95.50, 0.20, 1.70, 0.00, 1.90, 0.00, 0.00, 1.40, 34.20, 21.10, 0.50, 0.00, 1.70, 0.00, 1.90, 18.60, 2.10 and 2.90, for cases 1 to 20 in order). The outlier for X_1 was introduced by Kennedy and Cade (1996) with the idea that this would upset tests based on randomization of observations.

The question that can now be asked is which of the three analyses that have been used has the best properties in the situation being considered. This has been investigated by a simulation experiment similar to a number that have been used with earlier examples. What was done was to take the 20 errors e listed above in a random order and construct alternative sets of data by calculating Y values from the equation $Y = \beta_1 X_1 + \beta_2 X_2 + e$. These sets of data were then analysed with the three methods used on the original data. The values used for β_1 and β_2 were (0,0), (2,0), (4,0), (0,5), (0,10), (2,5) and (4,10), with 5000 sets of data generated and analysed for each of the seven pairs. The percentage of significant results for t_1, t_2 and F was then determined for each pair, with 99 randomizations of observations or residuals.

The simulation results are shown in Table 8.7. It can be seen that:

(a) Randomizing observations gave the best performance in terms of controlling the percentage of significant results close to 5% when the null hypothesis was true. There were 35 000 tests altogether under these conditions, for which randomizing observations gave 4.74% significant results, randomizing residuals gave 5.46% significant results, and using the t- and F-distributions gave 3.91% significant results.

(b) In terms of power when the null hypothesis was not true, using the t- and F-distributions gave the best results overall, although sometimes this method gave the lowest power. There were 70 000 tests carried out under these conditions. Of these, 43.1% were significant by randomizing observations, 37.08% were significant randomizing residuals, and 58.88% were significant using the t- and F-distributions.

To examine the effect of correlation between X variables, the simulation experiment was also carried out using the artificial data shown in Table 8.8. In this case the second X variable was calculated as $0.5X_1 + X_2$ using the values in Table 8.6 for X_1 and X_2. The result is that for the data in Table 8.8 the correlation between the two X variables is 0.972.

The results from this modified simulation experiment are summarized in Table 8.9. Overall, when the null hypothesis was true randomizing observations gave 4.77% significant test results while randomizing residuals gave 4.81% significant test results. Both of these percentages are close to the desired 5%. However, using t- and F-tables only gave 4.03% significant results. In terms of power when the null hypothesis was not true, using the t- and F-distribution was best (with 54.51% significant tests) followed by randomizing observations (with 48.7% significant tests), and randomizing residuals (with 45.34% significant tests).

Table 8.8 An artificial data set of 20 observations with highly correlated X variables

Case	Y	X_1	X_2	Case	Y	X_1	X_2
1	99.00	33.00	18.59	11	38.09	1.28	2.43
2	5.94	1.97	1.86	12	25.62	1.50	2.28
3	103.45	2.65	4.15	13	7.54	2.33	3.51
4	8.33	2.72	1.78	14	2.51	0.82	0.60
5	3.83	0.70	3.15	15	10.13	2.80	2.12
6	2.82	0.94	2.40	16	4.66	1.55	3.51
7	4.19	0.76	1.20	17	4.15	0.76	2.67
8	2.86	0.95	0.79	18	19.97	0.46	3.16
9	4.18	1.39	1.89	19	4.23	0.70	2.37
10	3.47	0.69	1.51	20	8.83	1.98	2.93

Table 8.9 Results of a simulation experiment to examine the properties of three methods for analysing multiple regression data. The tabulated values are the percentage of results from 5000 sets of data where the test statistics t_1 (for β_1), t_2 (for β_2) and F (for the whole equation) were significant at the 5% level. Percentages that are shown in italic (outside the range 4.4% to 5.6%) are significantly different at the 5% level from what is desired (also 5%). Results in bold type indicate that the testing method gave the lowest observed power when the null hypothesis was true

		Results from								
		Randomizing observations			Randomizing residuals			Using t- and F-tables		
β_1	β_2	t_1	t_2	F	t_1	t_2	F	t_1	t_2	F
0	0	5.3	5.2	5.6	4.9	4.4	*5.9*	4.4	*3.3*	*5.8*
2	0	11.6	*4.3*	**24.5**	12.5	4.7	35.2	**11.5**	*3.5*	68.4
4	0	29.2	4.9	91.2	29.7	4.7	**64.3**	**28.4**	*3.5*	100.0
0	5	*3.8*	15.6	68.4	4.5	16.7	**47.2**	*3.6*	**14.7**	100.0
0	10	*4.3*	45.9	96.5	4.6	46.3	**95.9**	4.1	**42.4**	100.0
2	5	**11.9**	17.1	93.6	13.0	16.9	**80.3**	11.9	**14.5**	100.0
4	10	**26.8**	50.5	**99.0**	30.3	46.4	100.0	28.8	**42.5**	100.0

In truth, it must be recognized that using the t- and F-distributions for testing has worked best for these simulations because this gave the highest power on average, while tending to give a significant result even less than 5% of the time when the null hypothesis was true. Randomization of observations has on balance performed better than randomizing residuals, but the results are really quite mixed, with each method of randomization appearing to be the best for some conditions. A more extensive simulation study is clearly needed to compare these and other methods more fully, particularly on highly non-normal data.

The randomization analyses on the original data for this example were performed with the RT package (Manly, 1996a). The simulation experiment required a special computer program to be written.

8.7   BOOTSTRAPPING AND JACKKNIFING WITH REGRESSION

There has been considerable interest in recent years in the use of jackknifing and bootstrapping in the regression context. Jackknifing is carried out as discussed in Chapter 2 by removing the data points one by one to produce partial estimates of regression coefficients, from which pseudo-values can be calculated in the usual way. For a further discussion see Hinkley (1983) and Weber (1986).

Bootstrapping can be done either by resampling the regression residuals or resampling the Y values and their associated X values. If the residuals are resampled this should give results that are similar to those obtained by ter Braak's (1992) method of randomizing residuals, which is a type of perfectly balanced bootstrapping because each of the residuals is used only once. However, bootstrapping the observations has the advantage of being potentially more robust to errors in the regression model itself (Weber, 1986). Of course, if the X values were fixed before Y values were observed (for example from a designed sampling plan) then it is difficult to justify generating new sets of data that could not have occurred by resampling the Y and X values. For a further discussion see Hall (1992a, Chapter 4) and Efron and Tibshirani (1993, Chapter 9). See also Hu and Zidek (1995) for the description of a method based on bootstrap resampling of linear combinations of residuals.

8.8   FURTHER READING

Maritz (1981) has reviewed the theory of randomization methods in the regression setting, and Brown and Maritz (1982) have described some 'rank method' alternatives to least squares for estimating regression relationships, with significance based on randomization arguments. Brown and Maritz also described how a method of restricted randomization can be used for inferences concerning the regression coefficient for one X variable when a second variable X variable also influences the response variable Y. Randomization tests for time and spatial correlation in regression residuals have recently been justified by Schmoyer (1994).

Randomization methods show great potential for use with robust methods for estimating regression equations, such as those reviewed by Montgomery and Peck (1982). The essence of these methods is that data points that deviate a good deal from the fitted regression are given less weight than points that are a good fit, with some form of iterative process being used to determine the final equation. The use of randomization with least absolute deviation regression has been examined by Cade and Richards (1996) both by simulations and in terms of biological examples, while De Angelis et al. (1993) have discussed bootstrapping in this context. Efron and Tibshirani (1993, p. 117) consider bootstrapping with least median of squares regression, which is another robust method that involves fitting the regression line that minimizes the median of the squared residuals. See also Crivelli et al.'s (1995) use of bootstrapping with ridge regression.

Plotnick (1989) and Kirby (1991a,b) provide Basic and Fortran programs for fitting simple and multiple regression equations when there are errors of measurement in the X variables as well as Y, with bootstrapping for estimating the standard errors of the estimated regression coefficients and other parameters. See also Booth and Hall's (1993) suggestion for a bootstrap method for finding a confidence region for a simple regression equation in this situation.

Edwards (1985) also gives an example of randomization with robust regression involving the calculation of a multiple regression equation in the usual way, the detection and removal of outliers in the data and the re-estimation of the equation on the reduced data. Classical methods are not appropriate for testing the final equation obtained by this process. However, it is straightforward to repeat the whole estimation procedure many times with permuted data to find randomization distributions for regression coefficients.

Mention must also be made of the usefulness of randomization for examining the properties of stepwise methods of multiple regression. With these methods, X variables are either (a) added to the regression equation one by one according to the extent to which they improve the amount of variation in Y explained, or (b) the multiple regression with all potential X variables included is considered first, and then variables are removed one by one until all those left make a significant contribution to accounting for the variation in Y. Many statistical packages allow this type of stepwise fitting of X variables, using several different procedures for determining whether variables should be added to or removed from the regression. They all suffer from the problem that the proportion of the variation in Y that is explained by the final equation will tend to be unrealistically high because of the search that has been carried out for the best choice of X variables. Essentially, the problem is that when there are many X variables at choice the chance of finding some combination that accounts for much of the variation in Y is high even when none of the X variables is related to Y (Bacon, 1977; Flack and Chang, 1987; Rencher and Pun, 1980; Hurvich and Tsai, 1990).

For example, Manly (1985, p. 172) investigated the use of the BMDP2R stepwise program (Dixon, 1981) in relating gene frequencies to environmental variables with McKechnie et al.'s (1975) data on *Euphdryas editha* that have been discussed in Examples 8.1–8.3. It was found that when stepwise regression was applied to random numbers rather than gene frequencies a regression equation 'significant at the 5% level' was obtained in seven out of ten trials.

In a situation like this, an alternative to testing the stepwise procedure with random numbers would be to apply it with the observed Y values randomly allocated to the X variables. This maintains the correlation structure for the X variables and has the advantage of taking into account any unusual aspects of the distribution of Y values. It could be, for instance, that one extreme Y value has a strong influence on how the procedure works. By randomly allocating the Y values a large number of times it is possible to determine the probability of obtaining a 'significant' regression by chance alone, and also to determine what apparent level of significance to require for the final equation from the stepwise procedure so that it is unlikely to have arisen by chance.

Bootstrapping may also help with model selection in regression when there is a need to choose predictor variables from a large number of candidates. Breiman (1992) and Breiman and Spector (1992) discuss this problem in terms of minimizing the errors involved in predicting future values of the dependent variable, and show that bootstrapping performs well in comparison with a more conventional approach.

Finally, the use of randomization and bootstrapping in non-standard regression situations must be mentioned. Two such situations are considered in Chapter 14. One involves testing for a relationship between the body size of ant species and latitude in Europe (Cushman *et al.*, 1993). The complication here is that some species occur at many different latitudes (the X variable), but the same size (the Y variable) has to be used for each of these. Therefore the regression data has a complicated structure, with some Y values that must be the same, and conventional regression tests are completely inappropriate.

The second regression situation that is considered in Chapter 14 involves the comparison of regression curves relating the weight of pigeon guillemots to their wingspan on Naked and Jackpot Islands in Prince William Sound, Alaska (Hayes, 1995). The complications here are due to data being collected from either one or two chicks in a nest, with observations on the chicks being made on different numbers of occasions. Again, the structure of the data means that conventional regression methods do not apply.

8.9   CHAPTER SUMMARY

- A randomization test of the regression coefficient for a simple linear regression involves comparing the observed coefficient with the distribution found by randomly pairing the Y values with the X values. This can be justified if (1) the observed (X, Y) pairs are a random sample from a population where X and Y are independent, (2) the X values are randomly assigned to sample units and the Y values are an observed treatment response, or (3) the mechanism generating the data makes the X and Y values potentially independent.

- An alternative to randomizing observations with simple linear regression involves randomizing the regression residuals. However, this is potentially more useful with multiple regression. A numerical example of simple linear regression gives very similar results from randomizing observations and residuals.

- The usual randomization test with simple linear regression is for the null hypothesis that the regression coefficient is zero. It can be modified to test for a non-zero coefficient quite easily. This then permits a randomization-based confidence interval for the true regression coefficient to be determined.
- With multiple regression, tests can be based on randomly reallocating the observed Y values to the corresponding sets of X values. The usual F-statistics and t-statistics can be used for these tests. However, alternative procedures involving the randomization of various types of residuals have been claimed to be better.
- A simulation study based on a particular numerical example was found to give very similar results from randomizing observations, randomizing residuals and using the t- and F-distributions. A second numerical example gave different significance levels for these three methods of analysis. In this case a simulation study showed that randomization of observations was on balance better than randomizing residuals, but the use of the t- and F-distributions was best of all. It is suggested that a more extensive simulation study is needed to examine the properties of the alternative methods of analysis, particularly with highly non-normal data.
- Bootstrapping can be used with regression, either with resampling of sample units or resampling of residuals. Jackknifing can also be used with regression.
- A number of extensions to the methods discussed in the chapter are briefly reviewed, particularly with robust regression, model fitting, and situations where the data have a complicated structure.

8.10 EXERCISES

8.1 The data below come from a study by Hogg *et al.* (1978) of alternative methods for estimating the throughfall of rain in a forest through gaps in the canopy and drip from branches and leaves:

X: 10.1 10.7 12.5 12.7 12.8 14.9 18.3 18.3 25.8 26.5 29.4 39.7

Y: 6.5 1.7 6.7 5.1 3.7 11.3 10.1 9.6 13.3 14.7 9.8 24.0

The variable X is the mean rainfall in millimetres measured from rain gauges placed outside a Douglas fir plantation. This mean is considered to measure accurately the magnitude of the 12 storms for which results are given. The variable Y is the mean rainfall in mm measured from 10 gauges placed at random positions within the fir plantation and relocated between rainfall events.

Table 8.10 Data on 'islands' of paremo vegetation: N = number of species, AR = area (thousands of square km), EL = elevation (thousands of m), DEc = distance from Ecuador (km) and DNI = distance to nearest other island (km)

'Island'	N	AR	EL	DEc	DNI
Chiles	36	0.33	1.26	36	14
Las Papas-Coconuco	30	0.50	1.17	234	13
Sumapaz	37	2.03	1.06	543	83
Tolima-Quindio	35	0.99	1.90	551	23
Paramillo	11	0.03	0.46	773	45
Cocuy	21	2.17	2.00	801	14
Pamplona	11	0.22	0.70	950	14
Cachira	13	0.14	0.74	958	5
Tama	17	0.05	0.61	995	29
Batallon	13	0.07	0.66	1065	55
Merida	29	1.80	1.50	1167	35
Perija	4	0.17	0.75	1182	75
Santa Marta	18	0.61	2.28	1238	75
Cende	15	0.07	0.55	1380	35

Find the regression line for predicting the throughfall measured in the forest from the mean rainfall in the outside gauges. Assess the significance of the regression coefficient by using the t-distribution and by randomization. Also, compare the 95% confidence limits for the true regression coefficient that are obtained using the t-distribution with the limits obtained by randomization.

8.2 To investigate whether organisms living on high mountains have insular patterns of distribution, Vuilleumier (1970) studied species richness for birds living in 'islands' of paramo vegetation in the Andes of Venezuela, Colombia and northern Ecuador. Part of the data from the study is shown in Table 8.10, where N is the total number of species on an 'island', AR is the area in thousands of square kilometres, EL is the elevation in thousands of metres, DEc is the distance from the assumed source area of Ecuador in kilometres and DNI is the distance from the nearest island of paramo vegetation in kilometres. Fit a multiple regression relating the number of species to the other four variables. Assess the significance of the regression coefficients and regression sums of squares using t- and F-tables. See how the results compare with the significance levels found by randomization.

9
Distance matrices and spatial data

9.1 TESTING FOR ASSOCIATION BETWEEN DISTANCE MATRICES

Many problems that involve the consideration of possible relationships between two or more variables can be thought of in terms of an association between distance matrices, often with one of the matrices relating to spatial distances. The following examples show the wide range of situations that can be approached in this way.

1. In Example 1.2 a matrix of distances between the continents based on the present-day earwig distribution (**A**) was compared with a matrix that reflects the present location of the continents (**B**), and also with a matrix that reflects assumed positions of the continents before continental drift (**C**). A comparison of the level of association between **A** and **B** with the level of association between **A** and **C** showed that the latter association is stronger. This suggests that the present-day distribution of earwigs reflects evolution in Gondwanaland rather than evolution with the continents in their present positions.

2. In studying evidence of disease contagion, information on n cases of a disease can be obtained. An $n \times n$ matrix of geographical distances between these cases can then be constructed and compared with another $n \times n$ matrix of time distances apart. If the disease is contagious then the cases that are close in space will have some tendency to also be close in time. Generally it can be expected that cases will tend to be clustered in space around areas of high population. They may also be clustered in time because of seasonal effects. However, in the absence of contagion the space and time clustering can be expected to be independent. Hence the hypothesis of no contagion becomes the hypothesis of no association between the elements of the spatial distance matrix and the corresponding elements of the time distance matrix. If it is desirable, the distance matrices can be simplified so that they consist of zeros for adjacent cases and ones for non-adjacent cases (with suitable definitions

of 'adjacent'), as was done by Knox (1964) in his study of childhood leukaemia in Northumberland and Durham. See Besag and Diggle (1977), Robertson and Fisher (1983), Marshall (1989), Robertson (1990), and Besag and Newell (1991) for further discussions of this type of medical application.

3. If an animal or plant population is located in n distinct colonies in a region, there may be some interest in studying the relationship between environmental conditions in the colonies and genetic or morphometric variation, with a positive association being regarded as evidence of adaptation (Sokal, 1979; Douglas and Endler, 1982; Dillon, 1984). One approach involves constructing a matrix of environmental distances between the colonies and comparing this with a matrix of genetic or morphometric distances. The construction of measures of distance for a situation like this is discussed by Manly (1985, p. 180). Inevitably the choice will be arbitrary to some extent. One of the measures commonly used in cluster analysis (Manly, 1994a, Chapter 5) can be used for environmental and morphometric distances. A range of measures are available for genetic distances. Two examples of this type of situation are discussed by Manly (1985, p. 182). The first (which is also the subject of Example 9.3, below) concerns a comparison between environmental and genetic distances for the 21 colonies of the butterfly *Euphydryas editha* studied by McKechnie *et al.* (1975) in California and Oregon. The second concerns a comparison between morphological distances (based on colour and banding frequencies) and habitat distances (0 for the same habitat type, 1 for a different habitat type) for 17 colonies of the snail *Cepaea nemoralis* studied by Cain and Sheppard (1950) in southern England.

4. Suppose that birds are ringed at a number of locations and at a later date n of the birds are recovered in another area that they have migrated to. Here the $n \times n$ matrix of distances between the recovered birds prior to migration can be compared with the same size matrix of distances between the birds on recovery. If the distances in the two matrices appear to match to some extent, then this is evidence of pattern transference. Besag and Diggle (1977) describe a comparison of this type for 84 blackbirds ringed in Britain in winter and recovered in northern Europe in a subsequent summer.

5. An area is divided up into n equal-size quadrats, and in each quadrat a count is made of the number of individuals present for a plant species. The question of interest is whether there is any evidence that similar counts tend to occur in clusters. In this case, an $n \times n$ matrix of differences between quadrat counts can be constructed and compared with a matrix of distances between the quadrats. If the matrices show a similar pattern then it appears that quadrats that have similar counts tend to be close together.

In a case like 3 it is necessary to rule out the possibility that an apparent relationship between biological and environmental distances occurs because close colonies tend to have similar gene and morphometric characteristics as a result of migration, and also tend to have similar environments because they are close. This suggests that in some situations an attempt should be made to control for the effects of geographical closeness. This is possible, as discussed in sections 9.5 and 9.6.

9.2 THE MANTEL TEST

Mantel's (1967) randomization test is the one that was used with Example 1.2 concerning the distribution of earwigs. It involves measuring the association between the elements in two matrices by a suitable statistic (such as the correlation between corresponding elements), and determining the significance of this by comparison with the distribution of the statistic found by randomly reallocating the order of the elements in one of the matrices.

Mantel's test is sometimes called 'quadratic assignment' as a result of its use in psychology (Hubert and Schultz, 1979). However, strictly speaking this description refers to the different but related problem of finding the ordering of the elements in one matrix in order to maximize or minimize the correlation between that matrix and a second matrix.

Although Mantel (1967) discussed more general situations, for the present purposes it can be assumed that the two matrices to be compared are symmetric, with zero diagonal terms, as follows:

$$
\mathbf{A} = \begin{bmatrix} 0 & a_{21} & \cdots & a_{n1} \\ a_{21} & 0 & \cdots & a_{n2} \\ \vdots & \vdots & \ddots & \vdots \\ a_{n1} & a_{n2} & \cdots & 0 \end{bmatrix}, \quad \mathbf{B} = \begin{bmatrix} 0 & b_{21} & \cdots & b_{n1} \\ b_{21} & 0 & \cdots & b_{n2} \\ \vdots & \vdots & \ddots & \vdots \\ b_{n1} & b_{n2} & \cdots & 0 \end{bmatrix}.
$$

Because of symmetry, the correlation between all the off-diagonal elements in these matrices is the same as the correlation between the $m = n(n-1)/2$ elements in the lower triangular parts only, this correlation being given by the equation

$$
r = \frac{\sum a_{ij}b_{ij} - \sum a_{ij} \sum b_{ij}/m}{\sqrt{\left[\left\{\sum a_{ij}^2 - \left(\sum a_{ij}\right)^2/m\right\}\left\{\sum b_{ij}^2 - \left(\sum b_{ij}\right)^2/m\right\}\right]}},
$$

where all the summations are over the lower triangular elements.

The only term in this equation that is altered by changing the order of the elements in one of the matrices is the sum of products $Z = \sum a_{ij}b_{ij}$. This is therefore an equivalent statistic to r for a randomization test. The regression coefficient of **A** distances on **B** distances and the regression coefficient for **B** distances on **A** distances are also equivalent statistics.

If the items for which distances are calculated are labelled $1, 2, \ldots, n$ then this matches the data order for the rows and columns in **A** and **B**. For the randomization test the item labels are randomly permuted for one of the matrices, **A**, say, and left in the order $1, 2, \ldots, n$ for **B**. For example, if a random permutation gives the order $5, 3, \ldots, 1$ then the rows and columns in the **A** matrix are reordered to

$$
\mathbf{A_R} = \begin{bmatrix} 0 & a_{35} & \ldots & a_{15} \\ a_{35} & 0 & \ldots & a_{13} \\ \vdots & \vdots & \ddots & \vdots \\ a_{15} & a_{13} & \ldots & 0 \end{bmatrix}.
$$

A randomized value of the test statistic is then calculated from $\mathbf{A_R}$ and **B**.

The justification for this test needs some further discussion. All of the examples mentioned in the last section involve observational data, and there is no sense in which the randomized order for the elements of **A** is part of an experimental design. In fact, generally the justification for Mantel's test must be on the grounds of either random sampling or something inherent in the nature of the situation being studied.

A random sampling argument can be made in cases where the n items for which distances are being calculated are a random sample from some larger population of potential items that might be studied. It can then be argued that if the **A** distances are independent of the **B** distances in the population then choosing the same n items for the **A** and **B** distances is equivalent to taking one random sample of n items for the **A** distances and an independent random sample of n items for the **B** distances. In that case the sampling process ensures that all permutations of the items for the **A** matrix are equally likely with respect to the ordering for the **B** matrix, and the randomization test is justified.

If the n items being studied are all items that are of interest then the random sampling argument cannot be used. Instead, it must be assumed that the null hypothesis of interest is that the mechanism that generates the **A** distances is independent of the mechanism that generates the **B** distances, so that in effect it is as if the two sets of distances are based on different sets of n items.

9.3   SAMPLING THE RANDOMIZATION DISTRIBUTION

It is possible to use formulae provided by Mantel (1967) and a normal approximation for the randomization distribution of Z to test the significance of an observed value. However, the normal approximation has been questioned by Mielke (1978) and Faust and Romney (1985) so that it is better to determine significance by making a direct comparison of the test statistic with the randomization distribution. There are $n!$ possible permutations for the order of n items, which means that it is feasible to determine the full randomization distribution for up to about nine items, because $9! = 362\,880$.

The discussion in section 5.3 concerning the number of random permutations required to estimate a significance level applies here just as well as with other randomization tests. Generally, 1000 randomizations is a realistic minimum for estimating a significance level of about 0.05 and 5000 randomizations is a realistic minimum for estimating a significance level of about 0.01.

Example 9.1: testing for a spatial correlation in counts of Carex arenaria   As an example of Mantel's test applied in a spatial situation, consider the counts of the plants *Carex arenaria* in a 12 × 8 contiguous grid that are shown in Table 9.1. These are part of a larger set given by Bartlett (1975, p. 74) that were originally collected by Greig-Smith in Newborough Warren in 1953.

Bartlett fitted an autoregressive type of model to the larger set of data, involving the counts in a square being related to the counts in neighbouring squares. For the present example, the question addressed is whether there is any evidence that close squares have similar counts. Note that this is not the

Table 9.1 Counts of *Carex arenaria* in 96 10 cm × 10 cm quadrats at Newborough Warren in 1953. The rows and columns shown here follow the field layout

Row	Column							
	1	2	3	4	5	6	7	8
1	0	1	1	1	0	1	1	2
2	0	1	0	0	1	1	1	0
3	1	1	1	0	0	1	1	0
4	0	0	1	1	1	1	1	0
5	0	1	1	0	1	4	1	1
6	0	1	0	0	0	2	0	1
7	0	1	0	1	2	0	3	0
8	0	0	0	0	0	0	1	2
9	0	0	0	0	0	1	2	0
10	0	0	0	0	0	0	1	0
11	0	0	0	0	0	2	0	0
12	0	0	0	0	2	0	0	0

same as the question of whether the plants occur randomly over the area studied, which implies that the counts within quadrats have a Poisson distribution as well as being independent of each other.

If there is no tendency for close quadrats to have similar counts then there will be no relationship between the absolute difference in quadrat counts for two quadrats and the distance between those quadrats. The first matrix for Mantel's test can therefore be formed from the absolute differences of counts, and the second matrix formed from the spatial distances between the quadrats, taking the rows and columns as being one unit apart, and measuring distances from the centre of quadrats.

There is an undesirable aspect of the method for constructing the second matrix because it does not take into account the fact that it is only very close quadrats that are likely to have similar counts. For this reason, using the reciprocals of distances for the second matrix makes more sense, because this emphasizes small distances at the expense of moderate and large distances (Mantel, 1967). An example may make this clearer. Consider first the quadrat in row 1 and column 1 compared with the quadrat in row 6 and column 6. The distance apart in row and column units is

$$\sqrt{\{(6 - 1)^2 + (6 - 1)^2\}} = 7.07$$

with reciprocal 0.14. Next, consider the comparison of the row 1 and column 1 quadrat with the row 12 and column 8 quadrat. Here the distance apart is 13.04, with reciprocal 0.08. In both cases the quadrats being compared are unlikely to have similar counts because of closeness, and the similar values for reciprocal distances are more sensible than the rather different values for distances. Of course, the reciprocal transformation is not the only one that can be used here, but it serves for the purpose of the example.

Using reciprocal distances for the second matrix means that a negative correlation between the two matrices gives evidence that close quadrats have similar counts. Hence Mantel's test should be one-sided, testing to see whether the correlation is significantly low.

With the observed ordering of quadrats, the correlation between the count difference matrix and the reciprocal distance matrix is -0.051 for the $96 \times 95/2 = 4560$ quadrat comparisons. This value is close to zero, but when 4999 randomized allocations of quadrats were made for the first matrix, only 74 gave a correlation as low or lower than -0.051. Counting the observed result as a value from the randomization distribution therefore gives an estimated significance level of 1.5% (75 out of 5000). There is evidence that close quadrats tend to have similar counts.

The low observed correlation indicates that the effect of spatial correlation is fairly small. This is confirmed by noting that the average absolute count difference between the 172 quadrats that are adjacent in the same row or column is 0.63, the average difference for the 154 diagonally touching quadrats is 0.61, and the average difference for the 4234 non-touching quadrats is 0.73.

The calculations for this example were carried out using the program RT (Manly, 1996a).

9.4 CONFIDENCE LIMITS FOR REGRESSION COEFFICIENTS

The method for obtaining confidence intervals for a regression coefficient that was discussed in section 8.4 can also be used to obtain a confidence interval for the matrix correlation with Mantel's test. It will be assumed for this purpose that the data are scaled so that the distances in both matrices have means of zero and unit variances.

In order to calculate the confidence interval it is necessary to assume that a linear regression model applies, with

$$a_{ij} = \alpha + \beta b_{ij} + \varepsilon_{ij},$$

where ε_{ij} is an error term that is independent of a_{ij} and b_{ij}. The scaling of the distances then ensures that the least squares estimator of α is zero and the estimator of β is $b = \sum a_{ij}b_{ij}/m$, which is also the correlation coefficient r.

If the true value of the regression coefficient is β then the matrix $\mathbf{C}$ with elements

$$c_{ij} = a_{ij} - \beta b_{ij} = \alpha + \varepsilon_{ij} \tag{9.1}$$

should show no relationship with the matrix $\mathbf{B}$. A 95% confidence interval for β therefore consists of all values of β for which a randomization test is not significant when comparing $\mathbf{C}$ and $\mathbf{B}$. The lower limit will be the value of β such that 2.5% of randomizations give a higher correlation than that observed between C and B. The upper limit will be the value of β such that 2.5% of randomizations give a lower correlation than that observed between $\mathbf{C}$ and $\mathbf{B}$.

This procedure works equally well with the roles of the $\mathbf{A}$ and $\mathbf{B}$ matrices reversed. If it is assumed that there is a regression

$$b_{ij} = \alpha + \beta a_{ij} + \varepsilon_{ij},$$

where ε_{ij} is independent of a_{ij} and b_{ij} then a $\mathbf{D}$ matrix can be constructed with elements

$$d_{ij} = b_{ij} - \beta a_{ij} = \alpha + \varepsilon_{ij}, \tag{9.2}$$

that are independent of a_{ij}. A 95% confidence interval for β is then given by the range of β values for which the correlation between the $\mathbf{D}$ and $\mathbf{A}$

distances is not significant on a two-sided test.

If the β value is the same for equations (9.1) and (9.2) then the correlation between the **B** and **C** distances will be the same as the correlation between the **A** and **D** distances, both being equal to

$$r(\beta) = \left\{ \sum a_{ij}b_{ij}/m - \beta \right\} / \sqrt{\left\{ 1 + \beta^2 - 2\beta \sum a_{ij}b_{ij}/m \right\}}.$$

From this point of view, it does not matter which of the two approaches are used to determine confidence limits. However, the estimated distribution of the correlation between **B** and **C** distances will not be the same as the estimated distribution of the correlation between **A** and **D** distances even when the same randomizations are used for the estimation. Therefore the limits obtained may depend to some extent on which approach is used.

Example 9.2: confidence limits for spatial correlation in Carex arenaria   Consider again the matrix **A** of *Carex arenaria* count differences between the 96 quadrats shown in Table 9.1 and the elements of the matrix **B** of the reciprocals of spatial distances between the centres of the quadrats. Recall from Example 9.1 that the observed correlation between the elements of these two 96 × 96 matrices is −0.051, which is significantly low at the 1.5% level. The present example is concerned with the determination of 95% confidence limits for the coefficient of the **B** distances, from the regression of **A** distances on **B** distances.

To determine confidence limits it is necessary to find the two values of β which, when used with equation (9.1), just give a significant correlation between the elements of the matrix **C** and the matrix **B** on a two-sided test. To this end, a range of trial β values were evaluated, approximating the full randomization distributions by 4999 randomizations and the observed data order. It was found that for values of β of about −0.093 or less in equation (9.1) the correlation between **C** and **B** is exceeded by 2.5% or less of the randomization distribution, while for values of β of −0.006 or more the correlation between **C** and **B** is greater than 2.5% or less of randomized values. Thus any β value within the range from about −0.093 to about −0.006 does not give a significant correlation between **C** and **B** distances on a two-sided test at the 5% level. Consequently, this is an approximate 95% confidence interval for the true regression coefficient.

9.5 THE MULTIPLE MANTEL TEST

One of the examples mentioned at the start of this chapter involved a matrix of genetic distances between colonies of an animal being related to a matrix of environmental distances between the same colonies, with a positive correlation possibly being regarded as evidence of adaptation. This is a reasonable type of analysis providing that the colonies concerned are far enough apart to be treated as being independent. However, close colonies will tend to have similar environments, so that environmental and spatial distances will often be positively related. Close colonies will also be more likely to exchange migrants than colonies that are far apart, so that genetic distances will often be positively related to spatial distances. Consequently, a positive association between environmental and genetic distances may simply be due to spatial effects.

This type of problem was considered by Smouse *et al.* (1986) and Manly (1986), with similar solutions being suggested based on the use of multiple regression. Thus, suppose that an $n \times n$ distance matrix $\mathbf{G}$ is to be related to two other matrices $\mathbf{S}$ and $\mathbf{E}$. For example, $\mathbf{G}$ might be the genetic distances, $\mathbf{S}$ the spatial distances and $\mathbf{E}$ the environmental distances between n colonies of an animal population. Let g_{ij}, s_{ij} and e_{ij} denote the distance between item i and item j in these three matrices, respectively. Then a regression relationship

$$g_{ij} = \beta_0 + \beta_1 s_{ij} + \beta_2 e_{ij} + \varepsilon_{ij} \tag{9.3}$$

can be assumed, where β_1 measures the relationship between g_{ij} and s_{ij}, after allowing for any effects of e_{ij}, β_2 measures the relationship between g_{ij} and e_{ij} after allowing for any effects of s_{ij}, and ε_{ij} is an independent error.

Equation (9.3) can be fitted by ordinary regression methods, as discussed in section 8.5. Clearly, though, it is not valid to apply the usual tests of significance based on the t-distribution and the F-distribution because the distances being used as data are not independent observations. This suggests that the significance of the regression coefficients and the overall fit of the equation to the data is better assessed with reference to randomization distributions. In particular, it seems reasonable to compare suitable test statistics with the distributions for those statistics that are generated when equation (9.3) is re-estimated many times, using data where the distances in the $\mathbf{G}$ matrix have been rearranged according to a random reordering of the n items that these distances relate to, keeping the distances in the $\mathbf{S}$ and $\mathbf{E}$ matrices fixed. The idea is that if the $\mathbf{G}$ distances are unrelated to the $\mathbf{S}$ and $\mathbf{E}$ distances then the observed data will look like a typical randomized set of data.

The discussions in the previous chapter concerning randomization in the usual multiple regression context suggest that the following statistics should be used for assessing the fitted equation: $t_i = b_i/SE(b_i)$, where b_i is the estimate of β_i, with the estimated standard error $SE(b_i)$; the F-values for the extra sum of squares accounted for by adding each of the variables in to the equation, as in Table 8.3; and the F-value for the overall significance of the equation, as in Table 8.2. Of course, if $\mathbf{S}$ and $\mathbf{E}$ are strongly related then the usual problems associated with determining the relative importance of correlated regression variables (discussed briefly in section 8.5) will occur.

The method that has just been described for relating one dependent matrix of distances to two other matrices generalizes in an obvious way to situations where the dependent matrix is related to three or more other matrices. Basically, the relationship is estimated by usual multiple regression methods, with testing based on randomizing the order of the elements in the dependent matrix.

9.6 OTHER APPROACHES WITH MORE THAN TWO MATRICES

Apart from the multiple regression approach that has been described in the last section, several other methods have been proposed for extending the randomization testing of distance matrices to situations where there are more than two matrices. In particular:

(a) Based on a result of Wolfe (1976, 1977) concerning the correlation between a variable Y and the difference $Y - Z$ between two other random variables, Dow and Cheverud (1985) suggested comparing the elements of the matrix **A** with the elements of the difference matrix **B** − **C** to see whether the correlation between the elements of **A** and **B** is significantly higher than the correlation between the elements of **A** and **C**.

(b) Hubert (1985) suggested comparing the elements of a matrix **A** with a second matrix consisting of the products of the elements of matrices **B** and **C** to test for a significant partial correlation between **B** and **C** conditional on **A**.

(c) Smouse et al. (1986) suggested comparing the residuals from the regression of the elements of a matrix **B** against the elements of a matrix **A**, with the residuals from the regression of the elements of a matrix **C** against the elements of **A**. A significant result on a Mantel test is then assumed to give evidence of a partial correlation between **B** and **C** conditional on **A**.

(d) Manly (1986) proposed a method for calculating confidence limits for simple and multiple regression coefficients when the elements of one matrix are regressed against the elements of one or more other matrices.

(e) Biondini et al. (1988) discussed the application of the tests for randomized block experimental designs, with an allowance for blocks being made by only randomizing treatments within blocks.

Some applications of these extensions are described by Sokal et al. (1986) for the study of the genetic structure of tribes on Yanomama Indians; by

Dow *et al.* (1987) using morphometric, geographic and linguistic differences between human populations of Bouganville; by Burgman (1987) using coefficients of similarity between plant species in quadrats, geographical distances, substrate similarity and east–west locations; by Legendre and Troussellier (1988) for relating differences in the distribution of aquatic heterotrophic bacteria to environmental differences; by Cheverud *et al.* (1989) for testing hypotheses of morphological integration of biological characters; by Tilley *et al.* (1990) using genetic, geographical and ethological differences for a salamander species; by Brown and Thorpe (1991), Thorpe and Brown (1991) and Brown *et al.* (1993) using body dimensions, altitude, climate and geographical differences for lizards in Tenerife; by Pierce *et al.* (1994) for studying geographical variation of squid in the north-east Atlantic Ocean; and by Verdú and García-Fayos (1994) for comparing the abundances of fruits and frugivorous birds taking into account the effects of autocorrelation in time between observations.

Oden and Sokal (1992) have carried out a simulation study of the methods (a) to (d) for situations where the distances being considered are for objects in space and spatial autocorrelation is present. They found all four methods to be upset by spatial correlation and concluded that 'we cannot unreservedly recommend the use of any of the methods with spatially autocorrelated data'. It seems, therefore, that these extensions to the basic Mantel randomization test must be used with caution in the situations where they would have been most useful.

The multiple regression extension to the Mantel test will always be valid in the sense that if the dependent matrix is independent of all the other matrices then the probability of a significant result will be equal to the significance level being used. However, there is a potential problem with spatial data because the effects of autocorrelation will not necessarily be removed just by including geographical distance in the regression equation. This is an important consideration with many applications in ecology because spatial autocorrelation can be expected to be present. It is discussed further in the following example, where a possible method for overcoming most of the effect of autocorrelation is proposed, based on the idea of using restricted randomizations.

Example 9.3: colonies of Euphydryas editha　This example is concerned with the problem of whether determining genetic differences between 21 colonies of the butterfly *Euphydryas editha* in California and Oregon are related to environmental differences, taking into account the fact that colonies that are geographically close can be expected to have similar environments and also to be relatively similar genetically because of past migration. Data from the same colonies have already been used for several examples in Chapter 8.

Using information provided by McKechnie *et al.* (1975), it is possible to construct a matrix **H** of Hexokinase (Hk) genetic differences, a matrix **G** of geographical distances and a matrix **E** of environmental distances between the 21 colonies of *E. editha*. First, the elements of **H** were calculated using the equation

$$h_{ij} = \sum_{r=1}^{3} |p_{ir} - p_{jr}|,$$

where p_{ir} is the estimated proportion of allele r in the ith colony, and there were three hexokinase alleles. Next, the geographical distance g_{ij} between colony i and colony j was set equal to the straight line distance between these colonies in units of approximately 111.2 km (one degree of latitude or longitude). Finally, the elements in **E** were taken to be the standardized Euclidean distances between the colonies based on 10 variables that describe the environment (the altitude, the annual precipitation and eight temperature variables). To be more precise, let x_{ik} denote the value of the kth environmental variable for the ith colony after

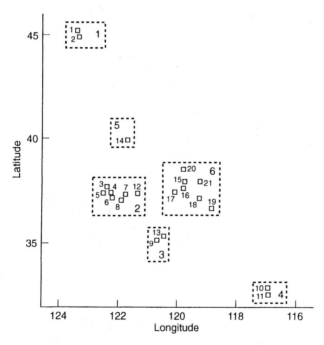

Figure 9.1 Locations of 21 colonies of the butterfly *Euphydryas editha*. The small squares represent the colonies and the large rectangles indicate the six groups of colonies used with a restricted randomization.

scaling so that the mean is zero and the standard deviation is one for the 21 colonies. Then the environmental distance from colony i to colony j was calculated to be

$$e_{ij} = \sqrt{\left\{ \sum_{k=1}^{10}(x_{ik} - x_{jk})^2 \right\}}.$$

The relative locations of the colonies are shown in Figure 9.1, and Figure 9.2 shows the distances plotted against each other. (The rectangles that

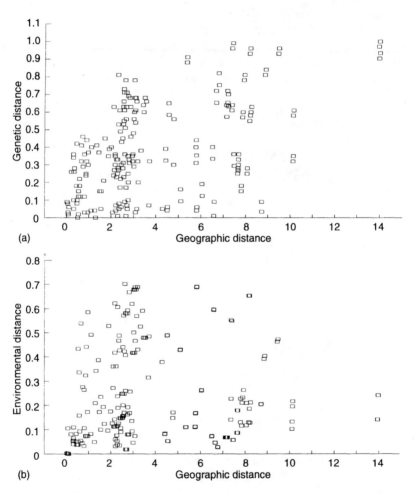

Figure 9.2 Plots of (a) genetic against geographic distances, (b) environmental against geographical distances, and (c) genetic against environmental distances for 21 colonies of *Euphydryas editha*.

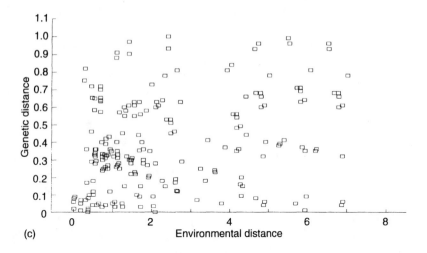

(c)

group colonies in Figure 9.1 will be explained shortly.) It appears from Figure 9.2 that there is little relationship between environmental and geographical distances or genetic and environmental distances, but a possible relationship between the genetic and geographical distances.

In examples like this it is usually worthwhile to begin an analysis by testing for associations between the matrices using simple Mantel tests. Here this was done using 4999 randomizations, from which the following results were obtained: the correlation between genetic and environmental distances is 0.29, which is significantly different from zero at the 0.64% level; the correlation between genetic and geographical distances is 0.49, which is significantly different from zero at the 0.02% level; and the correlation between environmental and geographical distances is 0.04, which is not at all significant. These results are consistent with the impressions from Figure 9.2 except for the significant relationship between genetic and environmental distances, which is presumably due to a fairly large number of cases where both the genetic and the environmental distance are close to zero.

When the genetic distances were regressed against the geographical and environmental distances the fitted regression equation was found to be

$$h_{ij} = 0.135 + 0.042g_{ij} + 0.036e_{ij}. \tag{9.4}$$

This gives a coefficient of multiple determination of $R^2 = 0.314$ and the following statistics that are relevant for assessing the importance of the geographical and environmental distances: $t_1 = 8.32$ (for testing the co-efficient of geographical distances); $t_2 = 4.74$ (for testing the coefficient of environmental distances); and $F = 47.48$ (for testing the overall fit of the equation). The randomization distributions of these statistics were approximated by producing and analysing 4999 other sets of data, keeping the **G** and **E** matrices fixed, and randomly permuting the order of the 21 colonies for the matrix **H**. In this way it was found that t_1 is significantly different from zero at the 0.02% level, t_2 is significantly different from zero at the 0.52% level, and F is significantly large at the 0.2% level. It appears, therefore, that there is evidence that genetic differences between the colonies are related to both their separation in space and their environmental differences.

It might seem from this analysis that the effects of spatial correlation (a tendency for close colonies to be similar in every respect just because they are close) is allowed for by including geographical distance in the regression equation. However, the situation is more complex than that because there is evidence that the residuals from the regression of genetic distances on geographical and environmental distances are autocorrelated. This evidence is found by measuring the difference in the residuals for all pairs of colonies, and then seeing whether these differences are related to the distances between the colonies.

The procedure is as follows. First the residuals are calculated as the differences between the observed genetic differences between colonies (i.e. the elements of matrix **H**) and the genetic differences that are predicted by equation (9.4). The residual difference between colony i and colony j can then be measured by

$$d_{ij} = \sqrt{\left\{ \sum_{\substack{k=1 \\ k \neq i,j}}^{21} (r_{ik} - r_{jk})^2 / 19 \right\}},\qquad(9.5)$$

where r_{ik} is the residual for the genetic difference between colony i and colony k. Obviously, d_{ij} will be small if two colonies have similar residuals with respect to other colonies, and large if their residuals are very different. Spatial correlation will show up as a tendency for close colonies to have similar residuals.

A plot of these residual differences against the geographical distances g_{ij} separating the locations is shown in Figure 9.3(a). There are some indications of spatial correlation in this plot. In particular, the residual distances are very small for the few colonies with geographical distances close to zero, and the residual distances are all less than about 0.4 for colonies with a geographical separation of less than two units. Furthermore, a regression of d_{ij} values against g_{ij} values gives the equation

$$d_{ij} = 0.202 + 0.018 g_{ij},$$

for which $R^2 = 0.141$, and a Mantel randomization test with 4999 randomizations shows that the coefficient of g_{ij} is significantly different

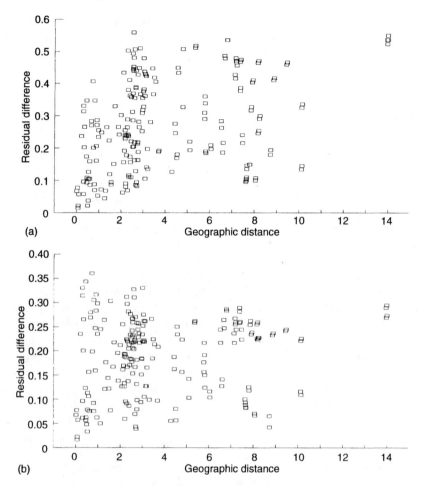

Figure 9.3 Plots of residual differences between colonies of *Euphydryas editha*: (a) based of residuals from the regression of genetic on geographical and environmental distances; (b) based on a regression with six groups of colonies.

from zero at the 0.02% level. It seems, therefore, that including geographical distances with environmental distances in the equation to account for genetic distances has not completely allowed for spatial autocorrelation, because this is still present in the residuals from the fitted equation.

The implication of these calculations is that randomizations that allow colonies that are close in the real world to become much further apart are not valid because they destroy an important feature of the original data (that the part of the genetic distance that is not accounted for by equation (9.4) tends to be similar for close colonies). It therefore seems possible that the significance of the coefficient of environmental distance in equation (9.4) may be an artefact of spatial autocorrelation rather than real evidence of an association.

A possible way around this difficulty involves applying some form of restricted randomization, as has been used, for example, by Sokal *et al.* (1987) to compare alternative non-null hypotheses related to the spatial structure of the graves in a Hungarian cemetery, by Biondini *et al.* (1988) with a randomized block experimental design, and by Sokal *et al.* (1989, 1990) to examine the association between differences between allele frequencies and differences between cranial measurements for language family groups in Europe.

With the *E. editha* data, restricted randomization can be used to ensure that colonies that are close in space with the real data stay close for randomized sets of data. One possibility here involves assigning each colony to a group of close colonies, and only randomizing within those groups. In addition, the regression equation that is fitted can include variables that take into account differences between all pairs of groups, as well as the geographical distance. In this way, a relationship between environmental and genetic distances can be tested for after allowing for overall differences between groups of colonies (large-scale spatial variation), as well as spatial differences within groups (small-scale spatial variation).

In order to apply this idea of restricted randomization, six groups of colonies were chosen, as indicated by the rectangles in Figure 9.1. The regression equation

$$h_{ij} = c_{v(i),v(j)} + w_1 g_{ij} + w_2 e_{ij}, \qquad (9.6)$$

was then fitted, where $c_{v(i),v(j)}$ is a parameter which depends on the groups that colonies i and j are in, and w_1 and w_2 are regression coefficients. To be more precise, $v(i)$ denotes the group that colony i is in and c_{rs} takes into account the overall genetic difference between colonies in group r and colonies in group s. It seems reasonable to assume that the parameter c is the same for all distances within groups of colonies by setting $c_{11} = c_{22} = \ldots = c_{66}$, and geographical distances are included in equation (9.6) to allow for the possibility of spatial correlation within groups.

When the equation was fitted to the data it was estimated to be

$$h_{ij} = c_{v(i),v(j)} + 0.027 g_{ij} + 0.041 e_{ij}, \qquad (9.7)$$

with $R^2 = 0.676$. Estimates of the $c_{v(i),v(j)}$ will not be given here because these are 'nuisance' parameters that are not important to the question of whether the genetic distances are related to environmental distances. Randomization testing for significant coefficients was carried out with values of h_{ij}

only randomized within groups of colonies, and with randomization distributions being approximated by 4999 randomly reordered sets of data plus the observed set. As a result it was found that the coefficient of genetic differences is not significantly different from zero at the 5% level ($p = 0.11$) but the coefficient of environmental differences is significantly different from zero at the 0.1% level ($p = 0.0006$). Thus it seems that genetic and environmental distances are related even after allowing for overall differences between different parts of the study area.

The residuals from the regression equation (9.7) were calculated and then used to determine residual distances between each of the 21 colonies using equation (9.5). Figure 9.3(b) shows these residual differences plotted against geographical distances. No relationship is apparent. A regression of the residual distances against the geographical distances gives the fitted equation

$$r_{ij} = 0.177 + 0.004g_{ij},$$

with $R^2 = 0.021$. A Mantel test with unrestricted randomizations shows that the regression coefficient is not quite significantly different from zero at the 5% level ($p = 0.075$), so that in this respect the model seems reasonable. Thus it appears that there really is evidence of some relationship between genetic and environmental distances that is not just due to spatial autocorrelation.

To check the validity of this conclusion a small simulation study was carried out. This involved maintaining the geographical distances and the environmental differences between the colonies as those for the original data, but generating new genetic differences in such a way that spatial correlation was present. The model used to generate the genetic distance between colony i and colony j was

$$h_{ij} = \beta_1 g_{ij} + \beta_2 e_{ij} + (a_i - a_j)^2,$$

where g_{ij} is the geographic distance, e_{ij} is the environmental distance, a_i and a_j are spatially correlated normally distributed 'errors', and β_1 and β_2 are constants that were varied in order to simulate data with different characteristics.

Three levels of spatial correlation were introduced. For the low level the correlation between a_i and a_j, was set at $R(i, j) = \exp(-4g_{ij})$, which means that this correlation was quite small (0.02) for two colonies that are one unit of distance apart. For the medium level of correlation, $R(i, j)$ was set at $\exp(-2g_{ij})$, which means that colonies had to be two units of distance apart in order for the correlation to be as low as 0.02. This seems to represent the situation with the real data fairly well. Finally, for the high level of correlation $R(i, j)$ was set at $\exp(-g_{ij})$, which means that two colonies had to be four units of distance apart in order for the correlation to be as low as 0.02. The standard deviation of a_i and a_j was always set at 0.55 and the required correlated normally distributed pseudo-random numbers were generated using the program provided by Bedall and Zimmermann (1976).

Six situations were simulated with each of the three levels of spatial correlation, with (β_1, β_2) given by (0,0), (0,0.1), (0,0.2), (0.05,0), (0.05,0.1) and (0.05,0.2). This was done in order to see whether randomization tests for a relationship between genetic and environmental distances:

(a) have the correct properties when there is no relationship $(\beta_2 = 0)$;
(b) have reasonable power to detect a relationship when one exists $(\beta_2 = 0.1$ or $0.2)$; and
(c) are adversely affected when the geographical distance between colonies has a direct influence on genetic differences $(\beta_1 = 0.05)$.

For each of the six sets of values for β_1 and β_2, 1000 sets of data were generated and analysed using unrestricted randomization, restricted randomization within three groups of colonies, restricted randomization within six groups of colonies, and restricted randomization within nine groups of colonies. The three groups of colonies consisted of 1 with 2, 10 with 11, and the other 17 colonies together, with the colony numbers as shown in Figure 9.1. The six groups of colonies were as shown in Figure 9.1. The nine groups of colonies consisted of groups 1, 3, 4 and 5 as shown in Figure 9.1, together with the groups consisting of the colonies with the numbers (3,4,5,6), (7,8,12), (15,16,17,21), (18,19) and (20).

Table 9.2 summarizes the results obtained from the experiment, in two parts. First the percentages of significant results from randomization tests are shown when these tests are carried out using unrestricted and restricted randomizations, and a 5% level of significance. The test statistics considered were F_1 (the F-value for the extra sum of squares accounted for by geographic distances in addition to the group difference variables, if any), F_2 (the F-value for the extra sum of squares accounted for by environmental differences in addition to group difference variables, if any, and geographical distances), F_{all} (the F-value for the variation accounted for by the full regression equation) and F_{res} (the F-value for the test for spatially correlated residuals using equation (9.5) to measure residual differences). Part (b) of the table shows the means and standard deviations of the estimates obtained for the regression coefficients β_1 and β_2. These are of interest in terms of the amount of bias that is introduced by spatial correlation in the data.

Only 19 randomizations were used with each set of data to determine whether the observed F-values were significant. Significance at the 5% level then required that the observed statistic was larger than any of the randomized values. This small number of randomizations was used because of the computation time required for matrix randomizations. The tests were still 'exact' in the sense discussed in section 1.3(a), and the effect of increasing the number of randomizations would just be to increase the power to detect non-zero values of β_1 and β_2. This is not a crucial consideration because the main purpose of the simulation was to see whether the probability of detecting a significant relationship between genetic and environmental differences was about 0.05 when there was in fact no such relationship.

From the results shown in part (a) of Table 9.2 it appears that:

(a) With a low level of spatial correlation ($D = 4$), and no relationship between genetic and environmental distances, a significant relationship between these two distances was detected with unrestricted random-ization (no groups) about 8% of the time for F_2, instead of the desired 5%. With a high level of spatial correlation ($D = 1$) the percentage of significant results increased to about 13%. The use of restricted randomization with six or nine groups improved the situation considerably, although 8% of results were still significant in some cases with the medium or high level of spatial correlation.

(b) The performance of the test for a relationship between genetic and environmental distances (F_2) is not affected greatly if geographical distances have a direct effect on genetic differences (i.e. if $\beta_2 = 0.05$ rather than 0).

(c) The power to detect a relationship between genetic and environmental differences reduces as randomizations are made more restricted.

(d) The test to detect spatial correlation in residual differences between colonies gave many significant results with unrestricted randomization. The number was much reduced with three groups of colonies, and became about 5% (the percentage expected by chance) with six groups of colonies. The percentage became higher again with nine groups of colonies, perhaps because of dependencies introduced into the residuals by the large number of regression coefficients being estimated in order to account for differences between groups.

From the means and standard deviations of the estimated regression coefficients, it also appears that:

(e) Spatial correlation shows up in strongly biased estimates of the coefficient of geographical distance (β_1).

(f) Although there are significant biases in estimates of the coefficient of environmental difference (β_2), the amount of bias is always small in absolute terms, and largely eliminated by using six groups of colonies.

Table 9.2 The results of a simulation experiment designed to examine the effects of spatial correlation on the Mantel randomization test. Part (a) shows the percentages of significant results for tests at the 5% level for 1000 sets of data (F_1, F-value for the extra sum of squares accounted for by the geographical distances; F_2, F-value for the extra sum of squares accounted for by the environmental distances; F_{all}, F-value for the variation accounted for by the whole regression equation; F_{res}, F-value from the test for correlated residuals). Percentages are shown in italic when it is desired that they should be 5% but they are significantly different from that value. Part (b) shows the means and standard deviations for the estimates of the regression coefficients β_1 and β_2 used to generate the data. Means of estimated regression coefficients are shown in italic when they are significantly different from the corresponding β value. The amount of serial correlation is dependent on the value of D, from low ($D = 4$) to high ($D = 1$), as explained in the text

(a) Percentages of significant test results

D	β_1	β_2	No groups F_1	F_2	F_{all}	F_{res}	3 groups F_1	F_2	F_{all}	F_{res}	6 groups F_1	F_2	F_{all}	F_{res}	9 groups F_1	F_2	F_{all}	F_{res}
4	0.00	0.0	11	8	14	11	12	5	10	4	8	6	8	4	8	4	8	6
4	0.00	0.1	12	55	54	11	30	51	57	3	21	37	42	4	25	13	28	8
4	0.00	0.2	13	91	90	12	55	89	91	5	42	69	69	5	38	30	53	9
4	0.05	0.0	40	9	39	11	31	10	19	4	12	5	9	4	8	6	8	6
4	0.05	0.1	42	56	67	14	59	56	69	6	29	30	39	4	23	15	26	7
4	0.05	0.2	44	91	92	14	79	89	94	4	49	70	75	5	43	26	46	8
2	0.00	0.0	12	8	15	10	22	8	15	6	9	8	10	4	8	5	8	7
2	0.00	0.1	16	65	62	15	46	52	61	9	24	42	45	5	23	17	32	5
2	0.00	0.2	17	92	92	15	75	92	95	7	47	70	73	5	43	33	55	5
2	0.05	0.0	46	11	44	15	46	8	29	8	18	4	9	5	8	7	7	5
2	0.05	0.1	52	61	74	17	67	57	73	5	36	36	47	5	33	15	34	6
2	0.05	0.2	51	91	93	16	89	89	95	3	61	68	77	4	53	31	58	6
1	0.00	0.0	25	13	29	27	34	14	26	12	12	8	9	4	9	7	9	7
1	0.00	0.1	26	74	76	28	66	66	78	14	36	47	55	4	35	17	34	10
1	0.00	0.2	24	93	93	27	84	91	97	13	60	75	79	6	56	47	67	7
1	0.05	0.0	61	14	61	26	61	15	44	10	16	7	12	3	11	8	13	8
1	0.05	0.1	68	73	87	26	86	67	88	12	50	53	64	5	34	18	39	10
1	0.05	0.2	69	95	97	22	96	93	97	9	71	79	86	6	61	40	71	5

Table 9.2 (*Cont.*)

(b) Means and standard deviations of estimated regression coefficients

D	β_1	β_2	No groups estimate of β_1 Mean	SD	No groups estimate of β_2 Mean	SD	3 groups estimates of β_1 Mean	SD	3 groups estimates of β_2 Mean	SD	6 groups estimates of β_1 Mean	SD	6 groups estimates of β_2 Mean	SD	9 groups estimates of β_1 Mean	SD	9 groups estimates of β_2 Mean	SD
4	0.00	0.0	0.01	0.01	0.01	0.06	0.03	0.08	0.01	0.06	0.04	0.18	0.01	0.08	0.05	0.45	0.01	0.09
4	0.00	0.1	0.01	0.05	0.11	0.05	0.04	0.09	0.11	0.07	0.05	0.19	0.10	0.08	0.11	0.41	0.10	0.08
4	0.00	0.2	0.01	0.05	0.21	0.06	0.03	0.08	0.21	0.06	0.05	0.22	0.20	0.08	0.08	0.42	0.20	0.09
4	0.05	0.0	0.06	0.05	0.01	0.06	0.08	0.08	0.01	0.06	0.11	0.21	0.00	0.07	0.16	0.39	0.00	0.09
4	0.05	0.1	0.06	0.05	0.11	0.06	0.08	0.09	0.11	0.06	0.09	0.19	0.10	0.08	0.13	0.42	0.09	0.09
4	0.05	0.2	0.06	0.06	0.21	0.06	0.09	0.10	0.21	0.07	0.10	0.20	0.21	0.07	0.13	0.41	0.20	0.10
2	0.00	0.0	0.01	0.05	0.01	0.06	0.06	0.10	0.01	0.07	0.08	0.20	0.00	0.09	0.09	0.39	0.00	0.09
2	0.00	0.1	0.02	0.05	0.12	0.06	0.06	0.11	0.11	0.07	0.05	0.20	0.10	0.08	0.06	0.39	0.10	0.10
2	0.00	0.2	0.02	0.05	0.21	0.06	0.05	0.10	0.21	0.07	0.06	0.20	0.21	0.08	0.12	0.39	0.20	0.09
2	0.05	0.0	0.06	0.05	0.02	0.06	0.10	0.09	0.01	0.06	0.12	0.21	0.01	0.07	0.15	0.40	0.00	0.09
2	0.05	0.1	0.07	0.06	0.11	0.07	0.10	0.09	0.11	0.07	0.13	0.20	0.10	0.08	0.14	0.37	0.10	0.10
2	0.05	0.2	0.07	0.06	0.21	0.06	0.12	0.10	0.21	0.06	0.13	0.22	0.21	0.07	0.16	0.40	0.21	0.09
1	0.00	0.0	0.03	0.06	0.02	0.06	0.08	0.10	0.01	0.07	0.08	0.18	0.00	0.07	0.12	0.34	0.01	0.09
1	0.00	0.1	0.03	0.06	0.12	0.06	0.08	0.10	0.11	0.06	0.09	0.22	0.11	0.07	0.13	0.34	0.10	0.09
1	0.00	0.2	0.03	0.06	0.21	0.06	0.08	0.11	0.20	0.06	0.08	0.21	0.21	0.07	0.10	0.36	0.20	0.08
1	0.05	0.0	0.08	0.06	0.02	0.07	0.13	0.11	0.01	0.07	0.12	0.20	0.00	0.07	0.14	0.32	0.01	0.09
1	0.05	0.1	0.08	0.06	0.12	0.07	0.12	0.10	0.12	0.07	0.11	0.19	0.11	0.08	0.15	0.35	0.10	0.09
1	0.05	0.2	0.08	0.06	0.22	0.07	0.13	0.11	0.21	0.07	0.13	0.21	0.21	0.07	0.18	0.36	0.20	0.09

Of course, a small simulation study like this based on the characteristics of one set of data is not sufficient to give complete confidence in the value of grouping and restricted randomization as a way of allowing for spatial correlation. Indeed, the simulation results indicate that spatial correlation is still likely to have some effect after these remedies are applied. However, with the *E. editha* data the coefficient of environmental distances was very highly significant. The simulations suggest that this significance level might be slightly exaggerated but not to the extent of casting doubt on the evidence for a relationship.

The randomization tests for this example were carried out using the program RT (Manly, 1996a). However, the simulation experiment was done using a computer program written just for that purpose.

9.7 FURTHER READING

Randomization tests on distance matrices are effected by transforming the distances. This fact was used in Example 9.1, where reciprocals of distances between quadrats were chosen in preference to the distances themselves for assessing the spatial correlation between counts of *Carex arenaria*. The effect can be quite pronounced, as was shown by Dietz (1983), who noted that the significance level for the simple correlation between anthropometric and genetic distances for tribes of Yanomama Indians (Spielman, 1973) changes from 20% to 32% if the anthropometric distances are squared. Clearly, care must be taken in deciding whether or not a transformation is appropriate.

A related matter concerns the situation where one of the distance matrices being used in a test is intended to represent the membership of certain specific groups. For example, consider the situation where the diet is recorded for 20 individual animals and from this information a 20 × 20 matrix of diet overlap values is constructed. Suppose that in addition the individuals are in three groups, and that there is interest in whether individuals in the same group tend to have a more similar diet than individuals in general. A second matrix 20 × 20 might then be constructed where the value in the *i*th row and *j*th column is 1 if individuals *i* and *j* are in the same group and 0 if they are in different groups. A Mantel test can then be used to test for a relationship between the two matrices. This test is an example of what is sometimes called a multi-response permutation procedure, as discussed more fully in section 12.3.

Luo and Fox (1996) have recently examined this situation in some detail. They found that the test has low power to detect diet similarities within groups when the sizes of those groups are not equal, and note that theoretical and simulation results indicate that this problem can be overcome by modifying the matrix that indicates group membership. Lou and Fox found that in their situation replacing the value of 1 for individuals in the same group by $1/n_i$, the reciprocal of the group size, gave better results than several alternative modifications, although theoretical results suggest that $1/(n_i - 1)$ may be better (Mielke, 1984; Zimmerman *et al.*, 1985).

The Mantel spatial correlogram is a useful tool for studying how correlation changes with distance for sample stations that are described by multivariate observations (Legendre and Fortin, 1989). This correlogram is a plot of the matrix correlation between a distance matrix based on the observations at the sampling station and a geographical distance matrix for which the element in the ith row and the jth column is 1 if station i and station j are in a particular distance class. The matrix correlation can be tested using Mantel's test to see whether objects within the distance class being considered are significantly related.

To conclude this chapter, it must be stressed that the randomization tests that have been discussed for comparing distance matrices will be only part of a comprehensive study of data. In most cases a significant test result on its own is of little value, and various other types of analysis should be considered as well. In particular, multidimensional scaling and cluster analysis may elucidate the nature of significant correlations.

Multidimensional scaling is a technique that can be used to produce a 'map' showing the positions of objects from a table of distances between those objects (Manly, 1994a, Chapter 11). For example, the genetic distances between colonies of *E. editha* that are shown in Table 9.3(a) can be used to construct a 'map' showing the relative positions of the colonies on this basis. The genetic 'map' can then be compared with the normal map of geographical positions (Figure 9.1), or an environmental 'map' obtained from multidimensional scaling of the differences shown in Table 9.3(b).

'Maps' from multidimensional scaling can be produced with one or more dimensions. With one dimension the objects are ordered along a line in the way that best represents the distances between them. With two dimensions a conventional 'map' is obtained. With three dimensions objects have 'heights' as well as positions in a plane. Graphical representation becomes difficult with four or more dimensions. A technique called Procrustes analysis (Seber, 1984, p. 253) can also be used to rotate and stretch the 'maps' obtained from two distance matrices in order to match them as closely as possible.

Cluster analysis can be carried out on a distance matrix in order to see which of the objects being considered are close together, and which are far apart (Manly, 1994a, Chapter 9). This is therefore another way to study the relationship between two distance matrices.

Table 9.3 Distance matrices for 21 colonies of *Euphydryas editha*. Lower triangular entries are shown (without the zero diagonals)

(a) Genetic distances

	1	2	3	4	5	6	7	8	9	10	11	12	13	14	15	16	17	18	19	20
2	0.03																			
3	0.64	0.61																		
4	0.28	0.25	0.36																	
5	0.30	0.27	0.34	0.02																
6	0.36	0.33	0.28	0.08	0.06															
7	0.18	0.15	0.46	0.10	0.12	0.18														
8	0.28	0.25	0.36	0.00	0.02	0.08	0.10													
9	0.35	0.32	0.29	0.07	0.05	0.01	0.17	0.07												
10	1.00	0.97	0.36	0.72	0.70	0.64	0.82	0.72	0.65											
11	0.93	0.90	0.29	0.65	0.63	0.57	0.75	0.65	0.58	0.09										
12	0.60	0.57	0.04	0.32	0.30	0.24	0.42	0.32	0.25	0.40	0.33									
13	0.61	0.58	0.03	0.33	0.31	0.25	0.43	0.33	0.26	0.39	0.32	0.01								
14	0.91	0.88	0.27	0.63	0.61	0.55	0.73	0.63	0.56	0.09	0.03	0.31	0.30							
15	0.81	0.78	0.17	0.53	0.51	0.45	0.63	0.53	0.46	0.19	0.12	0.21	0.20	0.10						
16	0.58	0.55	0.09	0.30	0.28	0.22	0.40	0.30	0.23	0.44	0.35	0.05	0.06	0.36	0.26					
17	0.63	0.60	0.05	0.35	0.33	0.27	0.45	0.35	0.28	0.40	0.31	0.04	0.04	0.32	0.22	0.05				
18	0.84	0.81	0.20	0.56	0.54	0.48	0.66	0.56	0.49	0.16	0.09	0.24	0.23	0.07	0.03	0.29	0.25			
19	0.96	0.93	0.32	0.68	0.66	0.60	0.78	0.68	0.61	0.04	0.06	0.36	0.35	0.05	0.15	0.41	0.37	0.12		
20	0.99	0.96	0.35	0.71	0.69	0.63	0.81	0.71	0.64	0.01	0.09	0.39	0.38	0.08	0.18	0.44	0.40	0.15	0.03	
21	0.96	0.93	0.32	0.68	0.66	0.60	0.78	0.68	0.61	0.04	0.06	0.36	0.35	0.05	0.15	0.41	0.37	0.12	0.00	0.03

Table 9.3 (*Cont.*)

(b) Geographical distances

	1	2	3	4	5	6	7	8	9	10	11	12	13	14	15	16	17	18	19	20
2	0.07																			
3	7.37	7.36																		
4	7.64	7.63	0.30																	
5	7.66	7.65	0.30	0.08																
6	7.68	7.67	0.34	0.04	0.10															
7	7.81	7.79	0.71	0.47	0.53	0.43														
8	8.07	8.05	0.80	0.50	0.53	0.46	0.31													
9	10.13	10.11	3.01	2.70	2.72	2.66	2.39	2.20												
10	13.97	13.94	7.46	7.16	7.20	7.12	6.77	6.66	4.56											
11	13.98	13.95	7.47	7.18	7.21	7.14	6.79	6.68	4.57	0.01										
12	7.86	7.84	1.11	0.90	0.97	0.87	0.44	0.66	2.27	6.50	6.52									
13	10.14	10.12	3.09	2.78	2.81	2.74	2.45	2.28	0.21	4.43	4.44	2.29								
14	5.36	5.34	2.33	2.53	2.57	2.55	2.55	2.85	4.78	8.71	8.73	2.53	4.78							
15	7.93	7.89	2.65	2.52	2.60	2.50	2.10	2.31	2.87	6.04	6.06	1.66	2.76	2.75						
16	8.19	8.16	2.60	2.44	2.51	2.41	1.99	2.17	2.55	5.78	5.80	1.55	2.44	2.96	0.32					
17	8.25	8.22	2.35	2.17	2.24	2.14	1.71	1.87	2.30	5.76	5.78	1.27	2.21	2.95	0.58	0.32				
18	8.88	8.84	3.20	3.00	3.07	2.97	2.54	2.65	2.39	5.09	5.11	2.10	2.23	3.69	0.95	0.74	0.87			
19	9.46	9.43	3.71	3.49	3.55	3.45	3.02	3.09	2.36	4.51	4.53	2.60	2.18	4.30	1.56	1.34	1.44	0.61		
20	7.40	7.36	2.74	2.67	2.74	2.65	2.30	2.56	3.41	6.58	6.59	1.90	3.32	2.34	0.58	0.90	1.11	1.49	2.07	
21	8.18	8.15	3.18	3.04	3.12	3.02	2.61	2.80	3.07	5.81	5.83	2.17	2.94	3.15	0.53	0.65	0.97	0.80	1.31	0.81

Table 9.3 (Cont.)

(c) Environmental distances

	1	2	3	4	5	6	7	8	9	10	11	12	13	14	15	16	17	18	19	20
2	1.05																			
3	2.26	1.40																		
4	1.78	0.87	0.55																	
5	1.78	0.87	0.55	0.01																
6	1.78	0.87	0.55	0.01	0.00															
7	2.26	1.30	0.45	0.51	0.50	0.50														
8	2.09	1.17	0.33	0.37	0.38	0.38	0.40													
9	1.95	1.02	0.46	0.19	0.19	0.19	0.34	0.31												
10	2.40	1.41	0.57	0.68	0.68	0.68	0.29	0.46	0.53											
11	2.40	1.41	0.57	0.68	0.68	0.68	0.27	0.47	0.53	0.03										
12	2.09	1.16	0.82	0.73	0.74	0.74	0.82	0.55	0.76	0.71	0.74									
13	2.16	1.32	0.73	0.79	0.80	0.80	0.90	0.53	0.81	0.81	0.84	0.38								
14	1.09	1.08	1.89	1.54	1.55	1.55	1.99	1.67	1.70	2.03	2.05	1.49	1.48							
15	2.63	2.32	2.60	2.46	2.47	2.47	2.72	2.37	2.57	2.60	2.63	1.93	1.91	1.62						
16	1.85	1.29	1.69	1.45	1.46	1.46	1.74	1.42	1.56	1.66	1.68	0.99	1.05	0.93	1.09					
17	2.11	1.28	1.24	1.11	1.12	1.12	1.24	0.96	1.15	1.09	1.12	0.49	0.60	1.35	1.58	0.65				
18	4.03	3.91	4.29	4.16	4.17	4.17	4.40	4.07	4.26	4.27	4.29	3.61	3.61	3.15	1.72	2.76	3.24			
19	4.72	4.62	4.85	4.78	4.79	4.79	5.01	4.66	4.88	4.87	4.90	4.22	4.17	3.78	2.34	3.42	3.87	0.93		
20	5.49	5.53	5.93	5.80	5.82	5.82	6.07	5.72	5.91	5.94	5.97	5.28	5.25	4.67	3.37	4.41	4.92	1.72	1.25	
21	6.53	6.53	6.88	6.78	6.79	6.79	7.01	6.68	6.88	6.87	6.90	6.22	6.20	5.70	4.33	5.39	5.86	2.65	2.09	1.06

9.8   CHAPTER SUMMARY

- Many hypotheses of interest in biology can be expressed in terms of whether there is an association between two matrices that measure different types of 'distance' between a set of objects.
- The Mantel randomization test can be used to assess whether or not an observed association between two distance matrices is significant. It involves comparing the observed value of a suitable statistic (such as the correlation between corresponding elements in the matrices) with the distribution that is generated by randomly permuting the object labels in one of the matrices.
- When the number of objects is too large to enable the full randomization distribution of a test statistic to be determined it is better to sample the randomization distribution than to assume that it can be approximated by a normal distribution.
- Confidence limits for a matrix regression coefficient can be determined as the range of values for which a Mantel randomization test fails to produce a significant result.
- The Mantel test relating one matrix to a second matrix can be generalized to the situation where a dependent matrix is related to two or more other matrices. The significance of matrix regression coefficients can then be determined by comparing conventional t-statistics and F-statistics with the distributions obtained by randomly reallocating the object labels for the dependent matrices.
- A number of other generalizations to the original Mantel two-matrix test have also been proposed. However, all generalizations are potentially affected by spatial autocorrelation when the situation being considered relates to objects in space.
- An example concerning the distribution of a butterfly in California and Oregon illustrates the complications that arise in ascertaining whether biological differences between colonies are related to environmental differences when spatial correlation exists. A way round these using spatial blocking and restricted randomizations is suggested.
- The results of matrix randomization tests depend upon how distances are measured, and any transformations that are made. In addition, when a matrix is an indicator of group membership the sizes of groups may need to be taken into account.
- It is stressed that matrix randomization tests should be part of a more comprehensive analysis that uses standard multivariate methods such as multidimensional scaling, Procrustes analysis and cluster analysis.

9.9 EXERCISES

The following exercises require the use of appropriate computer programs for matrix randomization tests, multidimensional scaling and cluster analysis. The randomization tests can be carried out using RT (Manly, 1996a). The other analyses are available in many popular statistical packages.

9.1 Using the distance matrices given in Tables 1.1 and 1.3, confirm that the matrix correlation between the coefficients of association of earwig species and 'jump' distances between continents before continental drift is −0.605, which is significantly low at about the 0.1% level. Using the method described in section 9.4, find an approximate 95% confidence interval for the 'true' regression coefficient. Using suitable software, carry out two-dimensional multidimensional scalings and cluster analysis on the two distance matrices to see if these throw light on the relationship between these matrices.

9.2 Confirm the results of the randomization tests described in Example 9.3 on the relationship between genetic, environmental and geographical distances between colonies of *E. editha*. Use multidimensional scaling to produce two-dimensional maps showing the positions of the colonies on the basis of genetic differences and environmental differences. See how these compare with the geographical map (Figure 9.1), and whether they help to explain the nature of the regression relationship between genetic distances and the other two distances.

9.3 Table 9.4(a) shows measures of the overlap in diet for 20 eastern horned lizards *Phrynosoma douglassi brevirostre*, as measured by Schoener's (1968) index. These overlap values were calculated from the data on 20 of the 45 lizards for which gut contents are provided by Linton *et al.* (1989), from a study by Powell and Russell (1984, 1985). For the purpose of the present example, the details of the calculation of the overlap values are not important, although more information is provided in section 14.4. What is important is to know that the values range from 0 (no prey in common) to 1 (exactly the same proportions of different prey used).

The lizards were sampled over a four-month period, and part (b) of Table 9.4 shows the sample time differences for the 20 lizards from 0 to 3 months. The lizards are also in two size classes (adult males and yearling females, and adult females) and part (c) of Table 9.4 shows a size difference matrix for the 20 lizards, with 1 indicating a different size and 0 indicating the same size. (Note that simple 0–1 indicator variables are used for size classes, rather than weighed values as discussed in section 9.7, because the numbers in the two classes (9 and 11 lizards) are similar.)

Use the methods of section 9.5 to investigate a regression of the diet overlap values against the sample time difference values and the size difference values to see to what extent differences in diet can be accounted for by time and size differences. There is no particular difficulty caused by only having zeros and ones in the size difference matrix. In a case like this the regression model simply allows the expected niche overlap values to differ by a fixed amount according to whether two lizards are in the same size class or not. Complete the analysis by producing a two-dimensional 'map' of the relationship between the lizards using multidimensional scaling and seeing how this relates to size and sample time differences.

Table 9.4 Matrices of (a) diet overlap values (based on gut contents), (b) time differences and (c) size differences for 20 eastern horned lizards.

(a) Diet overlap

	Lizard																		
Lizard	1	2	3	4	5	6	7	8	9	10	11	12	13	14	15	16	17	18	19
2	0.61																		
3	0.84	0.33																	
4	0.25	0.55	0.86																
5	0.34	0.53	0.86	0.18															
6	0.69	0.18	0.15	0.71	0.71														
7	0.67	0.57	0.86	0.59	0.53	0.71													
8	0.39	0.57	0.86	0.23	0.11	0.71	0.44												
9	0.36	0.54	0.86	0.20	0.08	0.71	0.50	0.06											
10	0.58	0.82	0.86	0.51	0.50	0.86	0.97	0.53	0.49										
11	0.62	0.57	0.86	0.54	0.48	0.71	0.05	0.39	0.45	0.92									
12	0.52	0.57	0.86	0.48	0.43	0.71	0.16	0.35	0.41	0.87	0.11								
13	0.51	0.83	0.87	0.43	0.34	0.87	0.88	0.44	0.39	0.44	0.83	0.78							
14	0.55	0.64	0.86	0.52	0.45	0.71	0.64	0.52	0.51	0.67	0.59	0.54	0.54						
15	0.49	0.94	0.94	0.73	0.68	0.94	0.94	0.75	0.75	0.77	0.89	0.78	0.60	0.67					
16	0.70	0.88	0.92	0.70	0.63	0.92	0.92	0.75	0.72	0.75	0.87	0.82	0.40	0.58	0.62				
17	0.67	0.57	0.86	0.59	0.53	0.71	0.00	0.44	0.50	0.97	0.05	0.16	0.88	0.64	0.94	0.92			
18	0.68	0.94	0.94	0.92	0.87	0.94	0.94	0.94	0.94	0.96	0.94	0.87	0.82	0.82	0.22	0.82	0.94		
19	0.65	0.94	0.98	0.72	0.67	0.98	0.98	0.77	0.74	0.73	0.93	0.82	0.44	0.64	0.54	0.12	0.98	0.76	
20	0.56	0.86	0.90	0.80	0.78	0.90	0.90	0.90	0.87	0.88	0.90	0.83	0.70	0.78	0.22	0.75	0.90	0.11	0.74

Table 9.4 (*Cont.*)

(b) Sample time differences

Lizard

Lizard	1	2	3	4	5	6	7	8	9	10	11	12	13	14	15	16	17	18	19
2	0																		
3	0	0																	
4	0	0	0																
5	0	0	0	0															
6	1	1	1	1	1														
7	1	1	1	1	1	0													
8	1	1	1	1	1	0	0												
9	1	1	1	1	1	0	0	0											
10	1	1	1	1	1	0	0	0	0										
11	2	2	2	2	2	1	1	1	1	1									
12	2	2	2	2	2	1	1	1	1	1	0								
13	2	2	2	2	2	1	1	1	1	1	0	0							
14	2	2	2	2	2	1	1	1	1	1	0	0	0						
15	2	2	2	2	2	1	1	1	1	1	0	0	0	0					
16	3	3	3	3	3	2	2	2	2	2	1	1	1	1	1				
17	3	3	3	3	3	2	2	2	2	2	1	1	1	1	1	0			
18	3	3	3	3	3	2	2	2	2	2	1	1	1	1	1	0	0		
19	3	3	3	3	3	2	2	2	2	2	1	1	1	1	1	0	0	0	
20	3	3	3	3	3	2	2	2	2	2	1	1	1	1	1	0	0	0	0

Table 9.4 (*Cont.*)
(c) Size differences

Lizard	1	2	3	4	5	6	7	8	9	10	11	12	13	14	15	16	17	18	19
1	0																		
2	1	1																	
3	1	1	0																
4	1	1	0	0															
5	0	0	1	1	1														
6	0	0	1	1	1	0													
7	1	1	0	0	0	1	1												
8	1	1	0	0	0	1	1	0											
9	1	1	0	0	0	1	1	0	0										
10	0	0	1	1	1	0	0	1	1	1									
11	0	0	1	1	1	0	0	1	1	1	0								
12	0	0	1	1	1	0	0	1	1	1	0	0							
13	0	0	1	1	1	0	0	1	1	1	0	0	0						
14	0	0	1	1	1	0	0	1	1	1	0	0	0	0					
15	0	0	1	1	1	0	0	1	1	1	0	0	0	0	0				
16	0	0	1	1	1	0	0	1	1	1	0	0	0	0	0	0			
17	1	1	0	0	0	1	1	0	0	0	1	1	1	1	1	1	1		
18	1	1	0	0	0	1	1	0	0	0	1	1	1	1	1	1	1	0	
19	1	1	0	0	0	1	1	0	0	0	1	1	1	1	1	1	1	0	0

10
Other analyses on spatial data

10.1 SPATIAL DATA ANALYSIS

The previous chapter was concerned with the analysis of various types of spatial data through the construction and comparison of distance matrices. The present chapter still involves spatial data. However, the concern is with approaches that do not explicitly involve distance matrices. No attempt will be made here to provide a comprehensive review of the large and growing area of statistics that is concerned with spatial data. Instead, a few particular situations will be used to illustrate the potential for computer-intensive methods in this area. Those interested in more information should consult a specialist text, such as that of Diggle (1983), Grieg-Smith (1983), Ripley (1981), or Haining (1990).

10.2 THE STUDY OF SPATIAL POINT PATTERNS

One major area of interest is in the detection of patterns in the positions of objects distributed over an area. For example, Figure 10.1 shows the positions of 71 pine saplings in a 10×10 metre square. A question here concerns whether the pines seem to be randomly placed, where a plausible alternative model involves the idea of some inhibition of small distances because of competition between trees.

There have been two general approaches adopted for answering this type of question. One approach consists of partitioning the area into equally sized square or rectangular quadrats and counting the numbers of objects within these. The second approach is based on measuring the distances between objects. In both cases a comparison can be made between observed statistics and distributions that are expected if points are placed independently at random over the area of interest.

Randomness implies a Poisson distribution for quadrat counts, and a chi-squared goodness of fit test can be carried out on this basis. However, this usually has low power against reasonable alternatives, and it is better to use

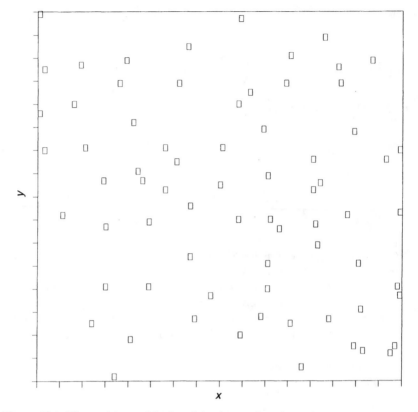

Figure 10.1 The positions of 71 Swedish pine saplings in a 10 × 10 m square. This figure is redrawn from Figure 8.7(a) of Ripley (1981) and is based on original data of Strand (1972).

an index such as the variance to mean ratio (Mead, 1974). Alternative approaches, which take into account the spatial positions of the quadrats, are a hierarchic analysis of variance (Grieg-Smith, 1952) for a 16 × 16 grid of quadrats or randomization tests (Mead, 1974; Galiano *et al.*, 1987; Dale and MacIsaac, 1989; Perry and Hewitt, 1991; Perry, 1995a, 1995b). Here only Mead's randomization test will be considered in any detail for quadrat count data. A Monte Carlo test for randomness based on distances between points has been considered already in Example 4.1. This is considered further in section 10.4.

10.3   MEAD'S RANDOMIZATION TEST

Mead (1974) described his randomization test in terms of quadrats along a line, but mentioned that the same approach can be used with a square grid.

1	2	5	6
3	4	7	8
9	10	13	14
11	12	15	16

Figure 10.2 An area divided into 16 quadrats in a 4×4 grid. The numbers in the quadrats are labels rather than quadrat counts.

Here the approach based on a square will be considered so as to enable the test to be used with the Swedish pine data displayed in Figure 10.1.

Consider an area divided into 16 quadrats in a 4×4 grid, as indicated in Figure 10.2. The question addressed by Mead's test is whether the division of the quadrats into the blocks of four (1,2,3,4), (5,6,7,8), (9,10,11,12), and (13,14,15,16) appears to be random. This can be tested by comparing the observed value of a suitable test statistic with the distribution of values that is obtained by randomly allocating the quadrat counts to the 16 positions. There are $16!/(4!)^4$, or approximately 6.3 million possible allocations, so that sampling the distribution is more or less essential.

A reasonable choice for the test statistic is the sum of squares between the blocks of four as a proportion of the total sum of squares. If T_i denotes the count for quadrat i, then the total sum of squares about the mean for the 16 counts is given by the equation

$$TSS = \sum_{i=1}^{16} T_i^2 - 16\bar{T}^2.$$

where $\bar{T}$ is the mean count. The sum of squares between the blocks of four is given by the equation

$$BSS = (T_1 + T_2 + T_3 + T_4)^2/4 + (T_5 + T_6 + T_7 + T_8)^2/4$$
$$+ (T_9 + T_{10} + T_{11} + T_{12})^2/4 + (T_{13} + T_{14} + T_{15} + T_{16})^2/4 - 16\bar{T}^2$$

The proposed test statistic is then $Q = BSS/TSS$.

The idea behind this test is that if some clustering occurs at the scale that is reflected by blocks of size four then this should be picked up by the test. Thus, to take an extreme case, suppose that the quadrat counts are as below:

$$
\begin{array}{cccc}
10 & 10 & 0 & 0 \\
10 & 10 & 0 & 0 \\
0 & 0 & 5 & 5 \\
0 & 0 & 5 & 5 \\
\end{array}
$$

Then there is no variation within the blocks of four, so that $BSS = TSS$ and $Q = 1$. This is a clear measure of the complete clustering in two of the blocks. At the other extreme is a case like the following one:

$$
\begin{array}{cccc}
5 & 10 & 5 & 10 \\
5 & 0 & 5 & 0 \\
5 & 10 & 5 & 10 \\
5 & 0 & 5 & 0
\end{array}
$$

Here there is no variation between the block totals, so that $BSS = 0$ and hence $Q = 0$. This is a clear measure of the pattern that is repeated from block to block. Obviously, real data cannot be expected to be as straightforward as these two examples. Nevertheless it is realistic to hope that the test will detect patterns.

Because patterns can make either high or low values of Q likely, it is appropriate to use a two-sided test to determine significance. Then an observed value will be significant at the 5% level if it is in either the bottom 2.5% or the top 2.5% of the randomization distribution. (However, see the remark on multiple tests at the end of Example 10.1 below.)

If there are several 4 × 4 grids of quadrats available in an area then a Q value can be calculated for each and the mean Q value used as an overall test statistic for the null hypothesis of randomness. Randomization can be done within each 4 × 4 grid and the significance of the mean Q value determined by comparison with the distribution that this randomization produces.

Upton (1984) has criticized Mead's test for quadrats along a line on the grounds that the starting point may be critical to the test statistic. The same difficulty occurs with the two-dimensional case being considered here. Thus suppose that the following pattern is found over an area.

```
...   .   .   .   .   .   .   .   .   .   .   .   ...
...   .   .   .   .   .   .   .   .   .   .   .   ...
... 0 4 4 0 0 4 4 0 0 4 4 ...
... 0 4 4 0 0 4 4 0 0 4 4 ...
... 4 0 0 4 4 0 0 4 4 0 0 ...
... 4 0 0 4 4 0 0 4 4 0 0 ...
...   .   .   .   .   .   .   .   .   .   .   .   ...
...   .   .   .   .   .   .   .   .   .   .   .   ...
```

Then the counts for Mead's test will all be of one of the following two types (or equivalents), repeated across the area, depending upon the quadrat where recording begins:

0	4	4	0		4	4	0	0
0	4	4	0		4	4	0	0
4	0	0	4		0	0	4	4
4	0	0	4		0	0	4	4

It follows that the Q statistic will be either 0 or 1, depending on the starting point. This example indicates that some care is necessary in interpreting significant values of the test statistic. The appropriate interpretation for the present example is that clustering and regularity both exist.

One of the advantages of Mead's test is that it can be carried out with a range of cluster sizes. For example, if quadrat counts are available for a 16×16 grid then this can be considered as a 4×4 array of blocks of 16 quadrats each. The test can then be carried out by randomizing the blocks of 16 quadrats within the 4×4 array. The 16×16 grid can also be thought of as four separate 8×8 grids. The test can be carried out by randomizing 2×2 blocks of quadrats within each of these 8×8 grids, using the mean of the four values of Q as a test statistic. Finally, the 16×16 grid can be thought of as 16 separate 4×4 grids. Within each of these the individual quadrats can be randomized, using the mean of the 16 Q values as a test statistic. In this way it becomes possible to test for patterns at three different levels of plot size. The following example will make the process clearer.

Example 10.1: Mead's test on counts of Swedish pine saplings Suppose that the area shown in Figure 10.1 is divided into 16 quadrats each of size 2.5×2.5 metres. The counts in the quadrats are then as shown below:

6	2	5	4
4	6	4	6
3	4	5	6
2	3	4	7

For these data Mead's test statistic is $Q = 0.389$. This value of Q plus 4999 values obtained by randomizations of the block counts to positions in the grid gave the estimated randomization distribution, for which it was found that 11.1% of Q values equal or exceed the one observed. It seems, therefore, that although the observed Q value is slightly high it is not a particularly unusual value to occur on the assumption of a random distribution of counts.

Next, suppose that each of the 2.5×2.5 metre quadrats is divided into four, so that the total area is divided into 64 quadrats of size 1.25×1.25 metres. The counts in these quadrats are then as follows:

```
1  0  0  1        1  1  1  0
3  2  0  1        2  1  1  2
2  1  2  0        2  0  1  0
0  1  3  1        1  1  3  2
(Q = 0.175)       (Q = 0.066)


1  1  1  1        1  2  2  1
0  1  1  1        0  2  1  2
0  1  0  2        2  1  1  2
0  1  1  0        0  1  1  3
(Q = 0.100)       (Q = 0.128)
```

The Q values for blocks of 16 quadrats are shown below the blocks, with mean $\bar{Q} = 0.117$. The observed statistic plus 4999 values obtained by randomizing counts independently within each of the 4×4 blocks was used to approximate the randomization distribution. It was found that the observed $\bar{Q}$ was equalled or exceeded by 89.2% of randomized values. Hence this is a low value, but within the range expected on the null hypothesis of no pattern in the data.

Finally, the study area can be divided into a 16×16 grid of quadrats of size 0.625×0.625 metres, and these can be considered in 4×4 blocks. For example, in the top left-hand corner of the area there are the quadrat counts shown below, there being 16 such blocks in all:

```
1  0  0  0
0  0  0  0
1  1  0  1
0  1  0  1
```

Each 4×4 block of quadrats provides a Q statistic. The mean is $\bar{Q} = 0.156$. The randomization distribution was approximated by this value plus 4999 other values obtained by randomizing independently within each 4×4 array. The observed value was exceeded by 90.7% of the randomization distribution so that again the observed statistic does not provide much evidence of non-randomness.

Overall, this analysis gives little indication of non-randomness, particularly if it is considered that since three tests are being conducted it is appropriate to control the significance level of each one to obtain an overall probability of only 0.05 (say) of declaring anything significant by chance. To achieve this, Bonferroni's inequality suggests that a significance level of $5/3 = 1.7\%$ should be used for each test.

The calculations for this example were carried out using the RT package (Manly, 1996a).

10.4   TESTS FOR RANDOMNESS BASED ON DISTANCES

Example 4.1 concerned a Monte Carlo test to compare mean nearest-neighbour distances for a set of points with the distribution of such distances that are obtained when points are allocated randomly over the study area. In particular, q_i was defined to be the mean distance from points to their ith nearest-neighbour, and the observed values of q_1 to q_{10} were compared to the distributions of these statistics that were obtained by allocating the same number of points to random positions. The test statistics q_1 to q_{10} are not, of course, the only ones that can be used in this type of situation. Some other possibilities are discussed by Ripley (1981, Chapter 8). In particular, the use of Ripley's K-function is becoming increasingly popular for providing test statistics, as discussed by Andersen (1992) and Haase (1995).

Perhaps the most important advantage of this type of Monte Carlo test is that it is not affected by boundary effects. The test can be used for points within any closed region providing that random points are equally likely to be placed anywhere within this region. In contrast to this situation, distributions that are quoted for use with tests on nearest-neighbour distances often ignore boundary effects and the dependence that exists between the distances, although corrections for these effects can be made (Ripley, 1981, p.153).

Example 10.2: nearest-neighbour distances for Swedish pine saplings   The nearest-neighbour test can be applied to the Swedish pine data shown in Figure 10.1. Here the q statistics are $q_1 = 1.27$, $q_2 = 1.78$, $q_3 = 2.08$, $q_4 = 2.34$, $q_5 = 2.59$, $q_6 = 2.88$, $q_7 = 3.18$, $q_8 = 3.43$, $q_9 = 3.65$ and $q_{10} = 3.92$, where the unit of distance ($10/16 = 0.625$ metres) is indicated on the border of the figure.

When the distribution of the q values was approximated by 499 random allocations of 71 points to the same area, plus the observed allocation, it was found that the observed values were the highest of the 500 for q_1 and q_2, the observed q_3 was equalled or exceeded by 22 (4.4%) of the 500, but q_4 to q_{10} were very typical values from the randomization distributions. Figure 10.3 shows the situation graphically, with the observed q values plotted together with the limits equalled or exceeded by 1% and 99% of the randomized values. There is a clear indication here of an absence of small distances between trees in comparison with what is expected from completely random positioning. Apparently there is an area of inhibition around each tree.

A similar conclusion was reached by Ripley (1981, p. 175), using different statistics on the same data. He suggested that an alternative to the random model might be a Strauss process, for which the probability of an object being at a point depends on the number of other objects within a distance r of this point. He found that the Strauss process with an inhibition distance

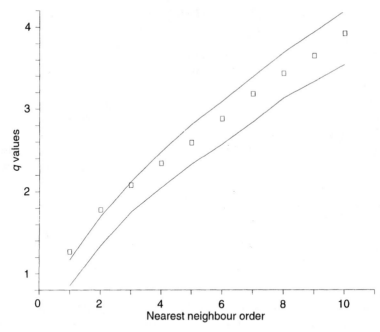

Figure 10.3 Swedish pine saplings: comparison between observed q-statistics ($\square$) and the lower and upper 1% points for the distribution found by randomly allocating 71 points to the study area (—).

of 70 cm (1.12 of the units being used here) gives a distinctly better fit to the data than the random model. A Fortran algorithm for generating data from the Strauss process is given by Ripley (1979).

A simple process is one for which there is a complete inhibition of objects less than a distance r apart, so that in effect each object covers a circular area of radius $r/2$. This is straightforward to simulate. Points are added to a region one by one in random positions. If the next point is closer than r to an existing point then it is not used, and another point is tried. An obvious upper limit to r is the minimum distance between two points observed in the data of interest.

Unfortunately, this simple model is not appropriate for the Swedish pine data. The closest observed distance between two trees is 0.36 units (23 cm). If this is taken as an inhibition distance round each point and data simulated accordingly, then the observed $g_1 = 1.27$ and $g_2 = 1.78$ are still larger than 499 simulated values. Even simulating with an inhibition distance of 0.76 (twice the minimum observed distance between trees) gave rather low g_2 values. The observed g_1 was equalled or exceeded by 13.2% of 499 simulated values, and g_2 was equalled or exceeded by 1.2% of simulated values. It seems that the conclusion must be that there was a tendency for the trees to be some distance apart, but the inhibiting process involved is more like the Strauss process discussed by Ripley than the simple inhibition process just considered.

This conclusion is not the same as the one reached from the analysis of the same data using Mead's randomization test (Example 10.1). However, with Mead's test on 64 quadrats the observed value of $\bar{Q}$ was exceeded by 89.2% of randomized values. Although this observed $\bar{Q}$ is not exceptionally low, it does indicate a tendency for the total counts to be rather similar in adjacent blocks of four quadrats. Similarly, with Mead's test on 256 quadrats the observed $\bar{Q}$ of 0.156 was exceeded by 90.7% of randomizations. Again this indicates a tendency for the total counts in adjacent blocks of four quadrats to be similar. Also, because Mead's test uses quadrat counts rather than information on the relative positions of all points it cannot be expected to be as sensitive as the Monte Carlo test for detecting spatial patterns.

The calculations for this example were done using the RT package (Manly, 1996a).

10.5 TESTING FOR AN ASSOCIATION BETWEEN TWO POINT PATTERNS

The spatial problems considered so far have only involved one set of points. However, there are some questions that involve two or more sets of points. For example, Figure 10.4 shows the positions of 173 newly emergent and

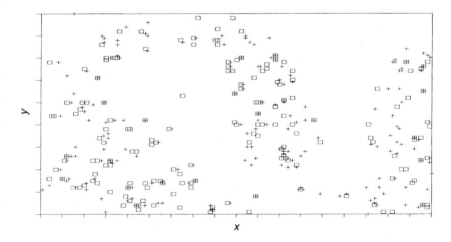

Figure 10.4 Positions of 173 newly emergent (□) and 195 one-year-old (+) brambles in a 9 × 4.5 m area.

195 one-year-old bramble canes, *Rubus fruticosus*, in a 9×4.5 m area, as given by Diggle and Milne (1983) and originally measured by Hutchings (1979). The open spaces suggest a tendency for the two types of bramble to cluster together, and there might be some interest in determining how likely this is to have occurred by chance alone, accepting that both types of bramble are likely to have a non-random distribution when considered alone.

As was the case with a single distribution, the relationship between two point patterns (A and B) can either be considered in terms of quadrat counts, or in terms of distances between points. Here the analysis of quadrat counts will be considered first.

10.6   THE BESAG–DIGGLE TEST

Besag and Diggle (1977) and Besag (1978) have discussed a generalization of Mead's (1974) test for the detection of an association between A and B counts at various levels of scale. To use this test, the area of interest is first divided into a 2×2 block of four quadrats, adding counts over smaller areas as necessary. The Spearman rank correlation between A and B counts is then used to measure their association. Assuming no ties, the observed value can be tested for significance using the usual tables for the Spearman rank correlation. With only four counts, a value of $+1$ or -1 occurs with probability $1/12 = 0.083$, so that the test is not very informative at this level of scale.

The second step is to divide the area of interest into four blocks, each consisting of a 2×2 array of quadrats. This does provide a useful test. A Spearman correlation coefficient can be calculated for each of the blocks and the average of these, $\bar{S}_4$, say, taken as a measure of the association between A counts and B counts within blocks of size four. Besag (1978) has calculated the exact distribution of this statistic on the null hypothesis that all of the 4! possible pairings of the A and B counts are equally likely within each block.

The procedure can be continued by dividing the area of interest into 16 blocks, each consisting of a 2×2 array, calculating Spearman's rank correlation coefficient for each block, and using the mean $\bar{S}_{16}$ as the measure of association between A and B counts. Besag (1978) states that the distribution of $\bar{S}_k$ under the null hypothesis that all pairings are equally likely within blocks is well approximated by a normal distribution with mean zero and variance $1/(3k)$, for $k \geq 16$. Continuing in the same way, 64, 256, or more blocks of 2×2 quadrats can be analysed to determine the evidence, if any, for association between the A and B counts at different scales of area.

Tied values may have to be handled in calculating Spearman's rank correlation. Also, with small quadrat sizes there may be cases where either all the A counts or all the B counts, or both, are zero within some 2 × 2 blocks. Besag and Diggle (1977) suggest breaking ties by randomly ordering equal values because this allows the exact distribution for $\bar{S}_4$, and the normal approximation for the distribution of $\bar{S}_k, k \geq 16$, to be used. However, it might be preferable to calculate the correlation using average ranks, ignoring data for blocks where there is no variation in either the A or B counts. The null hypothesis distribution of the mean correlation for all blocks used will then have to be determined by computer randomization.

The null hypothesis for the Besag–Diggle test is that within a 4 × 4 block of quadrats all of the 4! possible allocations of the A counts to the B counts are equally likely. However, there is some question about the reasonableness of this in cases where both the A and B counts have non-random distributions. An example will clarify the problem.

Suppose that the distribution of A counts is as follows for four contiguous 2 × 2 blocks of quadrats:

	Block		
1	2	3	4
0 25	25 0	0 10	10 0
0 25	25 0	0 10	10 0

Suppose also that, because of the mechanism that generates the counts, the cells with positive counts always occur in fours, as they do here for the four 25s and the four 10s. Then clearly, although it may be true that within one block all allocations of quadrat counts can be considered to be equally likely, this may not be so for two or more blocks considered together. For example, the configuration shown below may be impossible because of the way that the positive counts are split off from the blocks of four in which they must occur.

	Block		
1	2	3	4
25 0	25 0	10 0	10 0
25 0	25 0	10 0	10 0

This will not matter if the B counts can be considered to be randomly allocated and the A counts as fixed. However, if both A and B counts have a clear spatial structure then it is difficult to justify randomizing either of them. Put another way, if neighbouring quadrat counts are correlated for both the A and B points then it is not valid to allow any randomizations that change distances between quadrat counts.

A way around this difficulty involves separating the blocks of quadrat being tested with margins consisting of quadrats that are not being used. It may then be a reasonable assumption that randomization can be carried out independently in each block.

10.7   TESTS USING DISTANCES BETWEEN POINTS

The problem concerning how to randomize for testing the independence of two processes was addressed by Lotwick and Silverman (1982) for situations where the positions of individual points are known. They suggested that a point pattern in a rectangular region can be converted to a pattern over a larger area by simply copying the pattern from the original region to similar sized regions above, below, to the left and to the right, and then copying the copies as far away from the original region as required. A randomization test for independence between two patterns then involves comparing the test statistic observed for the points over the original region with the distribution of this statistic that is obtained when the rectangular 'window' for the A points is randomly shifted over an enlarged region for the B points.

As Lotwick and Silverman note, the need to reproduce one of the point patterns over the edge of the region studied in an artificial way is an unfortunate aspect of this procedure. It can be avoided by taking the rectangular window for the A points to be smaller than the total area covered and calculating a test statistic over this smaller area. The distribution of the test statistic can then be determined by randomly placing this small window within the larger area a large number of times. In this case the positions of A points outside the small window are ignored and the choice of the positioning of the small window within the larger region of A points is arbitrary.

Another idea involves considering a circular region and arguing that if two point patterns within the region are independent then this means that they have a random orientation with respect to each other. Therefore a distribution that can be used to assess a test statistic is the one that is obtained from randomly rotating one of the sets of points about the centre point of the region. A considerable merit with this idea is that the distribution can be determined as accurately as desired by rotating one of the sets of points about the centre of the study area from zero to 360 degrees in suitable small increments. The following example should make this idea clearer.

Example 10.3: testing for an association with brambles   Figure 10.5 shows the positions of newly emergent and one-year-old brambles in two circular regions extracted from the 9.0×4.5 m rectangular region of Figure 10.4. The left-hand circular region has a centre 2.25 m horizontally and 2.25 m vertically from the bottom left-hand corner, with a radius of 2.25 m. The right-hand circle has the same radius, with a centre 6.75 m horizontally and 2.25 m vertically from the bottom left-hand corner. Between them these two circular areas occupy most of the 9 m $\times$ 4.5 m rectangular region.

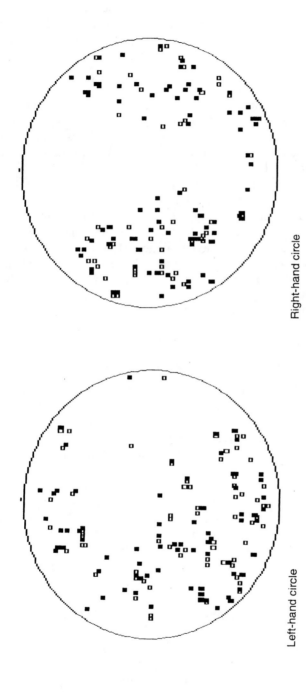

Left-hand circle   Right-hand circle

Figure 10.5 Circular test regions from the rectangular region shown in Figure 10.4: (□) newly emergent brambles; (■) one-year-old brambles.

There are various test statistics that can be considered for measuring the matching between the two point patterns. Here the average distance from each point to its nearest-neighbour of the opposite type is used (newly emergent to one-year-old and vice versa), with small values indicating a positive association between the two types.

For the left-hand circle the value of this statistic is 0.1152 m for the real data. This is the minimum value that can be obtained to four decimal places, the same value being found for any rotation of the one-year-old plants within a range from about −0.025 to 0.035 degrees of the observed configuration. The probability of being within this narrow range by chance is about $0.06/360 = 0.00017$, so that it is most unlikely that this is an accident. A similar situation is found for the right-hand circle. The observed test statistic is 0.1488, which is the smallest possible value to four decimal places. This value is obtained for rotations from about −0.035 to +0.015 degrees of the one-year-old plants so that the probability of obtaining such a small value by chance alone is $0.05/360 = 0.00014$. These results demonstrate clearly that the positions of the newly emergent and the one-year-old brambles are closely related in space, as is indeed obvious from Figure 10.4.

See Diggle (1983, p. 113) for a further discussion of the fitting of models to these data.

10.8 TESTING FOR RANDOM MARKING

Romesburg (1989) has published a versatile Fortran program called ZORRO for assessing the association between different types of points in one-, two- or three-dimensional space, where this can be thought of as testing whether the markings indicating the types of points could have been randomly allocated. Using a nearest-neighbour test statistic, the null hypothesis that objects were placed in their positions by a random process that ignores the type is tested against the alternatives that (i) similar types of object tend to occur close together, and (ii) similar types of object tend to repel each other. The testing is done by comparing the sum of squared nearest-neighbour distances with the distribution of this statistic that is obtained by randomly assigning objects to the observed positions, ignoring type labels. An application of this program would be to see whether the pattern of newly emergent and one-year-old brambles shown in Figure 10.4 is typical of what could be expected if the brambles were randomly allocated to the observed positions without regard to their age.

One important application of ideas of this type is for the detection of spatial clustering of events such as cases of a disease using controls to take into account spatial variation in the rate of occurrence. For example, Diggle and Chetwynd (1991) discuss the example where there are 62 cases of childhood leukaemia and lymphoma in north Humberside, England, together with 141 control individuals selected at random from the birth register. The question of interest was whether the disease cases show spatial clustering in comparison with the distribution of the controls. In particular, the null hypothesis was that the locations of the 62 cases were a random selection from the locations for all 203 individuals.

To test the null hypothesis, Diggle and Chetwynd used what are called K-functions. For a set of points over a region, the K-function is defined to be

$$K(s) = E[\text{number of other events within distance } s \text{ of an event}]/\lambda,$$

where E indicates the expected (mean) value and λ is the intensity, which is the mean number of events per unit area. When the points are also labelled (such as 1 for cases and 2 for controls), K functions can be defined as

$$K_{ij}(s) = E[\text{number of (other) type } j \text{ points within}$$
$$\text{distance } s \text{ of a type } i \text{ point}]/\lambda_j,$$

where λ_j is the intensity for type j points. With these definitions $K_{ii}(s)$ is essentially just the same as $K(s)$ for type i points. Although it is not immediately apparent, it can be shown that $K_{ij}(s) = K_{ji}(s)$.

The null hypothesis of interest is that the labelling of cases is random, which implies that $K_{11}(s) = K_{22}(s) = K_{12}(s)$. For this reason, Diggle and Chetwynd proposed the use of $\hat{D}(s) = \hat{K}_{11}(s) - \hat{K}_{22}(s)$ for studying departures from the null hypothesis, where $\hat{K}_{ii}(s)$ is an estimate of $K_{ii}(s)$. Here $\lambda_1 D(s)$ can be interpreted as an estimate of the mean number of excess disease cases within distance s of a typical disease case.

For estimating $K_{11}(s)$ the equation

$$\hat{K}_{11}(s) = A \sum_{i=1}^{n_1} \sum_{j=1}^{n_1} w_{ij} \delta_{ij}(s)/\{n_1(1 - n_1)\}$$

can be used, where A is the area of the study region, n_1 is the number of type 1 points, and δ_{ij} is 1 if the distance between points i and j of type 1 is less than s, or is otherwise 0. The definition of w_{ij} is a little complicated. Imagine a circle centred at point i with a radius equal to the distance from point i to point j, so that point j is on the circumference. Let p_{ij} be the proportion of this circle within the study area. Then w_{ij} is defined to be $1/p_{ij}$ when $i \neq j$, and 0 when $i = j$. For estimating $K_{22}(s)$ the same equation is used with obvious modifications so that it is calculated from type 2 points instead of type 1 points.

Having calculated $\hat{D}(s)$ for suitable equally spaced values of $s_1, s_2, \ldots, s_m$, these values can be combined to produce an overall test statistic which can then be compared with the distribution that is generated by randomly permuting the labels on the observations. One such statistic suggested by

Diggle and Chetwynd is

$$D = \sum_{k=1}^{m} \hat{D}(s_k)/SD\{\hat{D}(s_k)\}$$

where $SD\{\hat{D}(s_k)\}$, the standard deviation of $\hat{D}(s_k)$, can be calculated using an equation that they provide. One alternative to using D for a randomization test would be to test all of the $\hat{D}(s_1)$, $\hat{D}(s_2)$, ..., $\hat{D}(s_m)$ with randomization tests, making an allowance for the multiple testing with the method described in section 6.8.

When Diggle and Chetwynd applied their method to the data on childhood leukaemia and lymphoma, using $m = 10$ values for $\hat{D}(s)$, they found some suggestion of clustering of the cases. However, the test statistic of $D = 1.84$ was exceeded by 14 of 100 randomized values, so the evidence is very slight. A one-sided test was required in this case because large values of $\hat{D}(s)$ indicate that cases tend to have more other cases close to them than is expected from the distribution of the controls.

See Andersen (1992) for another examples of the use of K-functions, where this concerns the distribution of the plant *Polygonum newberryi* 'marked' by the presence of the aphid *Aphthargelia symphoricarpi*.

10.9 FURTHER READING

Some more examples of spatial data analysis using Monte Carlo methods are provided by Penttinen *et al.* (1992). A useful modern review of applications in ecology is also given by Legendre (1993).

Solow (1989a) has discussed the use of bootstrap methods with spatial sampling. As an example, he took the locations of 346 red oaks in Lancing Woods, Michigan, and estimating the number of trees per unit area by T-square sampling of 25 evenly spaced points. The distribution of the estimator was then approximated by taking 209 bootstrap samples of 25 from the obtained points. This was found to compare reasonably well with the distribution obtained by genuine repeated sampling of the study area. Also, bootstrap confidence intervals for the true tree density were similar to the limits found from genuine repeated sampling. As Solow notes, bootstrap determination of the level of sampling errors has much potential with spatial estimation problems.

In another paper, Solow (1989b) discusses a randomization test for independence between the successive n positions observed for an animal within its home range. The test statistic used is t^2/r^2, where t^2 is the mean squared distance between successive observations and r^2 is the mean squared distance from the centre of activity. The observed value is compared with the distribution obtained by randomly permuting the order of the observations.

See also the randomization test of Solow and Smith (1991) for clustering of the species in a community based on quadrat counts.

An important area of interest at the present time concerns the development of algorithms for searching for patterns in large geographical information systems containing data on the spatial location of items, information on the time when events occur, and other attributes of events. In this context Openshaw (1994) describes one approach which uses a search procedure and an associated Monte Carlo test procedure. No doubt methods of this type will receive increasing use in the future as a means of detecting patterns with a minimum of human input.

Another area with important potential applications concerns methods for detecting edges in mapped ecological data, such as the boundary between two vegetation types. It has been suggested, for example, that changes in such boundaries may be the first indication of global warming. One approach to this problem is based on using a randomization test to search for boundaries along which there is a high average rate of change for all the variables that are measured (Fortin, 1994). The process is called 'wombling' because it is based on Womble's (1951) algorithm for measuring rates of change. See also Williams (1996).

Finally, a type of data not yet mentioned is directional data where the items being considered are in the form of directions in a circle or sphere. Bootstrap methods in this area are reviewed by Fisher and Hall (1992).

10.10   CHAPTER SUMMARY

- There are two general approaches that are used for studying spatial patterns. One is based on partitioning the study area into rectangular or square quadrats and counting the number of objects of interest within each quadrat. The other approach is based on measuring the distances between objects.

- Mead's test is based on quadrat counts. As described here, it examines whether there is evidence of clustering within sets of four adjacent quadrats in a set of 16 quadrats. This test has been criticized because the test statistic can change from a high value to a low value depending on the location of quadrats when quadrat counts exhibit a pattern. However, the test can be used to examine non-randomness at a series of different scales.

- Tests based on distances between objects and their nearest neighbours are often used to test for complete spatial randomness. Ripley's K-functions have also been used in recent times as test statistics.

- The Besag–Diggle test is a generalization of Mead's test that is intended to test for correlation between the quadrat counts for two different types of object. With the Besag–Diggle test Spearman's correlation coefficient is tested for significance by comparison with the distribution obtained if all pairings of quadrat counts are equally likely within blocks of four quadrats. The validity of this test is questioned when both sets of quadrat counts are non-randomly distributed in space.

- One method for testing between the location of two point patterns involves comparing the value of an observed test statistic with the distribution obtained when a small rectangular window containing one of the sets of points is randomly positioned over the full area for the second set of points. Alternatively, if a circular region is considered then a measure of the association between two point patterns can be tested for significance in comparison with the distribution found from the random rotation of one point pattern with respect to the other one.

- A number of spatial problems can be reduced to a question of whether certain of the objects in a study region that are 'marked' appear to be a random sample from all of the objects, e.g. whether the cases of a disease represent a random sample of the individuals in a population in some respect. This can be examined using test statistics based on Ripley's K-functions.

- A number of generalizations and extensions to the methods discussed in the chapter are briefly reviewed.

10.11   EXERCISES

10.1   Table 10.1 is part of a larger one given by Andrews and Herzberg (1985, p. 296), from an original study by Sayers *et al.* (1977). The data are the number of cases of fox rabies observed in a part of South Germany from January 1963 to December 1970, at different positions in an area divided into a 16 × 16 array of quadrats, with a total area of 66.5 × 66.5 km. Carry out Mead's test, described in section 10.3, to assess whether the spatial distribution of counts appears to be random. Carry the test out at three levels: (i) using the individual quadrat counts to give 16 blocks of 16 quadrats each, (ii) using totals of four adjacent quadrats to give four blocks of 16 units each and (iii) using totals of 16 adjacent quadrats to give one block of 16 units.

10.2   Table 10.2 contains X- and Y-values which are the coordinates of the newly emergent brambles on the left-hand side of Figure 10.4. Use the Monte Carlo test described in section 10.4 to assess whether the positions of the brambles seem to be random within the 4.5 × 4.5 m study area.

Table 10.1 Number of cases of fox rabies in a 16 × 16 array of quadrats

Column

Row	1	2	3	4	5	6	7	8	9	10	11	12	13	14	15	16
1	0	0	0	0	0	0	0	0	0	1	5	4	6	3	1	2
2	0	0	0	1	0	1	0	2	0	3	5	2	4	6	1	2
3	0	0	0	0	0	0	2	0	0	1	1	2	5	3	2	1
4	0	1	0	0	0	0	0	2	0	1	5	3	1	8	5	2
5	0	1	0	0	1	0	1	1	0	0	6	6	5	10	5	3
6	0	0	0	0	0	0	0	0	2	0	1	8	5	5	5	4
7	0	0	0	0	0	0	0	0	1	3	13	15	5	2	8	1
8	0	0	0	0	0	0	0	0	1	8	8	8	7	8	8	1
9	0	0	0	1	1	0	0	0	0	0	7	5	2	2	4	1
10	0	1	1	3	1	0	0	0	0	0	20	4	8	5	4	4
11	0	0	2	3	0	0	0	0	0	5	6	10	2	3	4	4
12	0	1	1	6	0	0	0	0	0	0	7	3	5	2	1	3
13	0	0	0	1	1	0	0	0	1	5	2	2	1	6	2	3
14	1	0	1	2	1	0	0	0	0	5	9	1	9	7	2	3
15	0	2	0	1	1	0	1	0	0	1	0	2	2	3	5	3
16	0	1	2	5	0	0	0	0	0	1	6	0	4	1	8	4

Table 10.2 Coordinates of the newly emergent brambles on the left-hand side of Figure 10.4

	X	Y		X	Y		X	Y		X	Y		X	Y
1	0.2	0.8	21	1.3	0.7	41	2.0	0.4	61	2.6	0.2	81	3.6	4.4
2	0.2	3.4	22	1.3	1.2	42	2.0	0.6	62	2.6	0.8	82	3.9	0.1
3	0.3	2.2	23	1.3	3.1	43	2.0	0.6	63	2.7	1.6	83	4.0	0.1
4	0.3	2.2	24	1.4	1.7	44	2.0	1.9	64	2.8	0.8	84	4.0	0.2
5	0.4	1.0	25	1.4	3.8	45	2.1	0.3	65	2.8	0.7	85	4.0	3.9
6	0.5	0.8	26	1.5	1.6	46	2.1	0.7	66	3.0	1.9	86	4.1	3.8
7	0.6	1.2	27	1.5	2.5	47	2.1	1.9	67	3.1	0.5	87	4.1	3.9
8	0.6	1.3	28	1.5	1.1	48	2.1	1.4	68	3.1	1.0	88	4.1	4.2
9	0.6	2.5	29	1.5	1.9	49	2.2	0.2	69	3.2	0.2	89	4.3	0.3
10	0.8	0.6	30	1.5	3.5	50	2.2	1.4	70	3.2	0.9	90	4.3	3.2
11	0.8	2.3	31	1.5	3.5	51	2.3	0.1	71	3.2	3.9	91	4.3	3.3
12	0.9	0.6	32	1.6	0.5	52	2.3	0.2	72	3.3	0.7	92	4.3	3.4
13	0.9	2.4	33	1.6	1.1	53	2.3	4.1	73	3.3	2.7	93	4.4	3.2
14	1.0	2.5	34	1.6	1.2	54	2.4	0.3	74	3.5	0.6	94	4.4	4.4
15	1.0	0.7	35	1.6	2.1	55	2.4	0.7	75	3.5	0.7	95	4.5	2.1
16	1.0	3.3	36	1.7	3.5	56	2.4	2.1	76	3.5	0.7	96	4.5	2.7
17	1.1	2.5	37	1.7	3.5	57	2.4	3.7	77	3.5	1.1	97	4.5	2.0
18	1.2	0.8	38	1.8	3.5	58	2.5	0.6	78	3.5	3.8	98	4.5	3.5
19	1.2	1.2	39	1.8	4.1	59	2.6	1.6	79	3.5	4.0			
20	1.3	0.9	40	1.9	0.2	60	2.6	1.7	80	3.6	0.8			

11

Time series

11.1 RANDOMIZATION AND TIME SERIES

A time series is a set of ordered observations, each of which has an associated observation time. Because of the ordering, observations are inherently not interchangeable unless the series is 'random', where this means that all the observations are independent values from the same distribution. It is therefore in principle only possible to test a series for time structure against the null hypothesis that there is no structure at all. Procedures of this type, which are called tests for randomness or tests for independence, are reviewed by Gibbons (1986) and Madansky (1988, Chapter 3). Connor (1986) discusses applications associated with fossil records and Chatfield (1989) gives a useful introductory text for time series analysis in general.

With the randomization version of tests, the significance of a test statistic is determined by comparing it with the distribution obtained by randomly reordering observations. With n observations there are $n!$ possible orderings, which means that the full randomization distribution can be determined reasonably easily for n up to about 8 ($8! = 40\,320$). Of course, as usual it is straightforward to estimate the randomization distribution by sampling it. From the nature of time series, the only justification for randomization testing is in the belief that the mechanism generating the data may be such as to make any observed value equally likely to have occurred at any position in the series.

Spatial data collected along one dimension look exactly like a time series. Therefore many of the tests for spatial data discussed in the previous two chapters have time series counterparts: Mantel's test can be used to compare time differences with differences in the values of a series (section 9.2), the one-dimensional version of Mead's test can be applied (section 10.3), and nearest-neighbour tests can be considered for time clustering and regularity (section 10.4). Tests for the association between spatial point patterns also have their analogues for testing for association between points in time (sections 10.5–10.8). Some of these connections between spatial and time series methods are discussed further below, but no attempt is made to do this in a comprehensive way.

Time series that are not random can exhibit many different types of pattern. However, most tests for randomness fall into one of three categories corresponding to alternative hypotheses of serial correlation (autocorrelation), trend and periodicity, although the distinction between the first two categories is sometimes blurred because positive serial correlation can give the appearance of trend.

11.2   RANDOMIZATION TESTS FOR SERIAL CORRELATION

In a non-random time series, observations that are a distance k apart in time will show a relationship, at least for some values of k. These relationships can be measured using serial correlation coefficients that are positive or negative according to whether the observations tend to be similar or different at the distance apart in question. In many cases the plausible alternative to randomness is positive autocorrelation between close observations.

When the observations in a time series are equally spaced at times $1, 2, \ldots, n$, the kth sample serial correlation can be estimated by

$$r_k = \sum_{i=1}^{n-k}(x_i - \bar{x})(x_{i+k} - \bar{x})/(n - k)/ \sum_{i=1}^{n}(x_i - \bar{x})^2/n$$

where $x_1, x_2, \ldots, x_n$ are the values for the series, with mean $\bar{x}$. For a random series r_k will approximately be a value from a normal distribution with mean $-1/(n - 1)$ and variance $1/n$ when n is large (Madansky, 1988, p. 102). Tests for significance can therefore be constructed on this basis, by comparing the statistics

$$z_k = \{r_k + 1/(n - 1)\}/\sqrt{(1/n)},$$

$k = 1, 2, \ldots$, with percentage points of the standard normal distribution. A randomization test can be carried out instead by comparing the observed serial correlations with the distributions found when the x values are randomly ordered a large number of times. See also the approach proposed by Delgado (1996).

If the serial correlations are tested at the same time for a range of values of k there may be a high probability of declaring some of them significant by chance alone. For this reason it is advisable to choose the significance level to use with each correlation in such a way that the probability of getting anything significant is small for a series that is really random. One way to find the appropriate significance level is to follow the approach defined in section 6.8: the randomization distributions of the serial correlations are determined, the minimum significance level observed for all the serial correlations being considered is found for each randomized set of data, and the significance level used for testing individual serial correlations is the minimum significance level that is exceeded for 95% of all randomized sets of data. In this way the probability of declaring any serial correlation significant by chance is 0.05 or less.

As an alternative to this approach, the probability of declaring results significant by chance can be controlled by using the Bonferroni inequality, which says that if k serial correlations are all tested using the $(100\alpha/k)\%$ level of significance, then there is a probability of α or less of declaring any of them significant by chance. Numerical examples indicate that this gives very similar results to the method based on randomization, which is entirely consistent with the finding of Manly and McAlevey (1987) that test results have to be very highly correlated before the use of the Bonferroni inequality becomes unsatisfactory. Other methods for multiple testing are discussed by Westfall and Young (1993). See also the description of Gates's (1991) method in section 6.9.

One of the simplest alternatives to randomness is a Markov process of the form $x_i = \tau x_{i-1} + \varepsilon_i$, where τ is a constant and the ε values are independent random variables with mean zero and constant variance. For this alternative the von Neumann ratio

$$v = \sum_{i=2}^{n}(x_i - x_{i-1})^2 / \sum_{i=1}^{n}(x_i - \bar{x})^2,$$

is a suitable test statistic (Madansky, 1988, p. 93), with a mean of 2 and a variance of $4(n-2)/(n^2-1)$ for a random series. When the ε_i values have normal distributions the distribution of v is also approximately normal for large n, and the exact distribution for small samples is available (Madansky, 1988, p. 94). With randomization testing, v is compared with the distribution obtained when the x values are randomly permuted and no assumption of normality is required.

Another way of looking at correlation in a time series is in terms of the idea that observations that are close in time will tend to be similar. A matrix of time differences between observations can then be constructed where the element in row i and column j is $a_{ij} = (t_i - t_j)^2$. Similarly, an 'x distance' matrix can be constructed with elements $b_{ij} = (x_i - x_j)^2$. The matrix correlation

$$r = \sum(a_{ij} - \bar{a})(b_{ij} - \bar{b}) / \sqrt{\left\{\sum(a_{ij} - \bar{a})^2 \sum(b_{ij} - \bar{b})^2\right\}}$$

can then be calculated, where the summations are over the $n(n-1)/2$ values of i and j, with $i < j$. Testing r against the distribution found when the x observations are randomly permuted is in that case an example of Mantel's test, as described in section 9.2. As noted in that section, an equivalent test statistic to r is $Z = \sum a_{ij}b_{ij}$. An advantage of this approach is that it does not require the x observations to be equally spaced in time.

From the point of view of Mantel's test, there is no reason why the measures of distance have to be squared differences. A reasonable alternative approach involves taking the time difference as 1 for adjacent observations and 0 for other differences. Then Z is proportional to the von Neumann ratio v. Because of the definition of the time difference matrix, small values of Z and v indicate a similarity between adjacent x values.

All of the statistics mentioned so far for testing the randomness of a time series can be used with ranked data. That is to say, the original data values x can be replaced with their rank orders in the list from the smallest x to the largest one. This may result in tests that are more robust than those based on normal distribution theory, but still about as powerful (Gibbons, 1986; Madansky 1988).

Example 11.1: evolutionary trends in a Cretaceous foraminifer   As an example of the tests just described, consider Reyment's (1982) data on the mean diameters of megalospheric proloculi of the Cretaceous bolivinid foraminifer *Afrobolivina afra* from 92 levels in a borehole drilled at Gbekebo, Ondo State, Nigeria. Values for mean diameters at different depths are plotted against sample numbers in Figure 11.1(a). The samples are in chronological order, with sample 1 from the lowest depth of 3349 feet (the late Cretaceous age) and sample 92 from the highest depth of 2882 feet (the early Palaeocene age). Because of the sampling method used the samples are not evenly spaced in depth, and hence they are also not evenly spaced in terms of geological times. However, they will be treated as being approximately evenly spaced for this example.

The plot of the means indicates a positive correlation between the values for close samples. This is hardly surprising because of the continuity of the fossil record: any change in the series over a period of time must be from the value at the start of the period. Testing the mean series for serial correlation is therefore not a particularly useful analysis. However, there is some interest in knowing whether the differences in the mean between successive samples appear to be random. Figure 11.1(b) shows these differences with the change from sample $i - 1$ to sample i plotted against i. No trends are apparent, and it seems that this series could be random. In that case the mean series would be a random walk, which is the appropriate null model for testing evolutionary series (Bookstein, 1987).

The observed serial correlations for the difference series are shown in Table 11.1, together with their percentage significance levels on two-sided tests, as estimated using 4999 randomizations and the observed series to approximate the randomization distribution. There is only evidence of a non-zero value for the first autocorrelation. Also, the von Neumann ratio

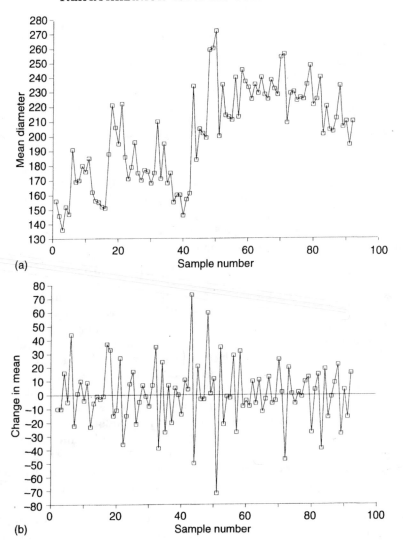

Figure 11.1 Plots of (a) mean diameters (μm) of proloculi in 92 samples of *Afrobolivina afra* at different depths, and (b) 91 differences between means for successive samples.

of $v = 2.862$ is greater than any of 4999 values obtained by randomization, and therefore has a significance level of 0.02%.

For a random series, the statistics

$$z_k = \{r_k + 1/(n-1)\}/\sqrt{(1/n)} = (r_k + 0.011)/0.105$$

Table 11.1 Serial correlations with significance levels obtained by randomization

k									
1	2	3	4	5	6	7	8	9	10
r_k									
−0.44	0.10	−0.10	0.07	−0.14	−0.00	0.13	−0.18	0.10	−0.17
Sig. (%)									
0.02	30.88	32.16	51.76	19.64	99.48	21.40	8.92	35.64	11.48

will approximately be random values from the standard normal distribution. The values for these statistics are shown in Table 11.2, with their corresponding significance levels from the normal distribution. The significance levels here are in fairly good agreement with the levels from randomization. For the von Neumann ratio the statistic

$$z = (v - 2)/\sqrt{\{4(n - 2)/(n^2 - 1)\}} = (v - 2)/0.2062$$

can to be compared with the standard normal distribution. With the observed $v = 2.862$ this gives $z = 4.179$, with a significance level of 0.002%. Again, this is consistent with the result from randomization.

Because 10 serial correlations are being tested at the same time it is appropriate to make some allowance for this when considering the evidence for non-randomness that is provided overall. The Bonferroni inequality suggests using the $(5\%)/10 = 0.5\%$ level for each test in order to have no more than a 0.05 probability of declaring anything significant by chance. On this basis, r_1 still clearly gives evidence of non-randomness.

The negative correlation between adjacent values in the difference series seems at first sight to be difficult to account for, and it is interesting to find that the same order of negative correlation was found by Bookstein (1987) in a similar type of difference series from fossil data. However, a simple explanation is that the correlation is due to sampling errors. Manly (1985, p. 349) has shown that when there are sampling errors the covariance between successive differences in a series is equal to minus the variance of

Table 11.2 Significance levels for serial correlations from the normal approximation

k									
1	2	3	4	5	6	7	8	9	10
z_k									
−4.09	1.06	−0.85	0.77	−1.23	0.10	1.34	−1.61	1.06	−1.51
Sig. (%)									
0.00	29.04	39.66	44.04	21.92	91.65	18.02	10.75	29.04	13.00

these errors. A simple extension of the same argument shows that for a random walk the correlation between two adjacent differences is

$$r = -\text{Var}(\varepsilon)/\{\text{Var}(\delta) + 2\text{Var}(\varepsilon)\},$$

where $\text{Var}(\varepsilon)$ is the sampling error variance and $\text{Var}(\delta)$ is the variance of the change δ in the true population mean between two-sample times. This shows that if $\text{Var}(\varepsilon)$ is much larger than $\text{Var}(\delta)$ then $r \approx -0.5$. It seems, therefore, that the *Afrobolivina afra* series may be a random walk with superimposed sampling errors.

The RT package (Manly, 1996a) was used to carry out the randomization tests on serial correlations and the von Neumann ratio for this example.

11.3 RANDOMIZATION TESTS FOR TREND

Trend in a time series is usually thought of as consisting of a broad long-term tendency to move in a certain direction. Tests for trend should therefore be sensitive to this type of effect as distinct from being sensitive to a similarity between values that are close in time. However, in making this distinction it must be recognized that high positive serial correlation will often produce series that have the appearance of trending in one direction for a long period of time.

One approach in this area has been promoted by Edgington (1995, Chapter 10). He discusses what is called the 'goodness of fit trend test' for cases where a very specific prediction can be made about the nature of a trend. For example, in one case (his Example 10.1) it was predicted before a training-testing experiment was carried out that the mean measured response of subjects would have a minimum when the time between training and testing is four days. Experimental results gave the following results:

Time	30 min	1 day	3 days	5 days	7 days	10 days
Mean	45.5	40.4	33.1	31.9	36.2	43.1

A randomization test with the null hypothesis of no trend gave a significance level of 0.7% with a test statistic designed to be sensitive to the alternative hypothesis that the data match the predicted U-shaped time trend.

The basis of Edgington's test is that trend values can be predicted for k observation times on the basis of the alternative to randomness that is being entertained. The test statistic used is

$$T = n_1(\bar{x}_1 - E_1)^2 + n_2(\bar{x}_2 - E_2)^2 + \ldots + n_k(\bar{x}_k - E_k)^2,$$

where n_i, $\bar{x}_i$, and E_i are the sample size, the observed mean and the expected mean, respectively, at the ith observation time. Significantly small values of this statistic are evidence against the null hypothesis of randomness in favour of the alternative of trend.

Edgington was concerned exclusively with experimental data that are usually thought of in terms of analysis of variance rather than time series analysis. However, time series are involved and therefore the approach is appropriately mentioned here. With non-experimental data the need to specify trend values exactly is a major drawback. Edgington also discusses a 'correlation trend test' where the test statistic is the correlation between the observed data values at different times and coefficients that represent the expected direction of a trend. Edgington's (1995) book should be consulted for more details about this and the goodness of fit trend test.

In many cases the alternative to the null hypothesis of randomness will be an approximately linear trend. In that case an obvious test statistic is the regression coefficient for the linear regression of the series values against their observation times. This can be compared with the distribution for this statistic obtained by randomization, as has been discussed in section 8.1.

A number of non-parametric randomization tests have been proposed to detect patterns in observations (including trends). The description 'non-parametric' is used because the numerical values in a series are not used directly. These tests can be applied with irregularly spaced series because a random series is random irrespective of the time of observations.

One such test is the runs above and below the median test, which involves replacing each value in a series by 1 if it is greater than or equal to the median and by 0 if it is below the median. The number of runs of the same value is then determined, and compared with the distribution expected with the 0s and 1s in a random order. For example, consider the following series, for which the median is 5: 1 2 5 4 3 6 7 9 8. Replacing the values with 0's and 1's as appropriate gives the new series: 0 0 0 0 0 1 1 1 1. There are only two runs, so this is the test statistic that requires to be compared with its randomization distribution. The trend in the initial series is reflected in the test statistic being the smallest possible value.

Given a series of r 0s and $n - r$ 1s there are $n!/\{r!(n - r)!\}$ possible orders, each with an associated runs count. For small series the full randomization distribution can be determined easily enough, and has been tabulated by Swed and Eisenhart (1943). For longer series the randomization distribution can either be sampled or a normal approximation for it can be used with mean

$$\mu = 2r(n - r)/n + 1,$$

and variance

$$\sigma^2 = 2r(n - r)\{2r(n - r) - n\}/\{n^2(n - 1)\}$$

(Gibbons, 1986, p. 556). A significantly low number of runs indicates a trend in the original series. A significantly high number of runs indicates a tendency for large values in the original series to be followed by small values and vice versa.

Another non-parametric test is the sign test. In this case the test statistic is the number of positive signs for the differences $x_2 - x_1$, $x_3 - x_2, \ldots, x_n - x_{n-1}$. If there are m differences after zeros have been eliminated then the distribution of the number of positive differences has mean $\mu = m/2$ and variance $\sigma^2 = m/12$ (Gibbons, 1986, p. 558) on the null hypothesis of randomness. The distribution approaches a normal distribution for moderate length series. A significantly low number of positive differences indicates a downward trend and a significantly high number indicates an upward trend.

The runs up and down test is also based on differences between successive terms in the original series. The test statistic is the observed number of 'runs' of positive or negative differences. For example, in the case of the series 1 2 5 4 3 6 7 9 8 the signs of the differences are $+ + - - + + + +$, and there are three runs. On the null hypothesis of randomness the mean and variance of the number of runs are $\mu = (2m + 1)/3$ and $\sigma^2 = (16m - 13)/90$, where m is the number of differences (Gibbons, 1986, p. 557). A table of the distribution is provided by Bradley (1968) among others. A normal approximation can be used for long series. A significantly small number of runs indicates trends, and a significantly large number indicates rapid oscillations.

In using the normal distribution to determine significance levels for the non-parametric tests just described it is desirable to make a continuity correction to allow for the fact that the test statistics are integers. For example, suppose that there are M runs above and below the median, which is less than the expected number μ. Then the probability of a value this far from μ is twice the integral of the approximating normal distribution from minus infinity to $M+\frac{1}{2}$, providing that $M+\frac{1}{2}$ is less than μ. The reason for taking the integral up to $M+\frac{1}{2}$ rather than M is to take into account the probability of getting exactly M runs, which is approximated by the area from $M-\frac{1}{2}$ to $M+\frac{1}{2}$ under the normal distribution. In a similar way, if M is greater than μ then twice the area from $M-\frac{1}{2}$ to infinity is the probability of M being this far from μ, providing that $M-\frac{1}{2}$ is greater than μ. If μ lies within the range from $M-\frac{1}{2}$ to $M+\frac{1}{2}$ then the probability of being this far or further from μ is exactly 1.

Example 11.2: testing for a trend in the extinction rates   Table 11.3 shows estimated extinction rates for marine genera from the late Permian period until the present. The times shown have a starting point of zero for the end of the Artinskian Stage, which is taken as being 263 million years ago. There are 48 geological stages covered, and for each of these a percentage

Table 11.3 Data on estimated percentages of marine genera becoming extinct in 48 geologic ages. The letter and number codes shown by the periods refer to the geologic stages in the Harland time-scale. The times shown are for the ends of geologic stages in millions of years before the present (MYBP). The percentage extinctions relate to the marine genera present during the stage. The 'mass extinctions' indicated are discussed in Example 11.4 and Exercise 11.3

	Period	Time (MYBP)	% extinct
Permian	1 A	265	22
	2 K	258	23
	3 G	253	61[a,b]
	4 D	248	60[b]
Triassic	5 S	243	45
	6 A	238	29
	7 L	231	23
	8 C	225	40
	9 N	219	28
	10 R	213	46[a,b]
Jurassic	11 H	201	7
	12 S	200	14
	13 P	194	26[a]
	14 T	188	21
	15 A	181	7
	16 B	175	22
	17 B	169	16
	18 C	163	19
	19 O	156	18
	20 K	150	15
	21 T	144	30[a,b]
Cretaceous	22 B	138	7
	23 V	131	14
	24 H	125	10
	25 B	119	11
	26 A	113	18
	27 A1	108	7
	28 A2	105	9
	29 A3	98	11
	30 C	91	26[a,b]
	31 T	88	13
	32 C	87	8
	33 S	83	11
	34 C	73	13
	35 M	65	48[a,b]

Table 11.3 (*Cont.*)

	Period	Time (MYBP)	% extinct
Tertiary	36 D	60	9
	37 T	55	6
	38 E1	50	7
	39 E2	42	13^{b}
	40 E3	38	$16^{a,b}$
	41 O1	33	6
	42 O2	25	5
	43 M1	21	4
	44 M1	16	3
	45 M2	11	$11^{a,b}$
	46 M3	5	6
	47 P	2	7
	48 R	0	2

[a] Times of mass extinctions according to Raup (1987)
[b] Times of mass extinctions based on deviations from a linear regression of logarithms of extinction rates on time

extinction rate has been read from Figure 1B of Raup (1987). The rates are defined as the percentages of families becoming extinct by the end of the stage out of those present during the stage, calculated from J.J. Sepkoski's compilation of stratigraphic ranges for about 28 000 marine genera. The question to be considered in this example is whether there is evidence of a significant trend in the rates. The plot shown in Figure 11.2 suggests that a trend towards lower extinction rates is indeed present.

The interpretation and analysis of data like these abounds with difficulties, and has created lengthy discussions in the scientific literature. There is no space here to completely review these discussions. However, two obvious points must be addressed briefly.

First, it is questionable to compare extinction rates for geological periods with different durations. However, Raup and Sepkoski (1984) argued against making an adjustment for durations on the grounds that (a) these durations are rather uncertain in many cases, (b) there is very little correlation between estimated durations and the extinction rates and (c) extinction may be an episodic process rather than a continuous one. In fact, the choice of the definition of the extinction rate seems to be relatively unimportant for many purposes (Raup and Boyajian, 1988).

Uncertainties about stage durations bring up the other point that needs mentioning. This is that there are several geologic time-scales in current use. The times shown in Table 11.3 are for the Harland scale as published by Raup and Jablonski (1986, Appendix). It seems unlikely that different

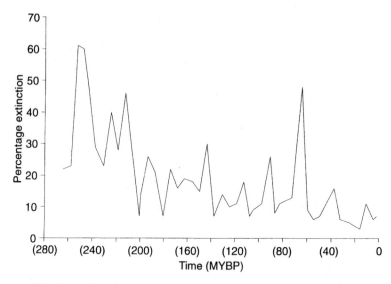

Figure 11.2 Plot of extinction rates against time (millions of years before the present) for marine genera.

scales will have much effect on tests for trend, but this has not been investigated for this example.

Four statistics have been used to test for randomness of the series of extinction rates: the regression coefficient for the extinction rates regressed on time measured as millions of years since 265 million years before the present, the number of runs above and below the median, the number of positive differences, and the number of runs up and down. The values obtained are as follows, with their estimated probability levels in parentheses: regression coefficient, -0.115 (probability of a value this far from zero, 0.0002 by randomization); number of positive differences, 23 (probability of a value this far from the randomization mean is 1.0 because no closer value is possible); runs above and below the median, 16 (probability of a value this far from the expected value is 0.008 by randomization and 0.013 from a normal approximation); runs up and down, 28 (probability of a value this far from the mean is 0.36 by randomization and 0.26 from a normal approximation).

The regression coefficient and the number of runs above and below the median have provided evidence of a trend here because they have taken into account the tendency for the higher observations to be at the start of the series and the lower observations to be at the end. The other two test statistics concentrate more on small-scale behaviour and have missed this important characteristic of the data.

11.4   RANDOMIZATION TESTS FOR PERIODICITY

One of the most interesting alternatives to randomness in a time series is often some form of periodicity, probably because explaining periodicity encourages intriguing speculations. Eleven year cycles suggest connections with sunspot numbers, a 26 million year cycle in the extinction rates of biological organisms suggests the possibility that the Sun has a companion star, and so on.

The conventional approach to testing for periodicity is based on modelling the series to be tested as a sum of sine and cosine terms at different frequencies, and testing to see whether there is significant variance associated with any of these frequencies. Some of the equations involved depend on whether n, the number of observations in the series, is odd or even. It is therefore convenient here to assume that n is even so that $m = n/2$ is an integer. Also it will be assumed initially that the observations are equally spaced at times $1, 2, \ldots, n$.

The model assumed takes the ith observation to be of the form

$$x_i = A(0) + \sum_{k=1}^{m-1}\{A(k)\cos(w_k i) + B(k)\sin(w_k i)\} + A(m)\cos(w_m i), \qquad (11.1)$$

where $w_k = 2\pi k/n$. Note that a $B(m)$ term is missing because $\sin(w_m i) = \sin(\pi i)$ is always zero. There are n unknown coefficients $A(0), A(1), \ldots, A(m), B(1), \ldots, B(m-1)$ on the right-hand side of equation (11.1). The n equations for the different values of x give n linear equations with n unknowns for these coefficients, and solving them can be shown to provide

$$A(0) = \bar{x},$$

$$A(k) = (2/n)\sum_{i=1}^{n} x_i \cos(w_k i),$$

$$B(k) = (2/n)\sum_{i=1}^{n} x_i \sin(w_k i),$$

for $k = 1, 2, \ldots, m-1$, and

$$A(m) = (1/n)\sum_{i=1}^{n} x_i(-1)^i.$$

Writing $S^2(k) = A^2(k) + B^2(k)$, it can also be shown that

$$n\left\{\sum_{k=1}^{m-1} S^2(k)/2 + A(m)^2\right\} = \sum_{i=1}^{n}(x_i - \bar{x})^2, \qquad (11.2)$$

which represents a partitioning of the total sum of squares about the mean of the time series into $m - 1$ components, representing variation associated with the frequencies $w_1 = 2\pi/n, w_2 = 4\pi/n, \ldots, w_{m-1} = 2\pi(m-1)/n$, and $A(m)^2$, which represents variation associated with a frequency of π.

To understand what this frequency representation of a time series means, consider as an example a series consisting of 100 daily observations on some process. Then the first frequency in the representation of equation (11.1) is $w_1 = 2\pi/100$, which is associated with the term

$$A(1)\cos(w_1 i) + B(1)\sin(w_1 i) = A(1)\cos(2\pi i/100) + B(1)\sin(2\pi i/100)$$

on the right-hand side of the equation. From the definition of the sine and cosine functions, this term involving $A(1)$ and $B(1)$ will be the same when $i = 1$ and $i = 101$, and generally this term will be the same for observations that are 100 days apart in the series. It represents a 100 day cycle that is just covered in the observed series. If such a cycle exists then it should account for some substantial amount of the variation in the x values so that $S^2(1)$ should make a relatively large contribution to the left-hand side of equation (11.2).

On the other hand, the last frequency in the representation of equation (11.1) is $w_m = 2\pi m/n = \pi$, which is associated with the term

$$A(m)\cos(w_m i) = A(m)\cos(\pi i).$$

This takes the values $-A(m), +A(m), -A(m), \ldots, +A(m)$ for observations on days $1, 2, 3, \ldots, 100$, respectively. It therefore represents a two-day cycle in the series. If such a cycle exists then it should show up in $A(m)^2$ making a relatively large contribution to the left-hand side of equation (11.2).

It has been demonstrated that the terms in equation (11.1) involving w_1 take account of a 100 day cycle, and the term involving w_m takes account of a two-day cycle. The other terms involving w_2 to w_{m-1} take into account cycles between these extremes. Thus

$$A(k)\cos(w_k i) + B(k)\sin(w_k i) = A(k)\cos(2\pi k i/100) + B(k)\sin(2\pi k i/100)$$

has the same value whenever $ki/100$ is an integer and represents a cycle of length $100/k$ days.

A plot of $nS^2(k)$ against w_k is called a periodogram (Ord, 1985), although this term is also used for plots of $nS^2(k)$ against the cycle length and plots of various multiples of $nS^2(k)$ against w_k or the cycle length.

A randomization test for peaks in the periodogram can be based directly on the $S^2(k)$ values, or on these values as a proportion of the total sum of squares of the x values about their mean. Thus let

$$p(k) = \begin{cases} S^2(k)/\sum_{i=1}^{n}(x_i - \bar{x})^2 & k < m, \\ A(m)^2/\sum_{i=1}^{n}(x_i - \bar{x})^2 & k = m. \end{cases}$$ (11.3)

Then the $p(k)$ values, with $\sum p(k) = 1$, estimate proportions of the variation in the series that are associated with different frequencies. High $p(k)$ values indicate important frequencies. Significance levels can be determined by comparing each $p(k)$ to the distribution found for this statistic from randomizing the order of the time series values. The $p(k)$ values are obviously equivalent statistics to the $S^2(k)$ values and $A(m)^2$ because the total sum of squares of x values remains constant for all randomizations.

It is advisable to take into account the multiple testing being carried out here when assessing the evidence for the existence of particular cycles. In principle this can be done using the same procedure as has been suggested for serial correlations in section 11.2, following the approach first discussed in section 6.7. That is, randomization can be used to find the appropriate significance level to use for individual test statistics in order that the probability of declaring any significant be chance alone is suitably small. However, this may cause some difficulties because of the need to store a large number of test statistics for a long time series. It may therefore be more convenient to use the Bonferroni inequality and assume that with m values of $p(k)$ to be tested it is appropriate to use the $(100\alpha/m)\%$ level with each test in order to have a probability of α or less of declaring anything significant by chance. Then for a series of 100 terms, with $50\,p(k)$ values to test, a realistic level for each test might be considered to be $(5/50)\% = 0.1\%$ with a probability of about 0.05 of declaring any frequency significant by chance.

Another approach involves using a statistic that tests the null hypothesis of randomness against the alternative that there is at least one periodic component. For example, consider the partial sums

$$u_j = \sum_{k=1}^{j} S^2(k)/\sum_{k=1}^{m-1} S^2(k).$$

On the null hypothesis that the time series being considered consists of independent random normal variates from the same distribution, each $S^2(k)$ for $k < m$ has an independent chi-squared distribution with two degrees of freedom, which means that the u_j values behave like the order statistics of a random sample of $m - 1$ observations from a uniform distribution on the range $(0,1)$. On this basis M.S. Bartlett proposed the use of the Kolmogorov–Smirnov test for an overall test of randomness. The randomization version of this involves calculating

$$D = \max\{D^+, D^-\}, \qquad\qquad (11.4)$$

where

$$D^+ = \max\{j/(m-1) - u_j\}$$

and

$$D^- = \max\{u_j - (j-1)/(m-1)\},$$

and comparing D with the distribution found when the observations in the original series are randomized. Essentially, D^+ is the maximum amount that the series of u values fall below what is expected, D^- is the maximum amount that the u values are above what is expected and D is the overall maximum deviation. A significantly large value of D indicates that at least one periodic component exists. The Kolmogorov–Smirnov test seems to have good power, but a number of alternative tests are available (Ord, 1985).

Example 11.3: testing for periodicity in wheat yields   Table 11.4 shows yields of grain from plot 2B of the Broadbalk field at Rothamsted Experimental Station for the years 1852 to 1925, as extracted from Table 5.1 of Andrews and Herzberg (1985). The series is also plotted in Figure 11.3. This example addresses the question of whether there is any evidence of periodicity in these yields.

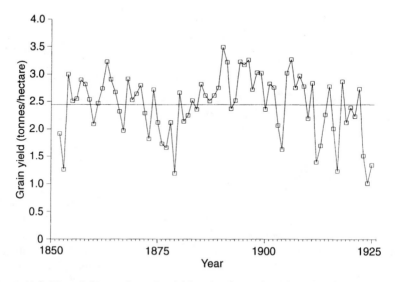

Figure 11.3 Plot of the yearly grain yield series shown in Table 11.3. The horizontal line on the plot is the average yield for the full period.

Table 11.4 Yearly grain yields from plot 2B of the Broadbank field at Rothamsted Experimental Station. During the period covered this plot was fertilized with farmyard manure only

Year	Yield	Year	Yield	Year	Yield
1852	1.92	1877	1.66	1902	2.76
1853	1.26	1878	2.12	1903	2.07
1854	3.00	1879	1.19	1904	1.63
1855	2.51	1880	2.66	1905	3.02
1856	2.55	1881	2.14	1906	3.27
1857	2.90	1882	2.25	1907	2.75
1858	2.82	1883	2.52	1908	2.97
1859	2.54	1884	2.36	1909	2.78
1860	2.09	1885	2.82	1910	2.19
1861	2.47	1886	2.61	1911	2.84
1862	2.74	1887	2.51	1912	1.39
1863	3.23	1888	2.61	1913	1.70
1864	2.91	1889	2.75	1914	2.26
1865	2.67	1890	3.49	1915	2.78
1866	2.32	1891	3.22	1916	2.01
1867	1.97	1892	2.37	1917	1.23
1868	2.92	1893	2.52	1918	2.87
1869	2.53	1894	3.23	1919	2.12
1870	2.64	1895	3.17	1920	2.39
1871	2.80	1896	3.26	1921	2.23
1872	2.29	1897	2.72	1922	2.73
1873	1.82	1898	3.03	1923	1.51
1874	2.72	1899	3.02	1924	1.01
1875	2.12	1900	2.36	1925	1.34
1876	1.73	1901	2.83		

For these data the Kolmogorov–Smirnov statistic of equation (11.4) is 0.3139. This is significantly large at the 0.12% level in comparison with the randomization distribution when it is approximated by the values obtained from 4999 randomizations of the data and the observed value. There is therefore clear evidence that this series is not random and it is worthwhile to consider the evidence for individual periodicities.

An analysis to detect periodicities is summarized in Table 11.5 and Figure 11.4. The table shows the periodicities $w(k)$, the corresponding cycle lengths in years, the proportions $p(k)$ of the total variance associated with the different periods, and the estimated significance levels associated with the $p(k)$ values. The estimated significance levels were determined by comparing the observed statistics with the randomization distributions

Table 11.5 Testing for periodicity in the Rothamsted grain yield series. Here $w(k) = 2\pi k/74$ is the period being considered, with a cycle length of $74/k$ years, and the $p(k)$ values are determined from equation (11.3). The significance levels are the percentages of values greater than or equal to that observed in the randomization distribution approximated by 4999 randomizations of the data and the observed data

k	w(k)	Cycle length	p(k)	Sig. level (%)
1	0.085	74.0	0.0877	3.48
2	0.170	37.0	0.1923	0.08
3	0.255	24.7	0.0678	7.92
4	0.340	18.5	0.0007	97.68
5	0.425	14.8	0.0082	73.86
6	0.509	12.3	0.0772	6.12
7	0.594	10.6	0.0355	27.46
8	0.679	9.3	0.0245	41.36
9	0.764	8.2	0.0320	31.26
10	0.849	7.4	0.0318	32.64
11	0.934	6.7	0.0165	56.06
12	1.019	6.2	0.0137	60.82
13	1.104	5.7	0.0074	77.04
14	1.189	5.3	0.0455	18.38
15	1.274	4.9	0.0107	69.20
16	1.359	4.6	0.0074	75.78
17	1.443	4.4	0.0114	67.20
18	1.528	4.1	0.0279	37.36
19	1.613	3.9	0.0301	34.74
20	1.698	3.7	0.0032	89.32
21	1.783	3.5	0.0114	66.62
22	1.868	3.4	0.0653	9.96
23	1.953	3.2	0.0185	51.44
24	2.038	3.1	0.0162	55.90
25	2.123	3.0	0.0039	86.84
26	2.208	2.8	0.0276	36.82
27	2.293	2.7	0.0050	83.62
28	2.377	2.6	0.0016	94.46
29	2.462	2.6	0.0154	58.42
30	2.547	2.5	0.0169	56.60
31	2.632	2.4	0.0225	44.84
32	2.717	2.3	0.0024	92.10
33	2.802	2.2	0.0002	99.04
34	2.887	2.2	0.0262	39.28
35	2.972	2.1	0.0155	57.48
36	3.057	2.1	0.0093	71.68
37	3.142	2.0	0.0104	38.52

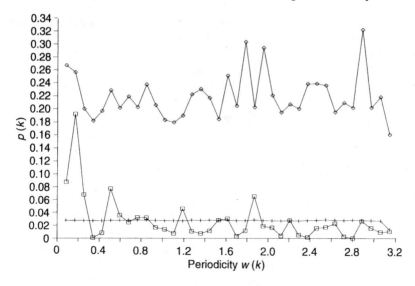

Figure 11.4 Periodogram for the Rothamstead grain yield series with the mean and maximum values obtained from approximating the randomization distribution. (□) estimate; (+) randomization mean; (◇) randomization maximum.

approximated by the same randomized orderings as were used with the Kolmogorov–Smirnov statistic. Figure 11.4 shows the sample periodogram with the mean and maximum values from the randomizations.

There are only two periodicities that can be considered at all seriously with this series. These are the first two, corresponding to cycles of length 74 and 37 years. In fact, the Bonferroni inequality suggests that to have only a 0.05 chance of declaring any periodicity significant by chance the appropriate level of significance to consider with individual tests is (5/37)% = 0.14% in this example. On this basis the only real evidence for periodicity is for the 37 year cycle.

Of course, common sense must be used in interpreting the results of this analysis. Only two 37 year cycles are covered in the data and this alone makes the assumption that the pattern shown would be repeated in a longer series rather questionable. It must also be born in mind that the null hypothesis that periodicity has been tested against is complete randomness. However, a series with no periodic components but positive serial correlation between close years can also give the type of pattern shown in Table 11.4. That is to say, the true situation might be that there is some mechanism that promotes serial correlation and it just happens that the trends induced by this mechanism have shown approximately 37-year movements. Furthermore, Fisher (1924, Table 14) has shown that 40% of the variation in yield can be attributed to rainfall by a linear regression. Rainfall was relatively high around the years 1879 and again around the years 1915 (Fisher, 1924, Figure 1), and these periods correspond to times of relatively low yields of grain. It seems therefore that the 37-year cycle that has been detected in grain yields may be merely a reflection of a cycle of about this length in the rainfall series.

The calculations for this example were carried out using the RT package (Manly, 1996a).

11.5 IRREGULARLY SPACED SERIES

Tests for periodicity are relatively straightforward when the series being considered is equally spaced in time. However, relatively little work has been done until recently on analyses for series that are either inherently unequally spaced, or equally spaced but with some values missing. See Ord (1988) for some general references.

The difficulty of dealing with unequally-spaced data led Raup and Sepkoski (1984) to adopt the randomization approach to testing for periodicity in fossil extinction rates over the past 250 million years that has been discussed briefly in Example 1.3. The process that they used was as follows.

The Raup–Sepkoski algorithm

1. A peak in the extinction series is defined as occurring at any point where the values on either side are both lower than the value at the point.
2. A perfectly periodic function is assumed to give the expected times of peaks, with a cycle of length C and a starting time t_0 corresponding to the start of the extinction series. That is, the expected times of peaks are $t_0, t_0 + C, t_0 + 2C$ and so on.
3. For each peak as defined in step 1, the distance to the nearest expected peak time is determined to give the 'error' in the observed peak time. The mean error is then subtracted from each expected peak time to find a better fit of the model to the data. This process is repeated several times to ensure the best possible fit.
4. The standard deviation of the errors from the best fitting function, $s(C)$, is used as the measure of goodness of fit of the model.
5. The extinction rates in the series are randomly allocated to geological ages and steps 2–4 are repeated. This is done many

---

times to approximate the randomization distribution of the goodness of fit statistic for the particular cycle length C being considered, and hence to determine whether the observed statistic is significantly low. A low statistic indicates a fit that is too good to be easily attributed to chance.

6. The steps 2–5 are repeated for cycle lengths of between 12 and 60 million years, at yearly intervals.

Using the above procedure, Raup and Sepkoski concluded that there was evidence of a 26 million year cycle because the goodness of fit for this cycle length is significantly low at the 0.01% level. This high level of significance was determined by doing 8000 randomizations for this particular cycle length as against 500 randomizations for the other non-significant lengths. Significance at the 5% level was also found for a 30 million year cycle.

Following the publication of Raup and Sepkoski's (1984) paper there was a good deal of discussion about the validity of this type of test. The objections were of two types. Some relate to the nature of the fossil data while others are concerned with the validity of the general procedure (Hoffman, 1985; Raup, 1985b, 1987; Jacobs, 1986; Sepkoski and Raup, 1986; Patterson and Smith, 1987; Quinn, 1987; Stigler and Wagner, 1987, 1988; Raup and Sepkoski, 1988; Stothers, 1989).

It was shown in Example 11.2 that there is evidence of a trend in extinction rates over the period studied by Raup and Sepkoski. This was noted also by Quinn (1987) who pointed out that this has undesirable effects on Raup and Sepkoski's definition of an extinction peak. He suggested that it is better to define peaks with reference to deviations from a linear regression of extinction rates on time and determine significance by randomizing these deviations.

Stigler and Wagner (1987, 1988) discuss several features of the Raup–Sepkoski testing procedure. They note that because a range of cycles is being considered, an appropriate test for overall significance is the minimum goodness of fit for all tested cycle lengths, with an appropriate adjustment to take into account the tendency for this statistic to increase (linearly) with the cycle length. They carried out a simulation experiment using the same geological time-scale as was used in Raup and Sepkoski's original paper. Rather surprisingly, they found that when the randomization test gives a significant result on a random series there is a strong tendency for the significant result to correspond to a 26 million year cycle. Furthermore, if the true model for the series is a moving average process due to some extinction events being recorded too early (the Signor–Lipps effect) then the tendency for the best fit to occur with the 26 million year cycle is enhanced. They conclude that the Raup–Sepkoski procedure is totally unreliable unless the periodicity in a series is very strong.

These problems are specific to the particular data being discussed, but they do emphasize the need for caution in determining test statistics, and the need to recognize that randomization tests may give a significant result not because the alternative hypothesis is true but because an entirely different non-random model is correct and the test statistic is somewhat sensitive to the patterns generated in the data by this other model.

11.6   TESTS ON TIMES OF OCCURRENCE

One of the criticisms of the Raup–Sepkoski procedure just reviewed is the definition of extinction peaks as occurring whenever an extinction percentage has lower values on either side. For one thing, these peaks are supposed to represent times of 'mass extinctions'. In reality such mass extinctions can presumably have occurred in adjacent geological ages, but the definition of peaks does not allow this.

Another problem is that the randomization procedure does not take into account the apparent downward trend in extinction rates in moving towards the present. If this trend is real then there is little sense in randomizing the extinction rates without knowing how a trend may affect tests for periodicity. These considerations suggest that a better testing procedure is one that establishes the times of events separately from the testing procedure, and tests whether the event times seem to be randomly chosen from the possible times.

One approach along these lines was used by Rampino and Stothers (1984a,b) in testing for periodicities in a variety of geologic series, and by Raup (1985a) in testing for periodicity in the times of magnetic reversals. The procedure is as follows.

The Rampino–Stothers algorithm

1. The times of n events are determined.
2. A perfectly periodic function is assumed to give the expected times of events, with a cycle of length C and a starting time t_0 corresponding to the start of the extinction series; that is, expected times of events are $t_0, t_0 + C, t_0 + 2C$, and so on.
3. For each event, the distance to the nearest expected event time is determined to give the 'error' in the observed peak time. The standard deviation of the errors for all events, $s(C)$, is used to measure the goodness of fit of the cycle.

4. A range of starting points for the series of expected event times are tried in order to find the one that gives the smallest goodness of fit statistic $s(C)$ for the observed event times. A 'residual index' $R(C)$ is then calculated (see below).

5. Values of $R(C)$ are determined for a suitable range of cycle lengths C and the maximum of these, R_{max}, is found.

6. The time intervals between events are put into a random order, and a new set of event times are determined, maintaining the times of the first and last events as fixed. For example, if the time between events 1 and 2 is 5 million years in the original data then this is equally likely to be the time between events 1 and 2, events 2 and 3, ..., events $n - 1$ and n, for the randomized data.

7. Steps 2–5 are carried out on the random data to get a randomized value for R_{max}. The randomization is repeated a large number of times to find the randomization distribution of this statistic. The significance of the individual $R(C)$ values for the observed data is determined with reference to the randomization distribution of R_{max}.

One advantage of the above procedure is that it is not sensitive to missing peaks because it only considers the extent to which known peaks agree with a period. It does, however, have two problems that have been pointed out by Lutz (1985). The first problem is with the residual index

$$R(C) = [C\sqrt{\{(n^2 - 1)/(12n^2)\}} - s(C)]/C.$$

This was originally suggested by Stothers (1979) as a means of normalizing the $s(C)$ values to have the same distribution for all cycle lengths. However, it does not achieve the desired result because of record length effects. The second problem, at least for Raup's (1985a) use of the test, is that trends in the frequency of magnetic reversals have unfortunate effects. In other words, the test for periodicity is made invalid because the series being tested is clearly non-random in another respect.

Quinn (1987) proposed another approach based on the times of occurrence of events. This involves considering the distribution of times between events in comparison with either the theoretical 'broken stick' distribution or in comparison with the distribution found by randomization. The Kolmogorov–Smirnov test is suggested as the method of comparison. Depending on the circumstances, event times can either be randomized freely within the time limits of the data or allocated to a finite number of allowable times (such as the ends of geological ages). This test has the disadvantage of being sensitive to missing events.

Whatever analysis is used, it must be stressed again that the existence of a significant cycle of a certain length in a time series does not necessarily mean that the cycle is a genuine deterministic component of the series. Significant 'pseudo-cycles' can easily arise with non-random series because of the serial correlation that they contain and the fact that the null hypothesis used in a randomization test is no more correct than the alternative hypothesis of periodicity.

11.7   DISCUSSION ON PROCEDURES FOR IRREGULAR SERIES

The last two sections have shown clearly the problems that are involved in determining periodicity in an unequally spaced series. Two points emerge as being important. First, it is not sensible to carry out a randomization test for periodicity when the series being considered is non-random because it contains a trend, unless this trend cannot effect the test statistics. Second, test statistics must be chosen in such a way that when a significant result is obtained by chance there is no bias towards indicating any particular cycle length as being responsible.

Noting these points, some sensible recommendations can be made for handling these types of data. If there is no trend in a series then the Raup and Sepkoski (1984) procedure described in section 11.5 is perfectly valid providing that the significance of the goodness of fit statistic for each cycle length is compared with the randomization distribution determined for this particular statistic. There is then a multiple testing problem, but that can be handled either using the Bonferroni inequality or the randomization method described in section 6.7. Another possibility when there is no trend involves interpolating the observed series to produce equally spaced data and then applying the standard periodogram methods discussed in section 11.4 (Fox, 1987). Significance levels can still be determined by randomizing the original data to the unequally spaced sample times and repeating exactly the same calculations as are used on the original data.

When a trend is present in a time series this can either be taken into account in the analysis or the problem can be reformulated to remove the influence of the trend. For example, deviations from a regression line (i.e. residuals) can be analysed (Connor, 1986; Quinn, 1987). Alternatively, many sets of data with a trend similar to that observed and appropriate superimposed random elements can be generated in order to approximate the distribution of test statistics when there is no periodicity. That is to say, a Monte Carlo test of significance can be used rather than a randomization test, as in Example 11.5 below.

A trend can be removed from consideration by formulating the problem in terms of the times of n key events such as m 'extinction peaks' or the most extreme m values in a series, where the definition of these events allows them to occur either at any times within the period being considered or at n out of m possible times. The Rampino and Stothers (1984a,b) and Raup (1985a) procedure described in the previous section is then valid providing that the test statistic used for each cycle length is assessed with respect to the randomization distribution of exactly the same statistic, with an allowance for multiple testing.

An alternative to this approach is to follow Quinn (1987) and consider the observed distribution of times between events in comparison with the randomization distribution. This comparison can either be on the basis of the Kolmogorov–Smirnov test or by a direct comparison between the full observed distribution of times between events and the range of distributions found by randomization.

A point that can be easily overlooked with procedures based on event times is that the definition of these times must make it possible for the m events to occur at any combination of the n possible times. In particular, defining events in terms of peaks in a time series is not valid because two peaks cannot occur at adjacent observation times.

Example 11.4: periodicity in the extinction record   Because the question of whether or not there is evidence of periodicity in the extinction record has caused so much controversy this will now be considered on the basis of Raup's (1987) genera data, which are shown in Table 11.3. As there seems to be a trend in the extinction rates (Example 11.2) it is simplest to base an analysis on the times at which mass extinctions are believed to have occurred, which for the moment will be accepted as being the eight suggested by Raup (1987, Figure 2), as indicated in Table 11.3.

Two possible approaches were suggested above for testing for periodicity using the times of m events, one based on determining the best fits of the times to a periodic model and the other based on considering the times between successive events. Here only the second (and simpler) approach will be used.

The seven times between the eight extinctions are shown in Table 11.6, in order from the smallest to largest. The upper, lower and two-tail percentages that the order statistics correspond to in comparison with the randomization distribution are also shown, where the randomization distribution was approximated by the observed data plus the results from randomly assigning the eight mass extinctions to the ends of the 48 possible geological stages a total of 4999 times. For each randomization the times between the eight events were also ordered from smallest to largest, so that the significance levels shown are for these order statistics. It is seen, for example that 1.2% of the randomized sets of data gave a minimum time between two events (the first-order statistic) of 19 or more. It seems,

Table 11.6 Order statistics for the times between eight mass extinctions, with upper, lower and two-tail significance levels

	Order statistic						
	1	2	3	4	5	6	7
Observed	19	26	27	27	40	50	53
Upper tail (%)	1.2	0.6	8.0	36.5	27.9	48.0	87.0
Lower tail (%)	99.4	99.4	93.6	67.4	75.8	60.7	14.1
Two-tail (%)	2.4	1.2	16.0	73.0	55.8	92.0	28.2

therefore, that the distance between the closest two events and the second closest distance for the observed data are larger than is expected for events occurring at random times.

Two points concerning the above percentage points need some discussion here. First, there is the question of how the two-tail significance levels were determined. Second, there is the question of how to take into account the multiple testing problem.

As has already been discussed in Example 4.1, making an allowance for tests being two-sided may not be as straightforward as it appears at first sight. The problem with the present example is that defining a two-sided significance level as the probability of a result as extreme as that observed does not produce a sensible answer because the randomization distributions of the order statistics are far from being symmetric. For example, the randomization mean for the first-order statistic is approximately 6.7 million years. The observed value is therefore $19 - 6.7 = 12.3$ million years above the mean. However, it is not possible to be this far below the mean because this would involve having a negative time between events. On these grounds it could be argued that the probability of a result as extreme as 12.3 million years from the mean is given by the upper tail value of 0.012 only. However, this ignores the fact that if the observed statistic has a very low value then the probability of being so low will be small, but the probability of being as far from the mean in either direction might be quite large. In fact defining the two-tail significance level in terms of the deviation from the randomization mean may make it impossible to get a significance result from an observed value that is below the mean.

A reasonable approach involves assuming that some transformation exists that makes the randomization distribution of an order statistic symmetric. In that case, the appropriate significance level to use for a two-sided test is twice the smaller of the upper and lower tail probability levels, which is the probability of a result as extreme as that observed on the transformed scale. In the present example this provides the two-tail significance levels for the seven order statistics that are shown in the bottom line of Table 11.6.

The simplest way to take account of the multiple testing involved in considering seven order statistics at one time involves using the Bonferroni inequality. This suggests that in order to have a probability of 0.05 or less of declaring any of the results significant by chance, the level to use with the individual order statistics is $(5/7)\% = 0.71\%$. On this basis none of the results is significant, and the evidence for non-randomness is not clear.

An alternative to using the Bonferroni inequality involves using the randomization approach discussed in sections 6.7 and 11.2. However, this is more complicated than using the Bonferroni inequality, and experience suggests that it tends to give rather similar results. This approach has therefore not been used here.

Another way to consider the results is in terms of the mean and standard deviation of the time between events. If eight events occur over a period of 265 million years then it is inevitable that the mean time between events will be approximately $265/8 = 33$ million years. This is the case for the present example, where this mean is 34.6 million years. It is the observed standard deviation of 13.2 million years which indicates non-randomness, with just 5.0% of randomizations giving a value this low. A one-sided test is appropriate here, because a low standard deviation is what indicates a tendency for the time between events to be equal to a particular value.

This analysis can be summed up by saying that it provides some limited evidence to suggest that the distribution of times between extinction events has a certain regularity. The significance level of about 5% on a one-sided test of the standard deviation is a reasonable indication of this, although the use of the Bonferroni inequality suggests that the order statistics for the times between events are not all that extreme.

If the existence of some regularity in the times between extinction events is accepted then it seems at first sight obvious that the best estimate of the cycle length is the observed mean. However, it can be argued that some of the events may not have been detected, in which case it is surely worthy of note that three of the observed intervals shown in the above table are about 26 million years and two more are about twice this length. In fact, if minor peaks at 175 MYBP and 113 MYBP are accepted as mass extinction (Table 11.3) then there are nine intervals between events, of which seven are close to 26 million years. This does then look remarkably like a cycle although, as has been emphasized before, this may just have arisen from some non-random but also not strictly periodic mechanism.

Unfortunately, there may be a flaw in this analysis due to the definition of mass extinction times. As was noted earlier, it is important that the definition of event times allows these times to occur at any combination of the possible n times. All the mass extinction times are at peaks in the extinction series given in Table 11.3. However, not all the peaks in the extinction series are classified as mass extinctions. The peaks that are recognized as being mass extinctions are supported by independent evidence (Sepkoski, 1986), but there must be some question about whether the procedure for identifying mass extinctions allows them to occur in two successive stages. Of some relevance here is the fact that the earliest recognized mass extinction in Table 11.3 is the Guadalupian at 253 MYBP. The following stage at 248 MYBP has almost the same percentage extinction, but is not counted separately by Raup (1987) because it is treated as relating to the same event as the previous stage (Sepkoski, 1986). It seems, therefore, that the definition of events may by itself have given rise to some non-randomness in the event times.

Given these circumstances, there is some interest in seeing how the analysis is changed if a different definition is used for mass extinctions. This idea is pursued further in Exercise 11.5 below.

The calculations in this example were carried out using the RT package (Manly, 1996a).

11.8   BOOTSTRAP AND MONTE CARLO TESTS

In some of the papers referenced above, randomization testing is incorrectly referred to as a 'bootstrap' procedure. The difference between these two methods is, of course, that in the context of the original Raup and Sepkoski (1984) test, randomization involves allocating the observed extinction rates to geological stages in a random order, but bootstrapping involves choosing the extinction rate to use for an individual stage from an infinite distribution where each of the observed rates is equally probable.

Quinn (1987) argued that bootstrapping is more realistic than randomization for testing significance with extinction data. However, this is debatable. The difference between the two techniques is in their justification. Randomization is based on the argument that the process producing the observed data is equally likely to produce them in any order, and it is of interest to see what proportion of the possible orders display a particular type of pattern. On the other hand, bootstrapping is justified by regarding the observed distribution in a time series as the best estimate of the distribution of values produced by the generating mechanism. Bootstrapping is therefore an alternative to assuming some particular theoretical distribution for series values (Kitchell et al., 1987).

Monte Carlo methods generally are of considerable value in the type of discussions that have surrounded the analysis of extinction data. For example, Stigler and Wagner (1987) generated 3000 series of 40 independent pseudo-random numbers to examine some aspects of the behaviour of Raup and Sepkoski's (1984) testing procedure, while Stothers (1989) examined the effect of errors in the dating of geological stages by generating series with the dates randomly perturbed. Monte Carlo tests that are not randomization tests seem to have been little used. However, there does seem some value in using a general Monte Carlo test in cases where the alternative hypothesis is something more complicated than complete randomness. One possibility mentioned in the last section was testing for periodicity and trend, against the null hypothesis of trend only, by comparing the observed data with data generated under the trend-only model. The following example shows the result of doing this with the extinction data in Table 11.3.

Example 11.5: periodicity in the extinction record reconsidered   Monte Carlo testing of the periodicity hypothesis was carried out on the logarithms of the extinction rates shown in Table 11.3 because the residuals from a linear regression of log extinction rates on time appear more normally distributed than the residuals from a regression using the rates themselves. The regression line was estimated in the usual way and the residuals determined. Interpolation was then used to produce an equally spaced time series of 48 'residuals' with a spacing of 5.64 million years. A periodogram of this series was calculated, and hence the $p(k)$ values of equation (11.3). These $p(k)$ values are shown in Figure 11.5. There is a peak

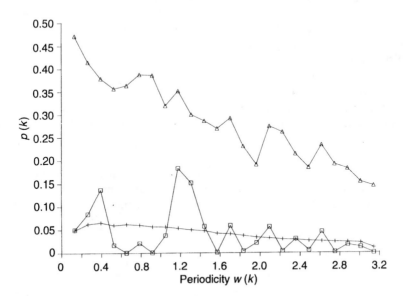

Figure 11.5 Periodogram for the marine genera extinction data, with the mean and maximum $p(k)$ values determined by simulating 1000 sets of extinction data with a trend similar to that observed. (□) observed; (+) simulation mean; (△) simulation maximum.

for the periodicity $w(9) = 1.18$, corresponding to a cycle of length 29.4 million years. The Kolmogorov–Smirnov test statistic defined in section 11.2 was also calculated for an overall test of the departure of the residuals from randomness. The value obtained was $v = 0.307$.

To determine the significance of the calculated statistics, 1000 sets of data were generated and analysed in exactly the same way as the real data. The generated data were obtained by taking the estimated trend values from the real data as the best estimates available of the true trend, and adding independent normally distributed random variables to these. These random variables were given means of zero and standard deviations equal to the residual standard deviation from the regression on the real data. The data thus generated look quite similar to the real data when plotted.

Figure 11.5 shows the mean values of $p(k)$ from the 1000 simulations, and also the maximum values. The observed values follow the randomization mean reasonably closely and the peak at $w(9) = 1.18$ is well within the total range of the simulations. This peak is equalled or exceeded by 2.7% of the simulated sets of data, while the observed value of $w(10)$ is equalled or exceeded by 4.2% of them. The Bonferroni inequality suggests that the appropriate level for testing individual $p(k)$ values is $(5/24)\% = 0.2\%$ in order to have a probability of 0.05 or less of declaring any of them significant by chance. Therefore the significance levels of 2.7% and 4.2% need not be taken too seriously. Furthermore, the observed Kolmogorov–Smirnov statistic is exceeded by 28.6% of the simulated values and is therefore not at all significant.

From a rather similar analysis, but without Monte Carlo testing, Fox (1987) concluded that there was evidence of a 26 million year cycle in the extinctions of genera. However, he did not make any allowance for multiple testing and analysed the data without the trend removed. Connor (1986) also did a periodogram analysis. He removed trend, determined significance levels by bootstrapping, and noted how multiple testing causes problems with interpreting results. He examined various sets of data and obtained fairly significant results for some data at some frequencies. However, because he was working with family data, and his bootstrapping was on the original fossil records rather than summary extinction rates, his study is not really comparable to the example that has been considered here.

11.9 FURTHER READING

Some rank-based methods for testing for randomness of time series against the alternative hypothesis of serial correlation are discussed by Hallin and Melard (1988), while Chiu (1989) has reviewed the literature on testing for periodicity and suggested a new approach.

Randomization has attracted considerable interest as a means of detecting density dependence in time series of the logarithm of the abundance of animal species, i.e. detecting whether the changes in a series are dependent on the level of the series. The 'null-model' in this case is a random walk, or possibly a random walk with trend, and the tests that have been proposed included those based on the correlation between changes and the level of the series, the range of the logarithms of abundances, and the tendency of the abundance series to move towards a particular range of values (Pollard *et al.*, 1987; den Boer and Reddingius, 1989; Reddingius and den Boer, 1989; den Boer, 1990; Crowley, 1992; Crowley and Johnson, 1992; Holyoak and Lawton, 1992; Holyoak, 1993; Holyoak and Crowley, 1993). Similar analyses have also been used with populations where the individuals pass through a series of development stages and there is interest in whether the survival through a stage is density dependent (Manly, 1990a, Chapter 7; Vickery, 1991). Tests for delayed density dependence have also been of interest (Holyoak, 1994). See also the bootstrap tests used by Kemp and Dennis (1993), Wolda and Dennis (1993) and Dennis and Taper (1994).

An important area for the application of time series methods is in the monitoring of important environmental variables to detect times of abrupt changes or trends. One such situation occurs when there are a number of sample stations in a region, with one or several variables measured at each station. Manly (1994b) describes a CUSUM method for detecting changes and trends in this context, with randomization used to assess the significance of the results either for one variable or for several related variables. This is based on a technique that was originally developed for comparing the marks on several examination papers taken by several candidates (Manly, 1988).

Bootstrapping has been used by Swanepoel and van Wyk (1986) to test for the equality of the power spectral density functions of two time series; by Veall (1987), Thombs and Schucany (1990), Breidt *et al.* (1995) and García-Jurado *et al.* (1995) to assess the uncertainty of forecasting the future values of a time series; by de Beer and Swanepoel (1989) to test for serial correlation using the Durbin–Watson statistic; by Hurvich *et al.* (1991) to estimate the variance of sample autocovariances; by Hubbard and Gilinsky (1992) for assessing the evidence for any mass extinctions in fossil records (as distinct from periodicity in mass extinctions); by Andrade and Proença (1992) to search for a change in the trend of an economic time series; and by Falck *et al.* (1995) to determine whether time series of animal abundance appear to be chaotic. Hurvich *et al.* (1991) also review and discuss the use of jackknifing with the estimation of autocovariances.

Most uses of bootstrapping with time series have involved resampling residuals from some suitable model. Recently, however, there has been interest in schemes that resample blocks of the original observations so as to avoid assuming a particular model and retain most of the correlation structure in bootstrapped data. Léger *et al.* (1992) discuss various implementations of this idea, using Canadian lynx trapping data from the Mackenzie river as an example. The choice of the block size causes some difficulty. Another possibility with a binary time series (i.e. a series of the form 0 0 1 1 1 0 0 ,..., with 1 indicating that a certain event occurs) involves resampling alternate runs of zeros and ones. Kim *et al.* (1993) provide several examples where this approach performs well. For recent developments see Politis and Romano (1994).

Finally, one technique that appears to have many potential uses is superposed epoch analysis. This is designed for the situation where there is interest in knowing whether certain key events are associated with extreme values of a time series. For example, Prager and Hoenig (1989) used this method to examine the relationship between elevated sea levels (often associated with El Niño events) and high recruitment success of chub mackerel *Scomber japonicus* off the coast of southern California. The basic idea is to compare the values of the time series when key events occur with the values for the time series immediately before and immediately after these events. A suitable test statistic for the observed data (such as the mean level of the time series for the key event times minus the 'background' mean for the preceding and following times) is assessed for significance in comparison with the distribution obtained when the key event times are chosen at random from the possible times. It is also possible to include in a test information on anti-key event years to see whether these are associated with low levels of the time series, in what Prager and Hoenig called a reflected-event analysis.

Prager and Hoenig (1989) found a significant relationship between high sea level years and high recruitment success before the collapse of the chub mackerel fishery in the late 1960s, but not for a longer time series extending up to 1983. In a later paper (Prager and Hoenig, 1992) they discuss the power of their randomization test with various test statistics. Superposed epoch analysis is better than some other techniques for assessing the impact of events such as randomized intervention analysis (Carpenter *et al.*, 1989) because it properly accounts for serial correlation in the time series being studied.

11.10 CHAPTER SUMMARY

- There are many methods available for testing the randomness of a time series, i.e. testing whether all values in the series are independently drawn from the same distribution. A randomization test involves comparing an observed test statistic with the distribution that is generated by randomly permuting the values in the time series.

- Spatial data collected along a line look exactly like a time series. Many methods of analysis that are proposed for spatial data have a time series counterpart.

- Tests for serial correlation can test the observed correlation coefficients directly, but in that case it may be appropriate to allow for multiple testing. Alternatively, the von Neumann ratio can be used for an overall test.

- A Mantel matrix randomization test can be used for the hypothesis that observations that are close in time tend to be similar in magnitude. This does not require observations to be equally spaced in time.

- Trend in a time series is a long-term tendency to move in one direction. A number of tests for randomness may have some power to detect this, including the 'runs above and below the median' test, the 'sign' test, the 'runs up and down' test, and a test based on regressing series values against time. These tests can all be carried out by randomization.

- Tests for periodicity can be based on modelling a time series as a sum of sine and cosine functions corresponding to cycles of different lengths. The variances associated with different cycle lengths, or these variances as a proportion of the total variance, can be tested directly by randomization, in which case an allowance for multiple testing may be considered necessary. The Kolmogorov–Smirnov test can also be adopted as an overall test for this situation.

- Most time series modelling assumes that observations are equally spaced in time. Several randomization procedures have been proposed for testing for periodicity with unequally spaced observations. A test due to Raup and Sepkoski (1984) is intended to detect a tendency for a series to have a maximum at regular intervals. This has been criticized for several shortcomings in the original application. A modified procedure proposed by Rampino and Stothers (1984a,b) avoids some of these shortcomings, but there was a problem with the test statistic used. Both the Raup and Sepkoski (1984) procedure and the Rampino and Stothers (1984a,b) procedure are valid for some sets of data, but some alternative randomization approaches are also possible, including some based on bootstrap and Monte Carlo methods.

- A number of generalizations and extensions to the methods discussed in the chapter are reviewed, including tests for density dependence in population sizes, CUSUM methods for environmental monitoring, bootstrap resampling of blocks of data and superposed epoch analysis, which is a randomization method for determining whether certain key events are associated with the extremes of a time series.

11.11 EXERCISES

11.1 The genetic composition of the scarlet tiger moth, *Panaxia dominula*, population at Cothill, in the Oxford area of England, has been studied intensively for many years by E.B. Ford and his associates (Ford, 1975). As part of this study, the population size (the total number of flying individuals produced) was estimated each year from 1941 to 1972, with the results shown in Table 11.7.

Use the methods discussed in section 11.2 to see whether the serial correlations and the von Neumann ratio are such as might arise if the series of percentage size changes is a random series. Compare the results of the randomization tests with what is found using large sample tests with the normal distribution.

11.2 Estimates of the percentage of *medionigra* genes in the Cothill population of the scarlet moth are as follows, in order from the 1940 to 1972. Use the regression and other tests that are described in section 11.3 to see if there is any evidence of a trend.

11.1 6.8 5.4 5.6 4.5 6.5 4.3 3.7 3.6 2.9 3.7 2.5 3.6 2.6 2.9 1.1 3.0
4.6 3.7 2.2 1.9 2.0 2.2 0.9 0.0 1.2 0.0 0.0 1.2 3.5 3.4 0.7 0.7

Table 11.7 Changes in the population size for a population of the scarlet tiger moth

Year	Approx. population size	Change (%)	Year	Approx. population size	Change (%)
1941	2300		1957	16000	45.5
1942	1600	−30.4	1958	15000	−6.3
1943	1000	−37.5	1959	7000	−53.3
1944	5500	450.0	1960	2500	−64.3
1945	4000	−27.3	1961	1400	−44.0
1946	7000	75.0	1962	216	−84.6
1947	6000	−14.3	1963	470	117.6
1948	3200	−46.7	1964	272	−42.1
1949	1700	−46.9	1965	625	129.8
1950	4100	141.2	1966	315	−49.6
1951	2250	−45.1	1967	406	28.9
1952	6000	166.7	1968	978	140.9
1953	8000	33.3	1969	5700	482.8
1954	11000	37.5	1970	4500	−21.1
1955	2000	−81.8	1971	7100	57.8
1956	11000	450.0	1972	3500	−50.7

11.3 Apply the periodogram tests that are described in section 11.4 to the data in Table 11.7 to see whether there is any evidence of cycles in the series of percentage size changes in the scarlet moth population.

11.4 It was noted in Example 11.4 that Raup's (1984) definition of the times of mass extinctions of marine genera may not have allowed these to occur at the ends of two adjacent geologic ages. There is therefore some value in seeing the effect of using an alternative definition of mass extinctions which ensures that this restriction does not occur. To this end, the approach of Quinn (1987) was followed, and extinction events were defined in terms of the residuals from a regression of the logarithm of extinction rates on time. (Taking logarithms seems to give a slightly better linear relationship than using the extinction rates themselves.) All points one residual standard deviation or more above the regression line were chosen as extinction events, these being the nine events that are indicated in Table 11.3, seven of which are the same as the ones suggested by Raup (1987). Repeat the analysis carried out in Example 11.4 using these nine events to see whether there is any evidence that the times between events are more constant than what is expected if events occur independently at random times.

12
Multivariate data

12.1 UNIVARIATE AND MULTIVARIATE TESTS

Multivariate analysis is the part of statistics that is concerned with interpreting results obtained by observing several related random variables simultaneously, with each of the variables being considered equally important, at least initially. Many of the standard univariate methods, such as those for comparing means of two or more samples, have multivariate generalizations. In addition, there some techniques that have no univariate counterpart (Manly, 1994a).

The possible justifications for randomization tests with multivariate data are the same as those that have been suggested for univariate data: (a) the experimental design involves the random allocation of the experimental units to the different treatments being compared, (b) random samples are taken from distributions that are identical if the null hypothesis being tested is true or (c) the null hypothesis implies that the mechanism generating the data makes all of the possible orderings equally likely to arise. The only difference between univariate and multivariate tests is that in the latter case each randomized unit carries two or more measurements rather than just one.

12.2 SAMPLE MEANS AND COVARIANCE MATRICES

Many univariate tests are defined in terms of sample means and variances. The multivariate generalizations of these are sample mean vectors and sample covariance matrices, which are defined as follows. Suppose that there are p variables $X_1, X_2, \ldots, X_p$ measured on n items, so the x_{ij} denotes the value for variable X_j for the ith item. The mean $\bar{x}_j$ and the variance s_j^2 for X_j can be calculated in the usual way, the latter being the sum of squared deviations from the mean divided by $n - 1$. Also, the sample covariance between X_k and X_r is defined as

$$c_{kr} = \sum_{i=1}^{n}(x_{ir} - \bar{x}_r)(x_{ik} - \bar{x}_k)^2/(n - 1),$$

where this measures the extent to which the variables are linearly related. With these definitions, the mean vector and the covariance matrix for the sample are

$$
\bar{\mathbf{x}} = \begin{bmatrix} \bar{x}_1 \\ \bar{x}_2 \\ \vdots \\ \bar{x}_p \end{bmatrix}, \quad \text{and} \quad \mathbf{c} = \begin{bmatrix} s_1^2 & c_{12} & \cdots & c_{1p} \\ c_{21} & s_2^2 & \cdots & c_{2p} \\ \vdots & \vdots & \ddots & \vdots \\ c_{p1} & c_{p2} & \cdots & s_p^2 \end{bmatrix}.
$$

12.3   COMPARISON OF SAMPLE MEAN VECTORS

One of the most frequently occurring questions concerns whether the differences between two or more sample mean vectors can be accounted for in terms of what can be expected from a random allocation of units to samples. For example, consider the data in Table 6.4. Here the two samples correspond to colonies of the snail *Cepaea hortensis* with and without the presence of the snail *Cepaea nemoralis*. The variables are the percentages of different types of *C. hortensis* in the colonies and there is interest in knowing whether the two types of colony differ in their average values for one or more of these variables. That question has been addressed in Example 6.3 by calculating and testing five *t*-values, one for each variable, and making an allowance for multiple testing. An alternative multivariate approach involves considering a single test statistic that takes into account the sample mean differences for all variables at once.

Many test statistics can be used in a situation like this, and only a few of these will be mentioned here. One possibility is Hotelling's T^2 statistic. Let n_1 and n_2 be the sizes of two samples being compared, with the sample mean vectors $\bar{\mathbf{x}}_1$ and $\bar{\mathbf{x}}_2$, and the sample covariance matrices $\mathbf{C}_1$ and $\mathbf{C}_2$. Assuming that both samples come from distributions with the same covariance matrix, this can be estimated by the pooled sample covariance matrix

$$\mathbf{C} = \{(n_1 - 1)\mathbf{C}_1 + (n_2 - 1)\mathbf{C}_2\}/\{n_1 + n_2 - 2\}.$$

Then T^2 is defined as

$$T^2 = \{n_1 n_2/(n_1 + n_2)\}(\bar{\mathbf{x}}_1 - \bar{\mathbf{x}}_2)'\mathbf{C}^{-1}(\bar{\mathbf{x}}_1 - \bar{\mathbf{x}}_2).$$

This statistic has the advantage of being known to have good power to detect sample mean differences when the two samples are randomly chosen from normal distributions with a common covariance matrix. Also, it is unaffected by the scales used for the variables and takes account of the correlations that these variables possess. For large samples, values of T^2 can be tested against the chi-squared distribution with p degrees of freedom, with high values of the statistic giving evidence of population mean differences. For small samples, an F-test can be used (Manly, 1994a, p.40).

A better choice if randomization testing is being contemplated may be the statistic

$$J = (\bar{\mathbf{x}}_1 - \bar{\mathbf{x}}_2)'(\mathbf{C}_1/n_1 + \mathbf{C}_2/n_2)^{-1}(\bar{\mathbf{x}}_1 - \bar{\mathbf{x}}_2),$$

which was proposed by James (1954), and has approximately a chi-squared distribution with p degrees of freedom for large samples, even if the distributions being sampled have different covariance matrices. Large values of J indicate mean differences between the sources of the samples.

Romesburg (1985) has published a Fortran program for a randomization test using a residual sum of squares statistic and Wilk's lambda statistic, for cases where there are two or more samples to compare. The sum of squares statistic is the sum of squares for deviations of observations from sample means, added for all p variables and all m samples. That is to say, if s_{ij}^2 is the variance for variable X_i in sample j, then the statistic is

$$E = \sum_{i=1}^{p} \sum_{j=1}^{m} (n_j - 1)s_{ij}^2.$$

This depends on the scaling of the different variables, so that in most cases it will be desirable to make sure that these scales are comparable by, for example, making the variance the same for all variables when the m samples are lumped together. Small values of E indicate low variation within samples, and hence are associated with large differences between samples.

Wilk's lambda statistic is

$$L = |\mathbf{W}|/|\mathbf{T}|,$$

where $|\mathbf{W}|$ is the determinant of the within-sample sum of squares and cross products matrix, and $|\mathbf{T}|$ is the determinant of the total sum of squares and cross products matrix (Manly, 1994a, p. 49). Like T^2 and J, this statistic takes proper account of correlations between variables, is unaffected by the scale used to measure variables, and should have good properties for comparing means of random samples taken from normal distributions with a common covariance matrix. Small values of L indicate mean differences between the populations being compared. For samples from normal distributions with a common covariance matrix the statistic $-[n - 1 - \frac{1}{2}(p + m)]\log_e(L)$ can be tested against the chi-squared distribution with $p(m - 1)$ degrees of freedom to see whether it is significantly large.

The calculation of the statistics T^2, J and L involve matrix inversions that cannot be carried out if one or more of the X variables being considered can be expressed exactly in terms of the other X variables. When this type of relationship exists it is therefore necessary to remove one or more variables from the data before starting the analysis. For example, if $X_1 + X_2 + \ldots + X_p = C$, where C is a constant, for all n sample units, so that X_i can be expressed exactly as C minus the sum of the other variables, then one of the X variables must be removed. The choice does not matter because T^2, J and L will take the same value whichever variable is omitted.

Edgington (1995, Chapter 8) discussed three test statistics for randomization testing. First, he suggested coding the observations on each variable to have means of 0 and variances of 1 for all samples lumped together. The sum of the scores for all p variables on a sample unit can then be used as a composite measure of the 'size' of that unit. This converts the multivariate problem of comparing the sample means of the p variables into the univariate problem of comparing the sample means for the composite variable, which can be handled using a t-test or analysis of variance. This approach involves the implicit assumption that if samples show systematic differences then these differences will be in the same direction for each variable.

The second possibility suggested by Edgington involves carrying out a t-test (with two samples) or an analysis of variance (with two or more samples) separately for each of the p variables and using the sum of the t or F-values as a statistic measuring overall sample differences. With two samples, and two-sided tests on each variables, the appropriate sum for the t tests is of absolute t values. Unfortunately this is not equivalent to using the sum of F values so Edgington suggests using either $\sum \log|t|$ or $\sum \log(F)$. These log sums are equivalent for randomization testing because for any variable $\log(F) = 2\log|t|$. Large values correspond to large sample mean differences. The use of a sum of univariate test statistics as a multivariate test statistic for a randomization test appears to have been first proposed by Chung and Fraser (1958).

Edgington's third approach involves comparing the sum of the squared Euclidean distances from observations to their sample centroids (the E statistic given above) with the sum of the squared distances from sample centroids to the centroids for all observations. This is analogous to calculating the ratio of the sum of squares between samples to the sum of squares within samples in a one-factor analysis of variance. In effect the statistic produced is equivalent to E. Usually the scales of the p variables will need to be made comparable to ensure that they contribute about equally to this test.

One way to carry out a test involves formulating it in terms of distance matrices, and then carrying out a Mantel test (section 9.2) to see if there is a significant association between a 'sample similarity' matrix and a 'distance between observations' matrix. In the first of these matrices the distance between two-sample units will be 1 if they are in the same sample and 0 otherwise. In the second matrix the distance between two units can be taken as $\sum(\delta X_i)^2$, where δX_i is the difference between the two units for the variable X_i, with the sum being over the p variables. Other measures of distance based on the X values are also possible, such as $\sum |\delta X_i|$. There is interest in whether the correlation between the terms in the two matrices is significantly low, suggesting that observations tend to be similar within groups.

Tests of this type were first proposed by Mantel and Varland (1970), and are sometimes called multi-response permutation procedures (Mielke *et al.*, 1976; Mielke, 1978; Zimmerman *et al.*, 1985). Mantel and Varland suggested avoiding the problem of different scales being involved for the different X variables by replacing the data values with their ranks before calculating distances between individuals.

An important aspect of multi-response permutation procedures is that with unequal sample sizes they may be made more powerful by defining the sample similarity matrix so that the elements are not just 0 (for observations in different samples) or 1 (for observations in the same sample). Instead, it is better to use 0 for observations in different samples and $1/(n_i - 1)$ for observations in the same sample of size n_i (Mielke, 1984; Zimmerman *et al.*, 1985; Luo and Fox, 1996).

Unfortunately, no real comparisons seem to have been made of the performances of the test statistics that have just been described. It is therefore hard to make any recommendations about which ones to use. However, T^2, J or W are reasonable choices if the aim is to have a sample comparison that is primarily a function of differences between the means of the X values, with the scales for the different variables and the correlations between the variables being taken into account automatically. The sum of $\log(F)$ values also gives a comparison between samples based mainly on mean differences with scaling effects taken into account. This might be considered a reasonable statistic because of the simplicity of its calculation. On the other hand, a test based on the concept that individuals within groups should have closer X values than individuals in general is conveniently set up either in terms of a Mantel test on distance matrices or with the use of Romesburg's E statistic.

A randomization test will require either all possible allocations of the data to the samples or a random sampling of these allocations. For large sample sizes the number of possible allocations becomes extremely large, but if these are to be enumerated then the Fortran subroutines of Berry (1982) and Berry and Mielke (1984) may be helpful. An alternative to either the full or partial enumeration of randomizations involves approximating the randomization distribution in some way. Mardia (1971) discusses this approach for approximating the distribution of a statistic for comparing sample mean vectors using a beta distribution. Mielke *et al.* (1976) and Mielke (1978) also proposed a beta distribution approximation for use with the multi-response permutation procedure.

Example 12.1: Cepaea hortensis *in colonies with and without* C. nemoralis   In Example 6.3 an analysis of variance was carried out on each of the five shell type percentages for the snail *Cepaea hortensis* that are shown in Table 6.4. It was found that there is some evidence that the mean percentage of brown snails differs between colonies of both *C. hortensis* and *C. nemoralis* and colonies of *C. hortensis* alone, but no evidence of any difference for the four other shell types. The difference between the two-sample mean percentages of brown shells was significant at the 0.5% level on a two-sided randomization test, and this was still significant after making an allowance for multiple testing.

As noted above, there are many different multivariate statistics that can be used to compare the compositions of the two types of colony. Here the results are considered for the sum of squares statistic (E), Wilk's lambda statistic (L), the sum of $\log(F)$ values from analysis of variance on the five individual variables, and the Mantel test statistic for the association between a matrix of sample similarities and a matrix of shell type proportion distances.

Data will usually be standardized for the calculation of the E statistic so as to avoid the possibility of one or two of the p variables having an undue effect on the statistic. However, scaling was not done for the present example because the data are percentages that are comparable without scaling. The E value for the data is 75 671.4. When the randomization distribution of E was approximated by 4999 randomizations of the data and the observed value, it was found that 34.2% gave values less than or equal to 75 671.4. Hence on this test there is no evidence of a difference between the two types of colony of *C. hortensis*. A one-sided test is appropriate because only low values of E indicate a similarity between individuals in the same sample.

The observed value of Wilk's lambda statistic is $L = 0.9085$. It was found that 24.6% of 4999 randomized values plus the observed value were this low or lower, so again there is no evidence of any difference between the two types of colony. Because the five shell percentages add up to 100% for each colony, this test was carried out using the first four percentages only. For comparison it can be noted that the value of the chi-squared test statistic $-[n - 1 - \frac{1}{2}(p + m)]\log_e(L)$ is 5.37, with $n = 60$, $p = 4$, $m = 2$ and $L = 0.9085$. With $p(m - 1) = 5$ degrees of freedom this is significantly large at the 37.2% level.

As shown in Example 6.3, the t-values for comparing the means for the two types of colony on the five shell type percentages are -1.27, 0.06, 0.44, 0.90 and 2.11. The F-values to compare the means are these t-values squared, and hence the observed statistic $\sum \log_e(F)$ comes out at -5.46 (avoiding rounding errors). This was exceeded by 52.9% of values of this statistic found when the randomization distribution was approximated by the observed statistic plus 4999 randomized values. Hence this statistic gives little evidence of any difference between the two samples.

For the Mantel test a 60 by 60 matrix of sample similarities was constructed with $1/(n_i - 1)$ as the entry for two colonies of type i, and a 0 for two colonies of a different type, where n_i is the number of colonies of type i. A matrix of shell proportion differences was constructed with d_{ij} as the distance from colony i to colony j where

$$d_{ij} = \sum_{r=1}^{5} |p_{ri} - p_{rj}|/2$$

and p_{ri} denotes the proportion of shell type r in the ith colony. This distance measure is one of many that are commonly used with proportion data (Manly, 1994a, p. 67). It takes the value 1 when two colonies have no shell types in common and 0 if they have exactly the same proportions of each type.

If anything, a negative relationship between these matrices is expected, with a large sample similarity corresponding to a small proportion difference. The regression coefficient for proportion differences against sample similarities is -0.073. This has the correct sign, but is not significantly low. From 999 randomizations the percentage of values this low or lower was found to be 36.4%. Again there is no evidence of a difference between the two samples.

The inability of these multivariate tests to detect a difference between the two types of colony of C. hortensis when a univariate test gives a significant result for the percentage of brown shells can be explained easily enough in terms of this difference being swamped by small differences between the percentages for other shell types. This phenomenon is not uncommon and has nothing to do with tests being made by randomization.

Example 12.2: Cepaea nemoralis *shell frequencies in different habitats* For a second example, consider the percentages shown in Table 12.1 for 10 shell types of the snail *Cepaea nemoralis* found in samples taken from 17 colonies in six types of habitat. These data were collected by Cain and Sheppard (1950) to consider the question of whether there is any evidence that the frequency of different shell types varies with the habitat.

The tests that were used in the last example gave the following results when they were applied to these data, with 4999 randomizations in each case:

1. The observed value of the sum of squares statistic with unstandardized data is $E = 8521.0$ with a significance level of 0.08%.
2. Because the 10 shell percentages add to 100, only the first nine were used for calculating Wilk's lambda, which has an observed value of $L = 0.000\,463$. This was found to have a significance level of 14.0% by randomization. For comparison, the approximate chi-squared statistic is

$$-[n - 1 - \tfrac{1}{2}(p + m)]\log_e(L) = 65.26,$$

Table 12.1 Percentages of shells with different colours and banding for samples of *Cepaea nemoralis* in colonies from six habitat types in southern England (UB = unbanded; MB = mid-banded; FB = fully banded; OB = other banding types; B = banded)

Habitat	Colony	Yellow				Pink				Brown	
		UB	MB	FB	OB	UB	MB	FB	OB	UB	B
Downland beech	1	9.6	15.4	0.0	0.0	48.7	25.0	0.0	0.0	0.6	0.6
	2	10.9	16.0	0.0	0.0	26.3	4.5	0.0	0.0	36.5	5.8
Oakwood	1	1.2	4.7	1.2	1.2	5.8	25.6	4.7	1.2	34.9	19.8
	2	0.0	13.0	0.5	0.0	27.6	43.2	3.8	8.6	0.0	3.2
	3	1.5	1.0	0.7	0.5	23.2	28.4	10.7	23.8	10.2	0.0
Mixed deciduous woods	1	3.0	1.5	4.5	0.0	50.0	6.1	16.7	6.1	12.1	0.0
	2	0.4	3.1	14.8	12.6	6.7	6.7	24.2	17.9	10.8	2.7
	3	0.0	0.0	11.4	5.7	17.1	14.3	25.7	14.3	11.4	0.0
	4	1.0	4.0	18.2	14.1	17.2	3.0	8.1	10.1	23.2	1.0
	5	13.5	0.0	5.8	7.7	13.5	0.0	23.1	21.2	15.4	0.0
Hedgerows	1	9.5	4.8	47.6	14.3	0.0	0.0	9.5	9.5	0.0	4.8
	2	6.1	9.0	16.5	9.3	2.5	9.7	21.1	21.9	3.2	0.7
	3	16.1	6.5	16.1	19.4	12.9	3.2	16.1	9.7	0.0	0.0
	4	1.3	1.3	13.2	2.6	0.0	1.3	55.3	25.0	0.0	0.0
Downside long coarse grass	1	0.0	15.5	43.3	5.4	0.0	7.5	19.2	2.6	1.9	4.7
	2	0.0	7.1	13.5	21.9	12.3	3.9	21.3	20.0	0.0	0.0
Downside short turf	1	28.7	12.8	0.0	3.7	3.0	7.9	0.0	4.3	29.9	9.8

with $p(m - 1) = 45$ degrees of freedom, which has a significance level of about 2.5%.

3. The value of the $\sum \log_e(F)$ statistic is 10.12 for the observed data. This has a significance level of 0.04%.

4. A Mantel test based on comparing similarities between colonies for habitat types and distances based on the proportions of different types of shell as described in the previous example (with 4999 randomizations) gives a matrix regression coefficient of -0.30, which is significantly low at the 0.03% level.

To obtain some idea of why the results differ so much between tests, a small simulation study was conducted, along the lines of studies that have been described with some of the examples in earlier chapters. The procedure was as follows:

1. The percentages for the 17 samples were randomly assigned to the six samples from different types of habitat so that, for example, the percentages for the first downland beech colony (9.6, 15.4, , 0.6) could be placed in any of the six habitats. The ordering of the 10 percentages within a sample was not changed in this process.

2. Habitat effects of various sizes (none, small, medium and high) were added to the randomized data.

3. The four tests were carried out to see whether the difference between habitats were significantly large at the 5% level, with 99 randomizations for the tests where these were needed.

4. Steps 1–3 were repeated 1000 times with each of four different levels of habitat effects.

Table 12.2 contains an explanation of how habitat effects were superimposed on the randomized data sets, and also shows the percentages of significant results that were obtained from the generated sets of data. It turns out that the simulation results are rather helpful in explaining the results obtained from analysing the real data in Table 12.1.

To begin with it can be noted that Wilk's lambda test with significance determined using the chi-squared distribution has given far too many significant results when the null hypothesis was true (20.2% instead of 5%). The conclusion must be that the use of the chi-squared distribution for testing Wilk's lambda is not reliable for these data, and it is not at all surprising to find that for the data in Table 12.1 it gives a much higher level of significance than is found by randomization.

The second important point to note from the simulation results is that using Wilk's lambda as the test statistic with randomization resulted in far less power than was obtained with the other statistics. For example, with high habitat effects only 53.5% of the test results were significant using

Table 12.2 A simulation experiment on tests to compare multivariate means, based on the data in Table 12.1

(a) Habitat effects: Habitat effects were superimposed on the randomized data from Table 12.1 by adding the following amounts to the percentages in the samples assigned to the different habitats, and then rescaling to make the total percentage 100 for each sample. The value of Θ was set at 0 (no effects), 1 (small effects), 2 (medium effects) or 4 (high effects). Abbreviations for morph descriptions (UB, MB, etc. are given in Table 12.1

Habitat	Yellow				Pink				Brown	
	UB	MB	FB	OB	UB	MB	FB	OB	UB	B
Downland beech	0	0	0	0	0	0	5Θ	5Θ	10Θ	10Θ
Oakwood	2Θ	2Θ	Θ	Θ	0	0	4Θ	4Θ	8Θ	8Θ
Mixed deciduous woods	4Θ	4Θ	2Θ	2Θ	0	0	3Θ	3Θ	6Θ	6Θ
Hedgerows	6Θ	6Θ	3Θ	3Θ	0	0	2Θ	2Θ	4Θ	4Θ
Downside long grass	8Θ	8Θ	4Θ	4Θ	0	0	Θ	Θ	2Θ	2Θ
Downside short turf	10Θ	10Θ	5Θ	5Θ	0	0	0	0	0	0

(b) Percentages of significant test results: Percentages are for 1000 sets of generated data. The tests are WLR: Wilk's lambda by randomization; WLC: Wilk's lambda using the chi-squared distribution; SLF: sum of log(*F*); RE: Romesburg's *E*-statistic; and M: Mantel test

Habitat effects	WLR	WLC	SLF	RE	M
None	4.6	20.2	5.1	4.8	5.2
Small	9.2	33.6	13.2	9.1	13.01
Medium	21.5	59.5	49.4	26.2	47.61
High	53.5	91.3	98.1	96.3	99.8

Wilk's lambda, compared to 96.3% or more using the other statistics. On this basis the test results for the data in Table 12.1 (Wilk's lambda by randomization not significant, the other tests very significant) is easily explained: Wilk's lambda is less likely to detect any habitat differences that exist. It is therefore the significant results using the E, $\sum \log(F)$ and Mantel tests that should be taken seriously.

This small simulation experiment should not be taken to indicate that Wilk's lambda is necessarily less powerful than the other statistics. Indeed, some other simulations that are not reported here have suggested that using Wilk's lambda gives a more powerful test for detecting some types of difference between samples, and that its lack of power with data like that shown in Table 12.1 may be related to the fact that the variables are percentages that are constrained to add to 100. It is difficult to say more in the absence of results from a much more extensive and searching simulation study of the properties of multivariate tests.

The calculations for the original data for this example and the previous one were carried out using the RT package (Manly, 1996a). The simulation study required specially written computer programs.

12.4   CHI-SQUARED ANALYSES FOR COUNT DATA

The examples just considered involved comparing colonies of snails in terms of the percentages of snails observed in different colour and banding classes. Because the original data consist of counts in the different classes, it is natural to consider an alternative analysis of these counts based on a chi-squared type of statistic.

One approach involves calculating a chi-squared value within each sample, and using the total of these as an overall measure of within-sample consistency. A small total chi-squared value indicates similar proportions for cases within the same sample category while a large value indicates a high level of variation in these proportions. The randomization test involves assessing whether the statistic is significantly low in comparison with what is expected if the cases are allocated randomly to the sample categories (Manly, 1985, p. 161). The following example will clarify the procedure.

Example 12.3: Cepaea nemoralis *in different habitats (reconsidered)*
Table 12.3 shows the counts of the numbers of different shell types that were used to produce the percentages in Table 12.1. Recall that these data come from Cain and Sheppard's (1950) study of 17 colonies of the snail *Cepaea nemoralis* in southern England. As in Example 12.2, the question to be considered is whether there is any evidence that the 10 shell type proportions vary significantly between habitat types.

Table 12.3 Sample frequencies of *Cepaea nemoralis* that produce the percentages in Table 12.1 (UB = unbanded; MB = mid-banded; FB = fully banded; OB = other banding types; B = banded)

Habitat	Colony	Yellow				Pink				Brown	
		UB	MB	FB	OB	UB	MB	FB	OB	UB	B
Downland beech	1	15	24	0	0	76	39	0	0	1	1
	2	17	25	0	0	41	7	0	0	57	9
Oakwood	1	1	4	1	1	5	22	4	1	30	17
	2	0	24	1	0	51	80	7	16	0	6
	3	9	6	4	3	135	165	62	138	59	0
Mixed deciduous wood	1	2	1	3	0	33	4	11	4	8	0
	2	1	7	33	28	15	15	54	40	24	6
	3	0	0	4	2	6	5	9	5	4	0
	4	1	4	18	14	17	3	8	10	23	1
	5	7	0	3	4	7	0	12	11	8	0
Hedgerows	1	2	1	10	3	0	0	2	2	0	1
	2	17	25	46	26	7	27	59	61	9	2
	3	5	2	5	6	4	1	5	3	0	0
	4	1	1	10	2	0	1	42	19	0	0
Downside long coarse grass	1	0	66	185	23	0	32	82	11	8	20
	2	0	11	21	34	19	6	33	31	0	0
Downside short turf	1	47	21	0	6	5	13	0	7	49	16

Consider first the downland beech habitat. Here there are two colonies available, and six shell types are present. A chi-squared statistic can be calculated to see whether there is any evidence of a difference between the shell type proportions in the two colonies. This statistic takes the form $\sum (O - E)^2 / E$, where O denotes an observed frequency and E its expected value on the hypothesis of no difference between the colonies. Its value is 93.3 with 5 degrees of freedom. This is very highly significant when compared with tables of the chi-squared distribution, reflecting the fact that colonies generally have quite variable shell type proportions within habitat types.

If a chi-squared value is calculated for each of the six habitat types (making this zero for downside short turf with one colony only), then the sum of these values, 824.8, gives an overall measure of the consistency of shell type proportions within habitat classes. The question of whether the proportions differ between habitats can then be phrased in terms of whether this total is smaller than is expected from a random allocation of colonies to habitats, because if this is the case then the differences between habitats must be larger than is expected from a random allocation. A randomization test can therefore be conducted by randomly allocating two colonies to be 'downland beech', three colonies to be 'oakwood', and so on, and calculating the chi-squared value that this gives. Repeating the randomization a large number of times allows the randomization distribution of chi-squared to be estimated in the usual way. The significance level of the observed statistic is the proportion of the randomized values less than or equal to this value. Manly (1985, p. 161) found that every one of 1000 randomizations gave a higher chi-squared than 824.8. The estimated significance level for this example is therefore 1 in 1001 (including the observed data when approximating the randomization distribution), or about 0.1%. There is clear evidence of a difference between habitats. It was also found that after allowing for multiple testing the contribution to the total chi-squared was significantly low for yellow fully banded and pink fully banded snails. In other words, it is these shell types in particular that show most consistency within habitats.

The difference between this chi-squared analysis based on frequencies and the analyses of Example 12.2 based on percentages is essentially in the weighting of the data. The chi-squared test assumes that the variance of a sample frequency is proportional to its expected value. In effect the frequencies in a sample are weighted according to their sum. However, an analysis based on proportions (or percentages) gives equal weight to the proportions in all samples. Neither of these weightings is necessarily 'correct' for a set of data, particularly when randomization testing is used.

The problem is that the sample counts are affected by sources of variation other than purely sampling errors. The variance due to sampling errors may therefore not be particularly important in comparison with general variation between colonies within one habitat. In fact, Manly (1985, p. 210) gives a calculation which suggests that pure sampling variation is relatively unimportant when comparing shell type proportions in different colonies of snails providing that at least 15 snails are taken from each colony.

12.5 PRINCIPAL COMPONENTS ANALYSIS AND OTHER ONE-SAMPLE METHODS

Principal components analysis is one of the simplest of the standard multivariate methods. It involves taking a set of data involving p variables $X_1, X_2, \ldots, X_p$ and finding linear combinations of these variables $Z_1, Z_2, \ldots, Z_p$ (the principal components) that are uncorrelated and are ordered in their variances so that $\text{Var}(Z_1) \geq \text{Var}(Z_2) \geq \ldots \geq \text{Var}(Z_{p-1}) \geq \text{Var}(Z_p)$ (Manly, 1994a, Chapter 6). This ordering is the best possible in the sense that $\text{Var}(Z_1)$ is as large as possible, $\text{Var}(Z_2)$ is as large as possible, given that Z_2 must be uncorrelated with Z_1, and so on, with the further constraint that the sum of the variances of the Z variables equals the sum of the variances of the X variables. The analysis is usually carried out with the idea of replacing the original X variables with a smaller number of the Z variables. This can either be in order to use the reduced number of variables for descriptive or summary purposes, or merely to aid in understanding the structure of the data.

Because principal components analysis treats the data as consisting of a single sample, randomizing the order of observations does not change the calculations in any way. Therefore there is no point in randomization testing with this particular multivariate method. However, Monte Carlo methods do have applications with principal components. For example, Karr and Martin (1981) carried out an interesting comparison between the principal components found from sets of real data and the principal components found from random data sets with the same numbers of uncorrelated variables and the same numbers of cases. They found that random data will often give rise to principal components that appear to be meaningful. This type of comparison was investigated further by Stauffer *et al.* (1985), who also considered how the bootstrap can be used to estimate standard errors for the variances of principal components. An earlier paper pointing out the value of bootstrapping with principal components analysis is that of Diaconis and Efron (1983). Jackson (1993, 1995) and Mehlman *et al.* (1995) discuss the use of bootstrapping to decide how many components to interpret from an analysis. See also Milan and Whittaker (1995).

Monte Carlo and bootstrap approaches are of value with other multivariate methods that treat the data as consisting of a single sample.

With cluster analysis (Manly, 1994a, Chapter 9), an attempt is made to divide a sample of n individuals into two or more clusters of similar individuals, using the values for p variables measured on each individual. Here it is interesting to see how many 'clusters' are found when in fact the data are random (Gordon, 1981; Jain and Dubes, 1988; Morgan, 1984). Bootstrapping and randomization can be used to test for significant clustering with species presence and absence and species abundance data (Harper, 1978; Strauss, 1982; Nemec and Brinkhurst, 1988a,b; Vassiliou *et al.*, 1989). See also Peck *et al.*'s (1989) bootstrap method for putting confidence limits on the optimal number of clusters from a cluster analysis.

With factor analysis (Manly, 1994a, Chapter 7) the values for the p variables are related to a smaller number of 'factors' which, it is hoped, will illuminate the structure of the data. Here there is potential for seeing the performance of different methods on simulated data (Francis, 1974; Seber, 1984, p. 223) and in using bootstrapping to assess the accuracy of estimates.

With multidimensional scaling (Manly, 1994a, Chapter 11) the distances between n individuals are used to produce a 'map' showing how the individuals are related in one, two, three or more 'dimensions'. Simulation can be used to assess the way this procedure behaves on data either with or without structure (Morgan, 1984, p. 218).

Jackknifing also offers possibilities with multivariate data. For example, Riska (1985) used this approach for comparing correlation matrices of morphological traits on an aphid.

A technique that can be approached with randomization is canonical correlation analysis. This is a technique for examining the relationship between one set of variables $X_1, X_2, \ldots, X_p$, and another set of variables $Y_1, Y_2, \ldots, Y_q$, when each of n sample units has observations on all $p + q$ of these variables (Manly, 1994a, Chapter 10). It can be thought of as a generalization of multiple regression, where there are q dependent variables instead of one. A randomization approach to assessing the significance of relationships between the X and Y variables consists of comparing estimates obtained from a standard canonical correlation analysis with the distributions of these estimates found from randomly allocating Y values to the sample units, while keeping the X values fixed. It is necessary to keep the Y values for a sample unit together for this randomization so that the correlations between these Y variables are maintained as well as the correlations between the X variables.

Mention must also be made of the use of randomization tests with canonical correspondence analysis, which is a method for relating the distribution of species in terms of their abundances or presences and absences at different study sites to the environmental characteristics of the sites. In this case important statistics can be compared with their random-ization distributions as generated by randomly permuting the environmental data between sites, possibly using the computer program CANOCO (ter Braak, 1988; Rejmánkova et al., 1991).

An unusual application of bootstrapping with multivariate data is reported by Hopkins et al. (1991). They were interested in the extent to which different variables that are intended to measure physical activities and fitness of adults are in fact measuring similar underlying variables. To this end they compared the mean correlations of various groups of variables with other groups of variables and determined which mean correlations were significantly different from zero and from each other using standard errors calculated by bootstrap resampling of subjects. It is difficult to see what alternative approach could have been used in this situation, although the validity of the bootstrap method requires further investigation before it can be used with complete confidence.

12.6 DISCRIMINANT FUNCTION ANALYSIS

One place where randomization testing is of particular value is with discriminant function analysis. This is a technique for comparing two or more groups and it is useful to know how an analysis on real data compares with what is obtained when the cases in the data are randomly reallocated to groups.

The description 'discriminant function analysis' typically covers the calculation of a range of statistics, and is justified in several different theoretical frameworks (Manly, 1994a, Chapter 8). Solow (1990) considers this topic from the point of view of a randomization test to see whether an estimated discriminant function gives a significantly high number of correct classifications for a set of data in comparison with a randomization distribution. Here only one aspect will be considered, namely the significance of canonical discriminant functions. Briefly, these functions are described as follows.

Suppose that there are m samples available, and the cases in each sample have variables $X_1, X_2, \ldots, X_p$ measured on them. Then there are generally s canonical discriminant functions, where s is the smaller of p and $m - 1$. The first of these,

$$Z_1 = a_{11}X_1 + a_{12}X_2 + \ldots + a_{1p}X_p,$$

is chosen so that for this linear combination of the X variables the F-ratio in an analysis of variance (the between sample mean square divided by the within sample mean square) is as large as possible. That is to say it is chosen to maximize, in this sense, the variation between samples relative to the variation within samples. If there are several discriminant functions then Z_i is chosen to maximize the F-ratio from an analysis of variance, subject to there being no correlation between this function and Z_1 to Z_{i-1}. From this point of view, the functions are ordered in importance from Z_1 to Z_s. Note that the scaling of these functions is arbitrary in the sense that multiplying all the X coefficients by the same amount does not change the F-value in the analysis of variance.

The coefficients of the discriminant functions are the eigenvectors of the equation

$$(\mathbf{B} - \tau\mathbf{W})\mathbf{a} = \mathbf{0}, \tag{12.1}$$

where $\mathbf{W}$ is the within-sample matrix of sums of squares and cross products and $\mathbf{B}$ is the between-sample matrix of sums of squares and cross products (Manly, 1994a, p.110). The eigenvalue τ corresponding to one of these vectors is the ratio of the between-sample sum of squares to the within-sample sum of squares for the discriminant function concerned. If the eigenvalues in order of magnitude are $\tau_1, \tau_2, \ldots, \tau_s$, then a test that is sometimes suggested for whether the jth discriminant function varies significantly from sample to sample is based on comparing the statistic

$$C_j = \{n - 1 - \tfrac{1}{2}(p + m)\} \log_e(1 + \tau_j),$$

where n is the total number of observations, with the upper percentage points of the chi-squared distribution with $p + m - 2j$ degrees of freedom. This test is based on the assumption that the m samples are drawn randomly from normal distributions with the same covariance matrix. The sum of the C_j values is Wilk's lambda statistic defined in section 12.3.

Tests based on C_j have been criticized by Harris (1985, p.172) on the grounds that for small samples the ordering of the sample eigenvalues of equation (12.1) may not match the ordering of the population eigenvalues. However, this causes no difficulty when the significance levels of the eigenvalues (or, equivalently, the C_j values) are determined by finding the percentages of estimates from randomized data that are greater than or equal to the observed values. This type of calculation can be done on some of the standard statistical packages for computers, although computation times may be considerable.

Example 12.4: Cepaea nemoralis in various habitats (reconsidered)   The percentages of different shell types in 17 colonies of *Cepaea nemoralis* in six different habitats shown in Table 12.1 have already been the subject of Example 12.2, which was concerned with assessing whether there is any evidence that the mean percentages vary with the habitat. With the present example the same question is considered from the point of view of seeing how well the samples from different habitats can be separated by using discriminant function analysis on these percentages. This is an interesting example because the sample sizes are small and the assumption of normality that is usually made when a discriminant function analysis is carried out is very questionable. Nevertheless, the question of how well the habitats can be separated using the shell type percentages is important.

The discriminant function analysis was carried out on the first nine percentages only because the tenth percentage can be expressed as 100 minus the sum of these. On the face of it, very good discrimination is obtained, with the eigenvalues of equation (12.1) being 25.88, 5.54, 2.62, 2.05 and 0.11. A plot of the values of Z_2 against the Z_1 is shown in Figure 12.1, with the habitats indicated. The grouping is quite clear. The C_j test statistics corresponding to the eigenvalues are found to be 27.98 with 13 degrees of freedom, 15.96 with 11 degrees of freedom, 10.94 with 9 degrees of freedom, 9.48 with 7 degrees of freedom, and 0.89 with 5 degrees of freedom. The first discriminant function is therefore significant at about the 1% level, but none of the other functions show much significance.

However, a randomization test gives very little evidence of anything

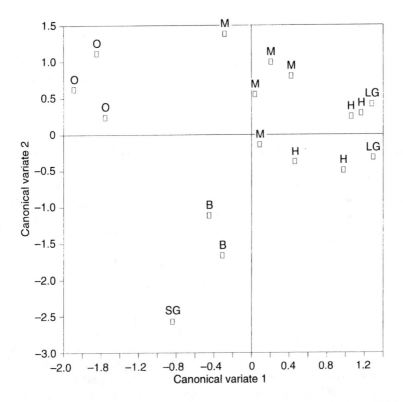

Figure 12.1 Plots of Z_2 against Z_1 for the observed *Cepaea nemoralis* data. Habitat types: (B) downland beech; (O) oakwood; (M) mixed deciduous wood; (H) hedgerows; (LG) downside long coarse grass; (SG) downside short turf.

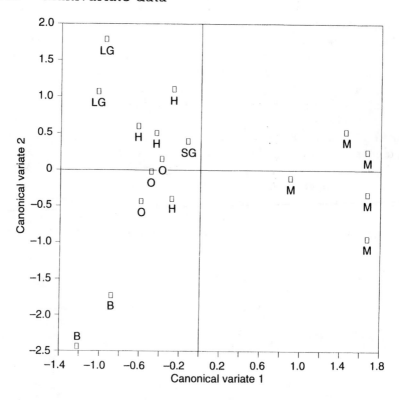

Figure 12.2 As for Figure 12.1 but for a random allocation of colonies to habitats.

being significant. The randomization distributions of the eigenvalues were approximated with 999 random allocations of the colonies to the habitats, maintaining the observed numbers of colonies for each habitat. On this basis,the significance levels estimated were 30.6%, 25.4%, 9.4%, 0.3% and 71.2%, respectively, for the eigenvalues in order. Here all that is apparent is that with randomized data a fourth eigenvalue as large as 2.05 is rather unlikely.

This result is made understandable by considering Figure 12.2, which shows a plot of Z_2 against Z_1 for a 'typical' set of randomized data, for which the first two eigenvalues are 26.95 and 4.43. The same good grouping of habitats seen in Figure 12.1 is also seen here. Clearly, discriminant function analysis is able to find groupings when none that have any real meaning can exist. That the chance grouping indicated in Figure 12.2 is by no means unusual is seen from the fact that the largest values observed for the first two eigenvalues in the randomized data were 1435.0 and 55.0, both of which are much larger than the observed values.

This outcome of randomization testing might be viewed with some alarm in view of the highly significant differences between the habitats found from other randomization tests in Examples 12.2 and 12.3. However, what must be remembered is that the outcome of a test is a product of both the data and the testing procedure. In the present case it simply seems that the discriminant function analysis is so good at finding groups when none exist that it is of little value for analysing the data. If the sample sizes were larger then presumably this problem would not occur.

In fact, the result of the discriminant function analysis mirrors what was found in Example 12.2. In that example the observed value of Wilk's lambda statistic (the sum of the C_j values from discriminant function analysis) was found to be significantly large at about the 2.5% level on an approximate chi-squared test, but significantly large at the 14% level by randomization.

12.7 FURTHER READING

In one of his early papers on bootstrapping, Efron (1979b) used the estimation of error rates in discriminant function analysis as one of the examples. Further work on this idea has been done by McLachlan (1980). Bootstrapping has also been used with correspondence analysis (Greenacre, 1984; Knox, 1989; Knox and Peet, 1989; Markus and Visser, 1992) and multidimensional scaling (Weinberg et al., 1984). Note that multidimensional scaling has already been briefly discussed in section 9.7 in relationship to the analysis of distance matrices.

Zhang and Boos (1992, 1993) discuss the use of bootstrapping to test for a significant difference between two or more sample covariance matrices. This is a potentially important application because of the known lack of robustness of tests based on the assumption of normality. However, difficulties with tests to compare variances that have been discussed in Chapters 6 and 7 suggest that the performance may not be very good with grossly non-normal data. See also Krzanowski's (1993) randomization test for comparing two correlation matrices.

There is a large literature on randomization tests for count data for which a test statistic is compared with the distribution for different possible sample counts subject to appropriate marginal constraints (Agresti, 1992; McDonald and Smith, 1995).

12.8 CHAPTER SUMMARY

- Multivariate analysis is concerned with the situation where each sample unit has several measurements associated with it. With randomization tests all observations associated with a unit stay together when they are moved between samples.

- A number of test statistics have been suggested for the comparison of two-sample mean vectors, including Hotelling's T^2 statistic, James's statistic that does not require covariance matrices to be equal for the two distributions being compared, Romesburg's residual sum of squares statistic, Wilk's lambda statistic, Edgington's difference between the means of two composite variables constructed from the original variables, Edgington's sum of the logarithms of absolute t-values or F-values, Edgington's sum of squared distances from observations to the centroid for the sample that they are in, and a statistic based on the use of a Mantel matrix randomization test. All of these statistics can be tested by randomizing the sample observations to samples. However, little is known about the merits of alternative statistics.

- The comparison of several types of randomization test on one set of data produced non-significant results in all cases. On another set of data different tests produced different conclusions. A simulation study showed that in the situation with the second set of data the different conclusions are due to Wilk's lambda test having low power. It is concluded that more extensive simulation studies are needed to examine the performance of competing tests.

- With count data it may be appropriate to test for sample differences using statistics based on chi-squared values. An example illustrating one such test is described.

- With a single sample randomization does not generally provide useful analyses. However, bootstrap and Monte Carlo methods are useful in conjunction with principal components analysis, cluster analysis, factor analysis, and multi-dimensional scaling for studying the effects of sampling errors.

- Randomization methods are potentially useful with canonical correlation analysis and canonical correspondence analysis.

- An example of discriminant function analysis demonstrates that conventional testing methods can be quite misleading with small samples.

- Some extensions and generalizations of the methods discussed in the chapter are briefly reviewed.

12.9   EXERCISES

Exercises 12.1 and 12.2 can be carried out using the package RT (Manly, 1996a). Exercise 12.3 requires the use of a computer program that includes discriminant function analysis as an option, together with the ability to run this analysis a large number of times with random allocations of the data to groups.

12.1 In earlier chapters some examples and exercises have involved data from a study by Powell and Russell (1984, 1985) of the diet of eastern horned lizard *Phrynosoma douglassi brevirostre* at a site near Bow Island, Alberta. Table 12.4 shows the full data for adult males and yearling females, as provided in a paper by Linton *et al.* (1989). Samples of the lizards were taken in the months of June, July, August and September, and the body of the table shows gut contents in milligrams of dry biomass for each of the sampled lizards, with the following categories: (1) ants, (2) nonformicid Hymenoptera, (3) Homoptera, (4) Hemiptera, (5) Diptera, (6) Coleoptera, (7) Lepidoptera, (8) Orthoptera and (9) Arachnida. Using randomization,

Table 12.4 Prey consumption by adult male and yearling female lizards

Month	Prey category								
	1	2	3	4	5	6	7	8	9
June	242	0	0	86	0	205	0	190	18
	13	0	0	0	0	0	0	0	0
	105	0	0	10	0	0	0	0	129
July	20	0	0	0	0	0	0	0	25
	2	0	0	0	0	0	0	0	5
	245	0	0	0	0	34	0	0	0
	59	0	0	0	0	0	0	0	0
	8	0	0	0	0	0	0	0	0
August	515	0	0	0	0	90	0	0	0
	600	1	0	1	0	32	0	0	0
	488	0	0	0	90	256	0	0	0
	1889	17	0	0	0	209	0	142	5
	88	81	15	46	0	0	0	52	0
	82	200	0	52	0	332	0	0	7
	50	0	0	36	0	13	0	340	0
	233	144	26	0	113	75	50	0	0
	52	0	0	34	0	0	0	429	0
	40	96	0	0	0	117	0	376	0
September	21	180	0	12	0	44	0	0	0
	5	0	0	39	0	0	0	94	0
	6	22	0	0	0	31	69	60	0
	18	50	0	0	0	6	1	50	1
	44	0	0	0	0	0	0	0	0
	0	23	0	0	0	49	0	0	0

test for a significant difference between the gut contents found in different months using Wilk's lambda statistic, the sum of log(F) values, and the sum of squares statistic E.

12.2 Table 12.5 contains data on shell type frequencies for *Cepaea nemoralis* from samples of 20 or more snails collected by Clarke (1960, 1962) in southern England. Assess the evidence for variation related to habitat using the chi-squared type of analysis described in section 12.4.

12.3 Carry out a conventional discriminant function analysis on the lizard data from Exercise 12.1. Compare the significance levels obtained for the discriminant functions using the C_j statistics and the chi-squared distribution (as described in section 12.6) with the significance levels obtained by randomization.

Table 12.5 Shell type frequencies for *Cepaea nemoralis*

Habitat	1	2	3	4	5	6	7	8	9	10	11	12
Beechwood	1	0	0	0	0	8	0	1	12	1	0	0
	0	0	5	4	0	0	0	5	20	0	1	1
	0	0	2	0	0	9	0	15	21	0	1	27
	0	0	0	0	0	6	1	0	23	0	0	0
	0	0	10	15	0	0	0	4	20	0	0	21
	0	0	3	1	0	0	1	6	2	0	3	9
	3	0	0	0	0	8	0	9	2	0	35	47
	5	0	7	0	0	3	0	19	20	0	0	0
Other	0	0	0	0	0	13	0	9	28	1	0	0
deciduous woods	9	1	1	0	0	63	8	4	10	0	0	8
Fens	54	1	3	3	1	54	0	8	13	7	0	20
	5	1	3	0	2	14	1	13	4	2	0	0
	12	0	1	1	2	18	0	2	3	1	0	2
Hedgerows	1	1	0	15	0	2	0	1	3	2	0	4
and rough	16	7	9	0	19	11	1	0	0	6	0	0
herbage	13	4	4	0	9	0	1	0	0	1	0	0
	6	2	0	2	4	9	0	0	0	1	0	1
	18	0	8	0	1	34	0	5	0	0	1	4
	8	0	2	1	1	1	0	1	0	0	0	9
	2	0	0	0	0	24	0	0	1	3	0	0
Grasslands	2	0	5	1	0	1	0	4	7	0	0	5
	4	10	4	0	3	3	3	7	0	1	2	1
	6	6	10	0	0	0	0	0	0	0	0	0
	7	2	12	0	7	7	4	5	0	2	0	0
	2	4	5	0	0	3	4	19	0	5	2	4
	1	2	3	0	2	2	3	12	0	1	0	4
	13	1	15	0	6	0	0	0	0	0	0	0
	5	0	1	0	5	5	0	1	3	1	0	0

Shell types: (1) yellow fully banded; (2) yellow part banded; (3) yellow mid-banded; (4) yellow unbanded; (5) other yellows; (6) pink fully banded; (7) pink part-banded; (8) pink mid-banded; (9) pink unbanded; (10) other pinks; (11) brown banded; (12) brown unbanded

13
Survival and growth data

13.1 BOOTSTRAPPING SURVIVAL DATA

Bootstrapping has been proposed as a method of analysis in a number of situations related to survival data, particularly where there are complications like censoring. For example, a typical medical study might involve recording the survival times of patients following a serious operation. Censoring occurs with patients who are lost to the study before they die. For these patients all that is known is that they survived for at least some minimum time after the operation. Censoring also occurs because a study is terminated while patients are still alive. Generally it is assumed that the censoring process is independent of the survival process, although theory is available for handling situations where this is not the case.

One problem involves determining confidence limits for the underlying survival function or parameters that describe this function. Efron (1979a) suggested using bootstrapping in the context of daily records of deaths for a group of subject with the Kaplan and Meier (1958) estimator of the survival function. In this situation the estimated survival function is $\hat{S}(0) = 1$ at the start of day 1 (time $t = 0$), and then changes according to the equation

$$\hat{S}(t) = \hat{S}(t - 1)\{n(t) - d(t)\}/n(t), \tag{13.1}$$

where $n(t)$ is the number at risk at the start of day t and $d(t)$ is the number of these that die on day t. Essentially this means that the survival rate on day t is estimated as the proportion surviving, $\{n(t) - d(t)\}/n(t)$, and the survival function is estimated as the product of the daily survival rates. Records for individual patients can be bootstrap resampled for estimating the bias, standard error and confidence limits for $\hat{S}(t)$, for $t = 1, 2, \ldots$. Further developments and discussions of bootstrapping in this context are provided by Efron (1981b), Reid (1981) and Akritas (1986). In particular, Akritas (1986) discusses how Efron's (1981b) method of bootstrap sampling can be used to construct confidence bands for a survival function, i.e. bands within which the entire survival curve lies with a certain specified level of confidence.

More recently, Flint *et al.* (1995) have discussed the use of bootstrapping with the Kaplan–Meier estimator of the juvenile survival of young waterfowl. In this case the survival may not be independent for the juveniles within one brood, and there may be movement of the juveniles between broods. Flint *et al.* therefore suggested that the broods are resampled rather than the individual juveniles. These authors also considered the situation where all broods are not observed at equally spaced times. Then the Mayfield (1961, 1975) estimator can be used. Problems related to non-independence of juveniles within broods also occur with growth studies, as illustrated in Example 13.2, below.

A popular approach for the analysis of survival data at the present time involves the use of Cox's (1972) proportional hazards model. With this model it is assumed that the probability of surviving until at least age t for an individual that is described by the values $x_1, x_2, \ldots, x_p$ for variables $X_1, X_2, \ldots X_p$ is given by

$$S(t; x_1, x_2, \ldots x_p) = \exp\{-\lambda(t)\exp(\beta_0 + \beta_1 x_1 + \ldots + \beta_p x_p)\}, \qquad (13.2)$$

where $\lambda(t)$ is an unknown, positive, non-decreasing function of t and $\beta_1, \beta_2, \ldots \beta_p$ are parameters that can be estimated by maximum likelihood.

Bootstrapping in the context of Cox's model has been discussed by Burr (1994). She considered three resampling algorithms:

1. Bootstrap resample the records for the n individuals in the study.
2. Generate a random survival time Y_i for the ith individual in the study using the estimate of the survival function (13.2). Generate a random censoring time C_i for the ith individual from the Kaplan–Meier estimate of the censoring time distribution. Decide whether the ith individual is observed to die or is censored before this. Use the resulting data for all n individuals as a bootstrap sample.
3. Generate a random survival time Y_i for the ith individual. If this individual is censored in the original data set then use the same censoring time in the bootstrap sample. Otherwise, generate a random censoring time from the Kaplan–Meier estimate of the censoring time distribution conditional on this being greater than the observed sample time. Use the resulting data for all individuals as a bootstrap sample.

The differences between these algorithms are related to the extent to which the bootstrap samples reflect the original data. Algorithm 1 allows the distributions of the X variables and the censoring pattern to change. Algorithm 2 retains the values of the X variables for each individual, but allows the censoring pattern to change. Algorithm 3 retains the X values and also the censoring pattern as far as possible for each individual.

From a simulation study, Burr concluded that algorithm 3 should be avoided because of the erratic results that it produced. Algorithm 1 was unreliable for obtaining confidence limits for β parameters, but worked quite well with confidence limits for $S(t)$, and the time corresponding to a survival probability of 0.5. Algorithm 2 was recommended overall, with the percentile method of equation (3.4) for β parameters, and the percentile method of equation (3.3) for $S(t)$ and the time corresponding to a survival probability of 0.5.

Cox's proportional hazard model is a special case of a general class of generalized linear models that can be written in the form

$$Y = f(\beta_0 + \beta_1 X_1 + \ldots + \beta_p X_p) + \varepsilon,$$

where Y is an observation with expected value $\mu = f(\beta_0 + \beta_1 X_1 + \ldots + \beta_p X_p)$, and ε is an error term from a specified distribution, adjusted to have a mean of zero. Many commonly used models for survival data fall within this general class (McCullagh and Nelder, 1989, Chapter 13). In the absence of complications due to censoring, bootstrapping of data from these models can either involve resampling the individual cases, generating data from the fitted model retaining the same values of the X for the cases as in the original data, or by the resampling of residuals.

The resampling of residuals from generalized linear model is complicated by the fact that the residuals do not have a constant variance. However, this can be allowed for, as discussed by Moulton and Zeger (1991) and Hjorth (1994, p. 195). In practice it is simplest to resample the original cases whenever this is reasonable because the X values for these cases were not somehow fixed by the process used to generate the data.

13.2 BOOTSTRAPPING FOR VARIABLE SELECTION

A second situation that occurs with survival data is where it is believed that the probability of survival for a given amount of time for an individual may be a function of certain covariates $X_1, X_2, \ldots X_p$ that are measured on that individual. There is then interest in procedures for selecting the variables that are important. The following example suggests the type of approach that can be used.

Example 13.1: predicting the survival of liver patients   Diaconis and Efron (1983) used the variable selection problem as one of their examples in a popular article on the bootstrap method. This example, which is also discussed by Efron and Gong (1983), involved a group of 155 patients with acute and chronic hepatitis, of which 33 died and 122 survived. There were 19 variables measured on each patient, and an important question concerned which of these variables, if any, were good predictors of survival.

The analysis began with a three-part screening process for the variables in which the 19 variables were examined one at a time by fitting logistic regression equations of the form

$$\pi(x_j) = \exp(\beta_0 + \beta_1 x_j)/\{1 + \exp(\beta_0 + \beta_1 x_j)\}$$

where $\pi(x_j)$ is the probability of a patient dying as a function of the jth variable (Manly, 1994a, section 8.10). Variables were only retained if the estimated value of β_1 was significantly different from zero at the 5% level. This removed six variables. Next, the remaining 13 variables were selected one at a time to be added to a multiple logistic regression equation for predicting death until the improvement of fit that was obtained by adding another variable was not significantly large at the 10% level for any of the variables not already included. This removed eight more variables. Finally, a stepwise multiple logistic regression was run again on the five remaining variables, but requiring a significance level of 5% for any variable added. This removed another variable.

After the screening process which removed all except four variables had been carried out, the final fitted logistic equation was used to predict survival or death for each patient. Because there were 33 deaths out of 155 patients, the prediction was death for all patients with an estimated probability of death of more than $33/155 = 0.213$. There were complications with missing data for the four predictor variables for 22 of the patients. Of the remaining 133 patients for which the prediction rule could be used there were 21 wrong predictions, giving an apparent error rate of $21/133$, or 15.8%.

The question that Efron and his colleagues addressed by bootstrapping concerned the amount of bias that was likely to be involved in the apparent error rate, taking into account both the initial screening of variables and the fact that the logistic regression equation was chosen to give the best fit to the data.

The process that they used consisted of repeating all steps in the analysis on bootstrap samples. In this way they estimated that the apparent error rate was probably about 4.5% too low, giving a true error rate of about $15.8\% + 4.5\% = 20.3\%$. Furthermore, they discovered that the bootstrap samples displayed surprisingly large variation in the variables that remained after the initial screening. If nothing else, this demonstrated very clearly the very large element of chance involved in the choice of variables, suggesting that the four variables that were selected with the original set of data should not be taken too seriously.

This use of bootstrapping with survival data is similar to the application with stepwise regression that has been mentioned in section 8.8, where the process is carried out many times with the Y values in a random order to assess the probability of obtaining a 'significant' regression by chance alone. This suggests the use of randomization with the liver patient data to test whether the predictive power of the final fitted logistic regression equation is significantly better than expected by chance, which can be done by comparing the number of correct predictions with the distribution of the number of correct predictions obtained when 33 patients are randomly chosen to be the survivors.

13.3   BOOTSTRAPPING FOR MODEL SELECTION

Another situation where bootstrapping can assist concerns the selection of a model for survival data when the alternatives being considered do not have the property of being nested. A series of models $M_1, M_2, \ldots, M_q$ are nested if model i is obtained by setting some of the parameters in model M_{i+1} equal to zero. The models are often fitted by maximum likelihood, in which case standard theory allows a test for whether model M_{i+1} fits the data significantly better than model M_i (McCullagh and Nelder, 1989) by finding the reduction in the deviance that is obtained by adding the extra parameters in model M_{i+1} into the model, and comparing this value with the chi-squared distribution. The 'deviance' referred to here is minus twice the maximized log-likelihood.

The comparison of non-nested models is more complicated. Wahrendorf *et al.* (1987) suggested that the fit of two non-nested models with the same number of parameters can be examined by testing whether the difference in their deviances is significantly different from zero, using bootstrapping to generate the null distribution for the test statistic. They considered two examples. One concerned a dose–response experiment on carcinogenesis, where seven groups of mice were given different doses of a drug and the number of tumour-free days was observed for each mouse. The problem was to decide which function of the dose fitted the data best. For this example Wahrendorf *et al.* used non-parametric bootstrapping to generate a distribution for the deviance difference, with resampling of the records for individual mice within each group.

The second example that was used by Wahrendorf *et al.* involved the comparison of two models for the death rates of British male doctors from coronary heart disease, using age and smoking status as explanatory variables. They used parametric bootstrapping in this case, with bootstrap sets of data generated by using observed sample counts in different categories as mean values for random values from Poisson distributions.

Subsequently, Cole and McDonald (1989) developed this type of application of bootstrap testing to the choice of the link function to use in a generalized linear model, i.e. the function which determines how the probability of an event is related to a linear combination of the variables that are being used to account for this probability. They used as an example the age at menarche for girls in north-east England related to the age and the number of siblings that the girls had. Non-parametric bootstrapping was used, with resampling of the observations for individual girls.

Hall and Wilson (1991) have argued that these uses of bootstrap testing for whether one model fits better than another violate one of the important principles of such tests, because the bootstrapped sets of data are not generated with the null hypothesis true. The result is then that the proposed tests have very little power. They suggested that if two models are to be compared then one appropriate approach requires two bootstrap tests to be carried out. First, bootstrap sets of data are generated with model 1 true to see whether the goodness of fit of this model is satisfactory (i.e. the deviance is not significantly large). Next, bootstrap sets of data are generated with model 2 true to see whether the goodness of fit of this model is satisfactory. If one model gives a satisfactory fit but the other does not then the choice of model is clear. It would also be interesting to test the goodness of fit of model 2 using data generated from model 1 and vice versa in order to see how well the wrong model fits. See also Becher (1993) and Schork (1993).

13.4 GROUP COMPARISONS

A problem that often occurs with survival data is the comparison of the survival rates for several different groups to see whether these are significantly different, possibly with an allowance for the measured effects of certain other variables. This can be handled by including variables that allow for group differences in a parametric model, as discussed in section 13.1 in terms of the proportional hazards model of equation (13.2), and finding confidence limits for the coefficients of these variables by bootstrapping.

When the problem is simply to test whether the survival rate is significantly different for two or more groups of individuals, without the complication of any allowance for the effects of potential confounding variables, it seems reasonable to use a randomization test. This is the recommendation of Flint *et al.* (1995) in their discussion of the estimation of the juvenile survival of waterfowl with an allowance for brood mixing and dependent survival within broods. They define a certain test statistic D^2 that measures the overall survival difference between two groups of broods, and then test whether this is significantly large in comparison with the distribution of D^2 values that is generated by allocating broods at random to the two groups, keeping the number of broods in each group constant.

13.5 GROWTH DATA

Growth data are similar to survival data in the sense that they are often based on repeated observations on individuals, with the possibility of missing data because these individuals leave the study. The use of randomization methods in this and related areas is discussed by Zerbe (1979a,b), Zerbe and Walker (1977), Foutz *et al.* (1985) and Zerbe and Murphy (1986), Raz (1989) and Raz and Fein (1992). See also the paper by Moulton and Zeger (1989) on the use of bootstrapping in this context. The following two examples illustrate some of the potential of these methods.

Example 13.2: growth of Pigeon Guillemot chicks   As part of the study of the effects of the *Exxon Valdez* oil spill in Prince William Sound in 1989, the growth of Pigeon Guillemot chicks was studied on Naked Island and Jackpot Island in 1994 (Hayes, 1995). Parts of Naked Island were oiled, but none of Jackpot Island was. Also, the diet of the chicks was quite different on the two islands. A question of some interest therefore concerned whether the growth of the birds was the same on both islands. This can be tested by randomization. In addition, bootstrapping can be used to assess the level of sampling errors in estimated growth curves.

Nests containing one or two chicks were sampled over the summer. Because few of these chicks were of known age when they were first encountered, the wing length in millimetres was used as a measure of age. The weight of the chicks in grams was therefore related to the wing length. Figure 13.1 shows plots of logarithms of weight against logarithms of wing length separately for the two islands. Various relationships between these two variables were explored, and a quadratic relationship involving logarithms was chosen as a reasonable model for both islands. The growth curves shown in the figure are

$$\log_e(\text{weight}) = -6.548 + 4.660 \log_e(\text{wing length}) - 0.422\{\log_e(\text{wing length})\}^2,$$

for Jackpot Island, and

$$\log_e(\text{weight}) = -3.932 + 3.554 \log_e(\text{wing length}) - 0.309\{\log_e(\text{wing length})\}^2,$$

for Naked Island, where these were both fitted to the data by ordinary regression methods.

A complication with using bootstrapping to assess the level of sampling errors in these fitted curves is due to the correlation that can be expected between observations taken on the same brood. This occurs because two chicks from the same nest can be expected to provide relatively similar observations and in addition there are multiple observations on individual chicks because the nests were resampled several times. Under these circumstances it seems best to regard the individual nests as the sample units, and to bootstrap by resampling nests, in the same way as was suggested by Flint *et al.* (1995) for survival estimation. For example, there were 17 sampled nests on Naked Island, with corresponding sets of measurements. A bootstrap sample for this island is therefore obtained by

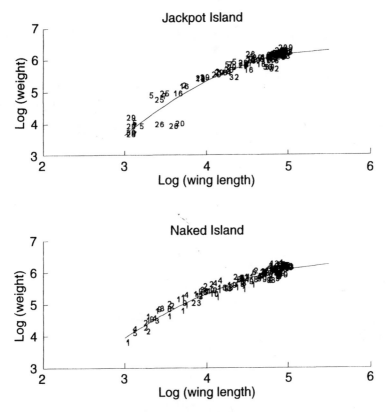

Figure 13.1 Body mass (g) related to wing length (mm) for pigeon guillemots on Jackpot and Naked Islands in Prince William Sound, Alaska. What is plotted is the nest number, from 1 to 17 on Jackpot Island (95 observations in total), and from 1 to 24 on Naked Island (144 observations in total). The multiple observations come from one or two chicks in a nest, with several measurement times for each chick. See the text for a description of the fitted growth curves.

randomly selecting 17 nests, with replacement, from the original 17 nests. All of the data from each selected nest is then included for the estimation of a bootstrap growth curve.

Figure 13.2 shows Efron's percentile limits for the fitted growth curves, as obtained by bootstrapping in the manner just described, using equation (3.3) to calculate the limits. The number of bootstrap samples used was 5000, and the confidence level used was 99%, where this level was chosen because the limits contain about 95% of all the growth curves estimated from the bootstrapped data. In a sense the limits therefore give a 95% confidence interval for the entire growth curve.

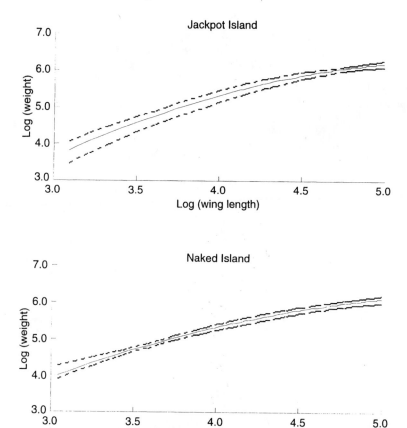

Figure 13.2 Bootstrap 99% percentile limits for the estimated growth curves for Jackpot and Naked Islands, based on 5000 bootstrap samples.

When Hall's 99% percentile limits were calculated using equation (3.4) they were found to be quite similar to Efron's percentile limits. Of course, the various other types of confidence limits that have been discussed in Chapter 3 could also have been used. It would be nice to know which of these various methods is best, particularly in terms of the problem of finding limits such that there is $100(1 - \alpha)$% confidence that a true growth curve lies entirely within them. However, this requires a study that has apparently not yet been carried out.

A randomization method for comparing the growth curves on the two islands can be based on the following algorithm.

Algorithm for comparing growth curves

1. Fit growth curves of the form

$$\log(\text{weight}) = b_0 + b_1 \log(\text{length}) + b_2 \{\log(\text{length})\}^2$$

separately to the data for each island. Also, fit the same curve to the combined data. Hence, calculate F_1, the F-statistic for the extra sum of squares accounted for by estimating separate equations rather than a combined equation, as discussed in section 8.5.
2. Take the 41 nests found on both islands and randomly assign 17 of these to Jackpot Island and the remaining 24 to Naked Island, to match the numbers in the original data. Allocate all of the data to the two islands according to the nests that they are associated with and calculate F_2, the F-statistic for the extra sum of squares accounted for by separate estimation of the growth functions.
3. Repeat step 2 a large number of times to generate further statistics $F_3, F_4, \ldots$ and declare F_1 to be significantly high at the $100\alpha\%$ level if it is among the largest $100\alpha\%$ of the set of F-statistics consisting of F_1 and all the randomized values.

This algorithm tests for a significant improvement in the fit of the growth curve model when estimated for the islands separately rather than combined, assuming that the model is reasonable in the first place.

When a test based on the above algorithm was run with 4999 randomizations, the extra sum of squares for the observed set of data was found to be significantly large at the 1.8% level of significance. There is therefore clear evidence of a difference in the growth curves for the two islands. According to the growth curves fitted for the islands separately, the chicks on Jackpot Island (unoiled) had lower weights than those on Naked Island (partly oiled) for short wing lengths. However, this was reversed for longer wing lengths. The extent to which these differences can be attributed to oiling is unclear.

The calculations for the bootstrap confidence limits shown in Figure 13.2 were carried out in a Fortran program written for the purpose. However, the randomization test was conveniently carried out in the spreadsheet Lotus 1-2-3 (Lotus Development Corporation, 1991). This needed two small macros to be written to randomize the data and to repeat the randomization 4999 times, but was otherwise a straightforward use of the Lotus regression and sorting facilities.

Example 13.3: growth and the sudden infant death syndrome   Williams *et al.* (1996) describe a comparison between growth curves for infants who died of the sudden infant death syndrome (SIDS) and randomly selected controls, using data from the New Zealand Cot Death Study that was set up in 1987 to explore the relationship between SIDS and possible risk factors (Mitchell *et al.*, 1991, 1992). The case–control study was carried out because in the past it has been suggested that some cases of SIDS are preceded by a period of low weight gains, and that careful monitoring of growth could prevent these deaths. However, the evidence for this claim has been equivocal.

Data were available for the birthweight and the weight at at least one later time for 309 cases (infants who died of SIDS) and 1491 controls. Separate growth curves were fitted for the four groups (male cases, female cases, male controls and female controls) using cubic splines (Efron and Tibshirani, 1993, p. 258), with the idea that the fitted curve for a group represents the growth for an average individual in the group. Bootstrapping was used to calculate confidence limits for the true mean growth curves, with 2500 resamplings of individuals with all of their data, rather than the resampling of the single data points. The procedure used was similar to that described in the previous example, with the idea of producing 95% limits in the sense that there is 95% confidence that a true average growth curve is entirely within the limits.

To compare the growth of cases and controls, growth 'velocities' were calculated from the fitted curves as the average change in weight per week, over a four week period. These velocities were then compared for cases and controls using a restricted randomization test, with separate tests for males and females. The procedure used involved dividing the male cases into four groups on the basis of race and whether they were bottle-fed (Maori bottle-fed, Maori not bottle-fed, non-Maori bottle-fed, non-Maori not bottle-fed). A similar grouping was used with the female controls. The difference between cases and controls in the growth velocity for a four-week period was then compared with the distribution of the difference obtained from 999 randomizations between cases and controls, maintaining the numbers in the race–bottle-fed classes equal to those in the real data.

The restricted randomization was used because of the belief that growth differences may exist related to race and bottle feeding. It was important that these differences should not affect the randomization test because it was desired to detect differences between cases and controls after allowing for race and bottle feeding effects.

Most of the randomization tests did not give a significant difference between cases and controls, and Williams *et al.* (1996) concluded that any average differences that might exist are too small to be of practical importance. Furthermore, weight gains for both groups were close to New Zealand norms for the first few months of life.

As was the case with the previous example, the complicated structure of the data with varying numbers of observations on individuals and confounding factors makes an analysis using conventional statistical methods quite complicated. By contrast, bootstrapping for obtaining confidence limits and randomization methods of testing are in principle straightforward. Nevertheless, it must be admitted that the bootstrap confidence intervals require further research before they can be used without reservations in studies of this type.

Finally, it can be noted that the idea of bootstrap resampling of individuals with all their associated data is not new. For example, Hjorth (1994, p. 250) discusses a similar procedure in a study of the relationship between two measures taken on human backbones involving 1083 observations on 94 individuals.

13.6   FURTHER READING

An approach to the analysis of survival data that involves bootstrapping and randomization which has not been mentioned so far in this chapter is through the development of survival trees (LeBlanc and Crowley, 1993). With this method the aim is to take a set of survival data with censoring and to divide the individuals involved into a number of distinct groups based on the values that they possess for variables $X_1, X_2, \ldots X_p$. For example, LeBlanc and Crowley (1993) started with a sample of 704 patients, of which approximately 14% had censored survival times. These patients were then divided into seven groups with median survival times varying from 5.8 months to 18.5 months, on the basis of a survival tree algorithm which involves splitting and combining groups according to specified rules. Bootstrapping was used as an aid in choosing the final number of groups by minimizing the expected number of misclassification errors. Randomization was used to determine whether there is a significant difference between the two parts of a group being considered for a split.

LeBlanc and Crowley's method is a development of the classification and regression tree (CART) methodology developed by Breiman *et al.* (1984). This is a type of computer-intensive method for producing groups of similar cases that can be used for many types of data other than those related to survival. For a further discussion of this methodology, and the role that bootstrapping can play, see Efron and Tibshirani (1993, Chapter 17).

Cubic spline curve-fitting was mentioned in Example 13.3 as the method used to relate the average weight of infants to their age. In that example bootstrapping was used for finding confidence limits for the true population curve, but not for choosing the amount of smoothing required. This latter application of bootstrapping is discussed by Efron and Tibshirani (1993, p. 258) in terms of attempting to minimize the error involved in predicting new values of the dependent variable.

Other applications of bootstrapping that have been proposed in relationship to survival studies are for comparing the survival of two groups when it is thought that one of these groups may initially have higher survival but this changes to lower survival later (O'Quigley and Pessione, 1991); finding confidence limits for the linear relative risk form of Cox's (1972) proportional hazards model and the linear hazards function model (Barlow and Sun, 1989; Aalen, 1993); determining the accuracy of life table functions (Golbeck, 1992); finding confidence bands for the median survival time as a function of covariates in Cox's model (Burr and Doss, 1993); finding confidence limits for a specific occurrence/exposure rate such as the probability of death due to cancer divided by the average lifetime of the individuals exposed (Babu *et al.*, 1992); finding confidence intervals for the ratio of such occurrence/exposure rates (Tu and Gross, 1995); and testing the goodness of fit of the Cox proportional hazards model (Burke and Yuen, 1995).

13.7 CHAPTER SUMMARY

- Bootstrapping has been proposed as a solution for various problems associated with survival data, including the determination of confidence limits for the survival function based on the Kaplan–Meier estimator, Cox's proportional hazards model and generalized linear models. Various alternative methods of resampling are discussed, with and without censoring.
- Bootstrapping has also been proposed for evaluating variable selection methods, the comparison of non-nested models, and the choice of the link function for a generalized linear model for survival data
- The survival of individuals in several groups can be compared either by using bootstrapping to test the significance of covariates that allow for group differences in a parametric model or by using randomization tests.
- Data on the growth of individuals is similar in many ways to survival data. The use of bootstrapping to assess the accuracy of growth curves and randomization tests to compare the growth of different groups is illustrated with an example comparing Pigeon Guillemot chicks on two islands in Prince William Sound, Alaska. A second example compares the growth of infants who died of the sudden infant death syndrome with normal infants.

• Various extensions and generalizations of the methods discussed in the chapter are briefly reviewed, including bootstrapping and randomization with the development of survival trees.

13.8 EXERCISES

The following exercises can be carried out using Lotus 1-2-3 (Lotus Development Corporation, 1991) although the spreadsheets will be quite complicated to set up. The alternative will be to write a special computer program, because software to do the calculations does not seem to be readily available.

13.1 Table 13.1 shows the results of a study in which 58 mule deer fawns were fitted with radio collars in the Piceance Basin in Colorado in December 1982. Ignoring the information on sex, estimate the Kaplan–Meier survival function using equation (13.1), regarding day 1 as the starting time for all the animals. Note that $n(t)$ increases when new fawns were fitted with radio collars on day t but otherwise equation (13.1) applies unchanged. Calculate 99% confidence limits based on (i) the standard bootstrap confidence limits of equation (3.1), (ii) the percentile confidence limits of equations (3.3), and (iii) the percentile limits of equation (3.4). (That is to say, calculate 99% limits for survival each day for from 1 to 191 days.) For each of the three methods for calculating confidence limits, estimate the coverage for the full survival curve, i.e. the probability that the true survival curve is completely within the stated limits. Choosing one of the sets of limits, plot the estimated survival function and the confidence limits for the true survival function.

13.2 Using the data from Table 13.1, carry out a randomization test to compare the survival to 50, 100 and 150 days for male and female mule deer fawns. There are various ways that this can be done. For example, the survival rates can be estimated using the Kaplan–Meier method and the observed differences compared with their randomization distributions.

Table 13.1 The results from a study in which 58 mule deer fawns were fitted with radio collars in the Piceance Basin in Colorado in December 1982, as given in Figure 9.1 of White and Garrott (1990), but with some simplification. Definitions: Animal, an arbitrary number; Sex, 0 for female, 1 for male; Entry day, the day when a collar was fitted; Death day, the day on which death was recorded for animals that survived; Exit day, the day on which radio failure occurred; and Days survived, the observed days from the entry day until death, radio failure or the end of the study on 15 June 1993

Animal	Sex	Entry day	Death day	Exit day	Days survived
1	0	1			191
2	0	1			191
3	0	1	115		114
4	0	1	68		67
5	0	1			191
6	0	1	13		12
7	1	1	12		11
8	1	1	95		94
9	1	1	136		135
10	1	1	88		87
11	1	1	96		95
12	1	1	73		72
13	0	2	100		98
14	0	2			190
15	0	2			190
16	1	2			190
17	1	2	131		129
18	1	2	39		37
19	1	2			190
20	1	2	22		20
21	1	2	101		99
22	1	2		161	159
23	0	3	57		54
24	0	3			189
25	0	3	115		112
26	0	3			189
27	1	3	13		10
28	1	3	143		140
29	1	3	11		8
30	1	3	10		7
31	0	4			188
32	0	4			188
33	0	4	68		64
34	0	4	110		106
35	0	4			188

Table 13.1 (*Cont.*)

Animal	Sex	Entry day	Death day	Exit day	Days survived
36	0	4			188
37	0	4	40		36
38	0	4	18		14
39	0	4			188
40	1	4	122		118
41	1	4			188
42	1	4	92		88
43	1	4	107		103
44	1	4	86		82
45	1	4			188
46	1	4	120		116
47	1	4		158	154
48	1	4	32		28
49	1	4			188
50	1	5			187
51	1	5	11		6
52	1	5			187
53	1	5	114		109
54	1	5	96		91
55	1	5	102		97
56	1	5	11		6
57	0	6	124		118
58	1	6	107		101

14

Non-standard situations

14.1 THE CONSTRUCTION OF TESTS IN NON-STANDARD SITUATIONS

In the previous chapters the various situations that have been discussed are more or less standard, and in most cases there are alternatives to using computer-intensive methods. However, one of the most useful things about computer-intensive methods is that they can often be devised for analysing data that do not fit into any of the usual categories. There is therefore some value in studying some unusual situations where this approach is valuable. Four such situations are considered in some detail in this chapter, and a number of others are briefly mentioned. The four situations considered in detail concern the Monte Carlo testing of whether species on islands occur randomly with respect to each other, randomization and bootstrap testing for changes in niche overlap with time for two sizes of lizard, the probing of multivariate data with random 'skewers', and the relationship between the size of ant species in Europe and the latitudes at which these species occur.

14.2 SPECIES CO-OCCURRENCES ON ISLANDS

There has been a considerable controversy in recent years concerning whether the effects of competition can be seen in records of species occurrences on islands. Questions concern whether when species A is present on an island, species B tends to be absent or, alternatively, whether species A and B tend to occur together.

The data to be considered can be represented by an array of zeros and ones, of the form shown in Table 14.1. This example is too small for a meaningful test, but it does allow the discussion of what the important considerations with data of this type are. Such an array of zeros and ones will be referred to as an occurrence matrix.

Notice that a row containing all zeros is not possible in the occurrence matrix. There may be species that could have been present but in fact are not, but the recording process tells us nothing about these. On the other hand, there could be small islands on which there are no species. In principle these can be recorded, although the definition of an 'island' may

Table 14.1 An example of presence-absence data, where a '1' indicates the presence and '0' the absence of a species on an island

Species	Island					Total
	1	2	3	4	5	
1	1	1	1	0	1	4
2	1	1	0	0	1	3
3	1	1	1	0	0	3
4	1	1	1	0	0	3
5	1	0	0	0	0	1
6	1	1	1	1	0	4
7	1	1	0	0	0	2
Total	7	6	4	1	2	20

cause some difficulties, and in practice only islands that are large enough to contain species are considered. Therefore, the only occurrence matrices that are possible are those with at least one occurrence in each row and each column.

As a rule some species are more common than others. Similarly, even islands that have the same area will usually be ecologically different, so that there is no reason to believe that they should have space for the same number of species, although in general the number of species can be expected to increase with island area (Williamson, 1985; Wilson, 1988). Therefore, the hypothesis that all species are equally likely to occur on any island is almost certainly false, and is not a useful null hypothesis. For this reason, Wilson (1987) suggested testing observed data by comparison with 'random' data subject to the constraints that the number of species on islands and the number of times each species occurs are exactly the same as those observed. This seems sensible, and will be accepted here, although other authors have assumed all islands equivalent (Wright and Biehl, 1982), that islands and species have occurrence probabilities (Diamond and Gilpin, 1982; Gilpin and Diamond, 1982), or that species can only occur for islands of a certain size (Connor and Simberloff, 1979). It has already been mentioned that this whole area is controversial, and this is reflected in the contents of the papers referenced above. Other relevant references on the same topic are Simberloff and Connor (1981), Connor and Simberloff (1983), Gilpin and Diamond (1984, 1987) and Jackson et al. (1992). More recent contributions have been concerned with the fitting and interpretation of a logistic model for the probability that species i is

Table 14.2 A nested occurrence matrix

Species	Island				Total
	1	2	3	4	
1	1	1	1	1	4
2	1	1	1	0	3
3	1	1	0	0	2
4	1	0	0	0	1
Total	4	3	2	1	10

on island k (Ryti and Gilpin, 1987), the use of new test statistics designed to detect competitive exclusion and species aggregation (Wilson, 1988; Roberts and Stone, 1990; Stone and Roberts, 1990, 1992; Wilson et al., 1992a; Wright and Reeves, 1992), and the examination of the relationship between species co-occurrences and their morphological similarity (Winston, 1995).

Fixing of row and column totals of an occurrence matrix does not by any means indicate what a 'random' occurrence matrix might be. However, fixing these totals can mean that there are very few or no alternatives to the patterns of zeros and ones that are observed. For example, Table 14.2 is a completely 'nested' matrix, where the islands and species can both be put in order in terms of their numbers of occurrences. This is the only possible matrix with these row and column totals. It therefore has no 'random' matrices to be compared with. In contrast, Table 14.3 is a 'chequerboard' matrix. This is not the only matrix with these row and column totals, but all the other possibilities can be obtained by just

Table 14.3 A chequerboard occurrence matrix

Species	Island				Total
	1	2	3	4	
1	1	0	1	0	2
2	0	1	0	1	2
3	1	0	1	0	2
4	0	1	0	1	2
Total	2	2	2	2	2

Table 14.4 Positions in an occurrence matrix that can be zero or one (indicated by question marks) when the row and column totals are fixed

Species	Island					Total
	1	2	3	4	5	
1	1	1	?	?	?	4
2	1	1	?	?	?	3
3	1	1	?	?	?	3
4	1	1	?	?	?	3
5	1	0	0	0	0	1
6	1	1	?	?	?	4
7	1	1	0	0	0	2
Total	7	6	4	1	2	20

reordering the rows and columns given here, with no effect on the pattern of co-occurrences of species.

With realistic data, the situation will usually be more promising than is indicated by the last two examples. Thus, consider again the 7×5 occurrence matrix in Table 14.1. Here, insisting on the row and column totals fixes some, but not all, of the zeros and ones, as indicated in Table 14.4. The first column has to be all 1s to give a total of 7. Row 5 is then fixed, and this fixes column 2 and then row 7. The positions indicated by question marks can then be either 0s or 1s.

Suppose now that a 'random' matrix is required in the sense that the remaining occurrences of species i (if any) are independent of the occurrences of the other species. This can be achieved by adding occurrences one at a time, taking into account the number of occurrences left of species and islands. For example, species 1 has two occurrences left, which must be on the islands 3, 4 and 5. The islands have 4, 1 and 2 spaces left, so one occurrence of species 1 can be allocated to island 3 with probability 4/7, island 4 with probability 1/7 and island 5 with probability 2/7. This allocation can be done by seeing whether a random integer in the range 1 to 7 is in the range 1 to 4, is equal to 5, or is 6 or 7.

Notice that if the random choice is to put species 1 on island 4, then the remaining elements in column 4 are all necessarily 0. Generally, whenever a random occurrence is entered into the occurrence matrix all the rows and columns must be checked to see if there are any other 0s and 1s that are thereby fixed. Failure to carry out this 'fill in' step appears to be one explanation for the 'hang-ups' in Connor and Simberloff's (1979) randomization procedure, whereby it becomes impossible to continue allocating species to islands without putting a species on an island more than once.

One algorithm to produce a random matrix therefore consists of the repetition of two steps. First, a species is put on one of the islands that it is not already on, with the probability of going to an island being proportional to the number of remaining spaces on that island. Second, any necessary 0s and 1s that are a consequence of the allocation are filled in the occurrence matrix. Another species is then allocated, necessary 0s and 1s are filled in, and so on. This algorithm seems to 'work' in the sense of not running out of spaces on islands before all species are allocated, at least when species are allocated in order of their abundance. That is to say, the species are ordered according to their total number of occurrences from greatest to least. The most abundant species is then allocated to islands, followed by the second most abundant (or equally abundant) species, and so on.

This is certainly not the only possible algorithm. One alternative is to fill up islands in the order of decreasing size, with species being chosen with probabilities proportional to the numbers of remaining occurrences. Numerical results suggest that this produces similar results to the algorithm just described (see below).

Other algorithms that could be used have been suggested by Wormald (1984) in the context of generating random graphs for graph theory problems, by Wilson (1987) in the context of the randomly allocating species to islands, by Vassiliou et al. (1989) in the context of testing the significance of clusters in cluster analysis on presence-absence data and by Snijders (1991) in the context of studying human social networks. All of the proposed algorithms produce 'random' occurrence matrices where there is no association between species because the probability of allocating a species to an island depends on the number of remaining spaces but not the species already present. It is most unlikely that any of them reflect at all closely the colonization process that took place to produce the real data. Nevertheless, they can be used as null models that will serve for assessing non-randomness in real data. It is quite conceivable that some of these algorithms will produce different distributions of occurrence matrices.

One of the sets of data that has been used in the controversy over how to detect competition between species on islands concerns 56 bird species on 28 islands of Vanuatu (the New Hebrides before 1980). The basic data matrix is shown in Table 14.5. For these data the distribution of the frequencies of co-occurrences of the 56 species that is obtained by the random allocation of species to islands was approximated by 999 of these allocations, plus the observed allocation. The algorithm used for random allocations was the first mentioned above, with allocations being from the most abundant to the least abundant species. The observed allocation was included with the simulated ones on the grounds of the null hypothesis

Table 14.5 The locations of 56 bird species on 28 islands of Vanuatu (1 = species present, 0 = species absent). The data in this form were kindly supplied by J.B. Wilson

Species	1	2	3	4	5	6	7	8	9	10	11	12	13	14	15	16	17	18	19	20	21	22	23	24	25	26	27	28
1	1	1	1	1	1	1	1	1	1	1	1	1	1	1	1	1	1	1	1	1	1	1	1	1	1	1	1	1
2	1	1	1	1	1	1	1	1	1	1	1	1	1	1	1	1	1	1	1	1	1	1	1	1	1	1	1	1
3	1	1	1	1	1	1	1	1	1	1	1	1	1	1	1	1	1	1	1	1	1	1	1	1	1	1	1	1
4	1	1	1	1	1	1	1	1	1	1	1	1	1	1	1	1	1	1	1	1	1	1	1	1	1	1	1	0
5	1	1	0	1	1	1	1	1	1	1	1	1	1	1	1	1	1	1	1	1	1	1	1	1	0	1	1	1
6	1	1	1	1	1	1	1	1	1	1	1	1	1	1	1	1	1	1	1	1	1	1	1	1	1	0	1	0
7	1	1	1	1	1	1	1	1	1	1	1	1	0	1	1	1	1	1	1	1	1	1	1	0	1	1	1	1
8	1	0	1	1	1	1	1	0	1	1	1	1	1	1	1	1	1	1	1	1	1	1	1	1	1	1	1	1
9	1	1	1	1	1	1	1	1	1	0	1	1	1	1	1	1	1	1	1	0	1	1	1	1	1	1	1	1
10	1	1	1	1	1	1	1	1	1	1	1	1	1	1	1	1	1	1	0	1	1	1	0	1	1	1	1	1
11	1	1	1	1	1	1	1	0	1	0	1	1	1	1	1	1	1	1	1	1	1	1	0	1	1	1	1	1
12	1	1	1	1	1	1	1	0	1	1	1	1	0	1	1	1	1	1	1	1	1	1	0	1	1	1	1	1
13	1	1	1	1	1	1	1	1	0	1	1	1	1	1	1	1	1	1	1	1	1	1	0	1	1	1	1	0
14	1	1	0	1	1	1	1	1	1	1	1	1	0	1	1	1	1	1	1	0	1	1	1	1	1	1	1	1
15	1	1	1	1	1	1	1	1	0	1	1	1	1	0	1	1	1	1	1	1	1	1	0	1	1	1	1	0
16	1	1	0	1	1	1	1	1	1	1	1	0	0	1	1	1	1	1	1	1	1	1	1	0	1	1	1	1
17	1	1	1	1	1	1	1	1	1	0	1	1	0	1	1	1	1	1	0	1	1	1	1	0	1	1	1	1
18	1	1	0	1	1	1	1	0	1	1	1	1	0	1	1	1	1	1	1	1	1	1	1	1	0	0	1	1
19	1	1	0	1	1	1	1	0	1	1	1	1	0	1	1	1	1	1	0	1	1	1	1	1	0	0	1	0
20	1	1	1	1	1	1	1	1	0	0	1	1	0	1	1	1	1	1	1	0	1	1	1	0	0	1	1	0
21	1	1	0	1	1	1	1	0	1	1	1	1	0	0	1	1	1	1	1	1	1	1	0	1	0	1	1	0
22	1	0	0	1	1	1	1	1	1	1	1	1	0	1	1	1	1	1	1	1	1	1	0	0	0	0	1	0
23	1	0	0	1	1	1	0	0	1	1	1	1	0	0	1	1	1	1	1	1	1	0	1	1	0	1	1	1
24	1	1	0	1	1	1	1	0	1	0	1	1	1	0	1	1	1	1	1	0	1	1	1	0	1	0	1	0
25	0	0	0	1	1	1	1	0	1	1	1	0	0	1	1	1	0	1	1	1	1	0	1	0	1	1	1	1
26	1	1	0	1	1	1	1	0	1	1	1	1	1	0	1	1	1	1	0	1	1	1	0	0	0	0	0	0
27	1	0	1	1	1	1	1	1	1	1	1	0	0	0	1	0	1	0	0	1	1	1	0	0	0	1	1	1
28	0	1	0	1	1	1	1	1	0	0	1	1	1	0	1	1	1	1	1	0	1	1	1	1	0	0	0	0
29	1	0	0	1	1	1	0	0	1	0	1	1	0	0	1	1	1	1	1	1	1	1	0	1	1	0	0	1
30	1	1	0	1	1	1	1	0	1	1	1	0	0	1	1	1	1	1	1	0	1	1	1	0	0	0	1	0
31	0	1	1	1	0	1	1	0	0	0	1	0	0	0	1	1	0	1	1	1	1	1	1	0	1	0	0	0
32	1	1	0	1	1	1	0	0	1	1	1	0	0	0	0	0	0	1	1	1	1	0	1	1	0	0	0	0
33	1	0	0	1	1	1	1	0	0	1	1	1	1	1	0	0	0	1	0	0	1	0	0	1	1	0	0	0
34	1	0	0	0	1	1	0	0	1	1	1	0	0	0	1	1	0	1	1	0	1	0	0	0	0	0	1	0
35	1	0	1	0	0	1	1	1	1	1	1	1	1	0	0	0	1	0	0	0	1	0	0	0	0	0	1	0
36	0	1	1	0	0	1	1	1	0	1	0	0	0	1	0	1	0	0	0	1	1	1	0	0	0	0	0	0
37	1	0	0	1	0	0	0	0	1	1	1	0	0	1	1	0	0	1	1	1	1	0	0	0	0	1	0	0
38	1	1	0	1	1	1	1	0	0	0	1	0	0	0	1	0	0	0	1	0	0	0	0	0	1	0	0	0
39	1	0	0	1	0	1	0	0	0	0	1	0	0	0	1	1	0	1	1	0	1	0	0	0	0	1	0	0
40	1	0	0	0	0	1	0	0	0	0	1	0	0	0	0	1	0	1	1	1	1	0	0	0	0	1	1	0
41	1	1	0	1	0	0	1	0	1	1	0	1	0	0	0	0	1	1	0	0	0	1	0	0	0	0	0	0
42	0	0	0	1	1	1	0	0	0	0	1	1	0	0	1	0	1	0	0	1	1	0	1	0	0	0	0	0
43	0	1	0	1	0	0	1	0	1	0	1	0	0	0	0	0	1	0	0	1	1	0	1	0	0	0	0	0
44	1	0	0	0	0	0	0	0	1	0	0	0	0	0	0	1	0	1	1	0	1	0	0	0	0	1	1	0
45	0	0	0	1	0	0	1	0	1	0	0	0	0	0	1	0	0	1	0	0	1	1	0	0	0	0	0	0
46	0	0	1	0	0	0	1	0	0	0	1	0	0	0	0	0	0	0	0	0	1	0	0	0	1	0	1	0
47	0	0	0	1	0	0	0	0	1	0	0	0	0	0	1	0	0	1	0	1	0	0	0	0	0	0	0	0
48	0	0	0	0	0	1	0	0	0	0	1	0	0	0	1	0	0	0	0	0	1	0	0	0	0	0	0	0
49	0	1	0	0	0	0	1	0	0	0	0	0	0	0	0	0	0	0	0	0	1	0	0	0	0	0	0	0
50	0	0	0	0	0	0	1	0	1	0	0	0	0	0	0	0	0	0	0	0	1	0	0	0	0	0	0	0
51	0	0	0	0	0	0	0	0	0	1	0	0	0	1	0	0	0	0	0	1	0	0	0	0	0	0	0	0
52	0	0	0	1	0	0	0	0	0	0	0	0	0	0	0	0	0	0	0	1	0	0	0	0	0	0	0	0
53	0	0	0	0	0	0	0	0	0	0	0	0	0	0	0	0	0	0	1	0	0	0	0	0	0	0	0	0
54	0	0	0	0	0	0	0	0	0	0	0	0	0	0	0	0	0	0	1	0	0	0	0	0	0	0	0	0
55	0	0	0	0	0	0	0	0	0	0	0	0	0	0	0	0	0	0	0	0	0	0	0	0	0	0	0	0
56	0	1	0	0	0	0	0	0	0	0	0	0	0	0	0	0	0	0	0	0	0	0	0	0	0	0	0	0

being tested, which is that the observed data were obtained by some random allocation that is equivalent to the computer algorithm.

Table 14.6 shows a comparison between the observed distribution of

Table 14.6 Comparison of the observed distribution of the co-occurrences of species pairs and simulated distributions. The lower percentage point is the percentage of times (out of 1000) that the value by random allocation is less than or equal to that observed. The upper percentage point is the percentage of times that a value by random allocation is greater than or equal to that observed. The randomization minimums, maximums and means are as determined from the 999 simulated distributions and the observed distribution

| Number of co-occurrences | Observed frequency | Percentage points | | Randomization | | |
		Lower	Upper	Minimum	Maximum	Mean
0	64	82.20	20.00	28	105	53.56
1	210	63.60	39.70	156	239	204.57
2	74	6.80	94.20	55	116	88.18
3	118	53.90	50.90	82	141	116.84
4	70	95.40	7.10	38	83	59.51
5	61	53.60	51.80	40	83	60.91
6	52	8.30	93.70	41	81	61.26
7	67	70.90	34.20	43	87	63.84
8	55	1.10	99.40	48	93	71.06
9	56	28.80	76.20	41	83	60.38
10	112	99.30	0.90	70	124	93.33
11	66	86.40	16.70	36	83	59.30
12	65	38.00	67.70	47	90	67.43
13	23	1.90	98.70	19	47	32.68
14	25	7.10	95.20	20	51	32.59
15	26	11.70	92.50	18	48	32.59
16	37	65.10	41.70	19	53	35.67
17	58	96.90	4.00	32	70	48.28
18	38	8.40	94.60	31	62	45.88
19	46	56.20	50.30	30	67	45.75
20	40	78.10	27.20	22	51	37.05
21	36	92.20	12.30	19	44	30.70
22	24	64.90	45.00	12	36	23.07
23	28	9.20	94.50	23	44	33.47
24	43	99.60	0.80	23	46	34.51
25	21	34.20	80.20	17	32	22.57
26	19	73.60	65.00	18	24	19.03
27	3	100.00	100.00	3	3	3.00
28	3	100.00	100.00	3	3	3.00

co-occurrences of pairs of species and the distributions obtained by simulation. Figure 14.1 shows the observed distribution with the upper and lower limits from simulation. There is a certain amount of evidence here which suggests that the observed distribution is not altogether typical of the simulated distributions. The frequencies of 8 and 13 co-occurrences are unusually low and the frequencies of 10 and 24 co-occurrences are unusually high. Obviously, though, there is a multiple testing problem here so that the percentage points for the individual numbers of co-occurrences should not be taken too seriously.

One way to compare the observed distribution with the simulated ones is by comparing means, standard deviations, skewness and kurtosis. This gives the results shown in Table 14.7. The mean number of co-occurrences is always the same, which is a result of the constant total number of occurrences. However, the observed standard deviation is larger than any of the simulated values and the kurtosis (the fourth moment about the mean divided by the squared variance) is in the bottom 2.4% tail of the estimated random distribution. The skewness (the third moment about the

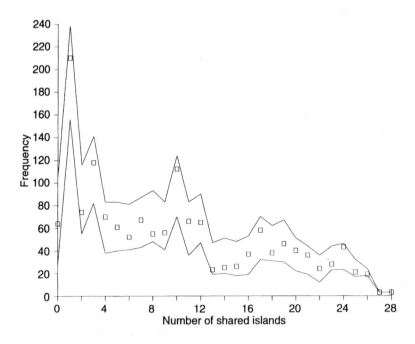

Figure 14.1 Comparison between the observed distribution of co-occurrences for Vanuatu birds (□) and the maximum and minimum frequencies seen in 999 simulated data plus the observed data (—).

Table 14.7 Comparison of the observed mean, standard deviation, skewness and kurtosis of the number of co-occurrences of pairs of species with the distributions of these statistics found by randomization

| | Observed | Percentage points | | Random | | |
		Lower	Upper	Minimum	Maximum	Mean
Mean	9.59	100.00	100.00	9.59	9.59	9.59
SD	7.55	100.00	0.10	7.38	7.55	7.46
Skewness	0.55	29.60	70.50	0.51	0.58	0.55
Kurtosis	2.14	2.40	97.70	2.12	2.22	2.17

mean divided by the third power of the standard deviation) is well within the simulated range.

Wilson (1987) followed Gilpin and Diamond (1982) in comparing distribution in terms of deviations between observed and expected numbers of co-occurrences, divided by their standard deviations. He reached the same conclusion as the one reached here: the observed occurrence matrix is not what is expected from the random allocation algorithm used to generate data.

One of the interesting aspects of this example is the difficulty in defining the appropriate null model of randomness. Even if the principle of fixing row and column totals at those observed is accepted as reasonable (which has been questioned), there is still no unambiguous way of defining a random allocation. It seems possible that alternative algorithms may give different results, although it has been found that the second algorithm mentioned above (filling up islands in order from the most to the least abundant, with species being chosen with probability proportional to the number of occurrences left) gives essentially the same results as the ones just described.

14.3 AN ALTERNATIVE GENERALIZED MONTE CARLO TEST

An alternative approach for the production of randomized species occurrence matrices involves starting from the observed occurrence matrix and changing this by a stepwise process of switching a randomly chosen pattern, as indicated in Table 14.8 (Connor and Simberloff, 1979; Roberts and Stone,1990). Switching species between islands in this manner maintains the total number of occurrences of each species and the total number of species on each island, and will eventually generate all possible

Table 14.8 Random switching to change an occurrence matrix. Changing pattern A to pattern B anywhere in an occurrence matrix will not change the row and column totals

	Pattern A		Pattern B	
	Island		Island	
Species	i	j	i	j
r	0	1	1	0
s	1	0	0	1

occurrence matrices with these constraints (Brualdi, 1980). It can therefore form the basis of a generalized Monte Carlo test, of the type that has been discussed in section 4.2 (Manly, 1995a).

The serial algorithm is as follows for testing whether an observed statistic that is calculated from an occurrence matrix is significantly large at the $100(k/n)\%$ level.

The serial algorithm

(a) Generate a random integer m with a uniform distribution between 1 and n. This gives the position of the observed occurrence matrix in a series of length n. The observed test statistic therefore becomes S_m.

(b) If $m > 1$, carry out $m - 1$ stepwise changes of species occurrences in the observed occurrence matrix as described in the previous section to obtain $m - 1$ occurrence matrices with their corresponding values $S_{m-1}, S_{m-2}, \ldots, S_1$, i.e. one change gives S_{m-1}, two changes give S_{m-2}, etc. These are considered to be backwards changes.

(c) If $m < n$, carry out $n - m$ stepwise changes of species occurrences in the observed occurrence matrix to obtain $n - m$ occurrence matrices with their corresponding values $S_{m+1}, S_{m+2}, \ldots, S_n$, i.e. one change gives S_{m+1}, two changes give S_{m+2}, etc. These are considered to be forwards changes.

(d) If S_m is one of the top k values for the test statistics ordered by magnitude then declare it significant at the $100(k/n)\%$ level.

In the absence of ties the above procedure is 'exact' in the sense that the probability of obtaining a significant result is k/n when the observed occurrence matrix is randomly chosen from the set of all possible occurrence matrices with the same row and column totals (Besag and Clifford, 1989; Manly, 1995a). If ties occur then these can be randomly broken. Alternatively, a conservative test can be constructed by placing the observed test statistic below any others with the same value in the ordered list. The probability of a significant result will then be less than k/n.

Besag and Clifford's (1989) alternative parallel method for a generalized Monte Carlo test is not considered here because it requires many more changes of the occurrence matrix than the serial method to generate the same number of test statistics, and appears generally to be less efficient than the serial method (Manly, 1993, p. 333).

It can be argued that the dependence of the serial method on a single choice of the position in the sequence where the observed test statistic will be situated is unsatisfactory. This limitation can be overcome by repeating the algorithm M times in order to obtain independent estimates $p_1, p_2, \ldots, p_M$ of the proportion of test statistics as large as or larger than that observed in a series of n statistics that includes the observed one. Then the mean of these estimates $\bar{p} = \sum p_i/M$ gives a combined estimate of the p-value for the observed test statistic which can be declared to be significant at the $100\alpha\%$ level if it is less than α. This replicated serial method has the advantage of providing information about the accuracy of the estimated p-value because a standard error for $\bar{p}$ can be estimated in the usual way by $\sqrt{\{\mathrm{Var}(p_i)/M\}}$.

Just deciding that an observed occurrence matrix appears to display interaction between species is not entirely satisfactory. In addition, it is desirable to identify the nature of at least the main interactions. An immediate problem that arises is the large number of possible interactions that can occur. For example, in a data set with 20 species there are $20 \times 19/2 = 190$ possible pairwise interactions. In other examples there might be as many as 50 species with 1225 possible pairwise interactions. To put it mildly, this generally rules out the possibility of estimating interaction parameters in some type of parametric model. However, progress is possible by asking which species exhibit non-random interaction with other species.

To this end, let e_{ij} denote the expected number of times that species i and j occur together on the same island if species occur independently of each other, with the row and column totals of the occurrence matrix fixed at those obtained for an observed matrix. Then the overall deviation from expected numbers for all species from an occurrence matrix can be measured by the mean square of deviations

$$S = \sum_{i=1}^{R} \sum_{j=1}^{R} (o_{ij} - e_{ij})^2 / R^2, \tag{14.1}$$

where o_{ij} is the observed number of co-occurrences and R is the number of species. This can also be written as

$$S = \sum_{i=1}^{R} v_i / R,$$

where

$$v_i = \sum_{j=1}^{R} (o_{ij} - e_{ij})^2 / R \tag{14.2}$$

is a measure of the extent to which numbers of co-occurrences of other species with species i differ from expectation. As $o_{ii} = e_{ii}$ (every species always occurs with itself) it matters little whether or not cases with $i = j$ are included or excluded from the sums in equations (14.1) and (14.2). For simplicity of expression they are included here.

The following testing scheme is based on these equations. First, S is tested to see whether it is significantly large and therefore gives overall evidence of deviations from expectations. This can either be done using one application of the serial method or by repeating the serial algorithm M times and combining the results as discussed at the end of the last section. Secondly, all the v_i values are individually tested using the serial method, with or without replication, with the same generated occurrence matrices as for the test on S. But, to make an allowance for the multiple testing of R species, v_i is required to be significantly large at the $100(\alpha/R)\%$ level before it is considered to provide evidence of interactions. This procedure for testing the v_i value for each species is based on the Bonferroni inequality which says that if R tests are each carried out at the $100(\alpha/R)\%$ level then the probability of declaring any of them significant by chance is α or less.

With many species this procedure requires a high level of significance before the result for an individual species gives evidence of interaction, which will only be possible with a large number of computer-generated test statistics. For example, with 50 species and a 5% level of significance v_i is required to be significantly large at the $5\%/50 = 0.1\%$ level before the result for species i is considered significant, which is just possible with 999 computer-generated statistics. However, the good thing about the serial method of testing is that even with a large occurrence matrix the computer time required to generate long series of statistics is not unreasonable.

In order to calculate the statistics S and v_i of equations (14.1) and (14.2), it is necessary to determine the expected numbers of co-occurrences, e_{ij}.

This does not appear to be simple to do analytically, and therefore the approach proposed here is to use the mean values obtained from the n sets of data generated with each application of the serial algorithm. This results in the deviations from expectation being generally somewhat less than would be obtained by using the mean values for all possible occurrence matrices. However, because all of the n statistics generated from one application of the serial algorithm will be affected in the same way it seems that this should not result in much loss of power in the test.

The generalized Monte Carlo test procedure was run on the data in Table 14.5 with 100 replicates of the serial test with series of length 10 000. The overall test statistic S from equation (14.1) was found to have a mean significance level of $\bar{p} = 0.0002$ (0.02%), with standard error 0.000 02 (0.002%), giving extremely strong evidence of non-randomness. Furthermore, a number of individual species had quite significant values for the v statistic of equation (14.2). However, the use and interpretation of the v statistics will not be considered in detail for this example.

See Manly (1995a) for further details about the use of the serial test on species co-occurrences, including some information about the number of random changes that are required in order to change an observed occurrence matrix into one with the appearance of a typical matrix, the interpretation of the p-value that the test produces and methods for examining non-random occurrence patterns for individual species. Manly (1996d) gives an example of the same method being used to detect non-random patterns in the associations between artefacts found in graves in Thailand.

It must be emphasized that the randomization tests that have been discussed here and in the previous section are not the only type of analysis that might be considered for analysing data on species occurrences on islands. Species relationships can be studied by the multivariate methods of cluster analysis and ordination (Manly, 1994a), and it may be illuminating to see whether islands that are close together tend to have similar species using, for example, the matrix randomization methods described in Chapter 9. However, these possibilities will not be pursued here.

14.4   EXAMINING TIME CHANGES IN NICHE OVERLAP

It is often useful to quantify the amount of overlap between two species in terms of their use of resources such as food or space. A number of indices have therefore been developed for this purpose (Hurlbert, 1978). In many cases the basic information available is the use by individuals of R resource categories, in which case Schoener's (1968) formula

$$O_{ij} = 1 - 0.5 \sum_{h=1}^{R} |p_{ih} - p_{jh}| \tag{14.3}$$

has the virtue of simplicity. Here, p_{ih} is the proportion of the resources used by individual i that are in category h and p_{jh} is the proportion in this category for individual j. The values of the index range from 0, when the individuals use different categories, to 1, when the individuals have exactly the same resource proportions. This is the only index that will be considered here. Note, however, that if the resource is essentially one-dimensional so that, for example, prey sizes are being considered, then it may be more appropriate to characterize each individual's use by a frequency distribution and determine the overlaps of the distributions. Two possibilities are then to assume normal distributions (MacArthur and Levins, 1967) or to assume Weibull distributions (Manly and Patterson, 1984).

One of the common questions with niche overlap studies is whether there are significant differences between individuals in different groups. These groups might be males and females, different species, individuals of different sizes or ages etc. In all cases what has to be ascertained is whether the overlap between individuals in different groups is significantly less than the overlap between individuals in the same group. In fact, this is not the question that is going to be addressed in this example, but it has been mentioned because it provides an obvious application for Mantel's matrix randomization test, which was discussed in section 9.2.

Another question concerns whether the niche overlap between two groups is constant under different conditions. One example concerns a set of data that has been considered several times in earlier chapters of this book. This is Powell and Russell's (1984, 1985) data on the diet compositions of two size classes of the eastern short-horned lizard *Phrynosoma douglassi brevirostre* in four months of the year at a site near Bow Island, Alberta. These data, which were published in full by Linton *et al.* (1989), are for 9 lizards sampled in June, 11 in July, 13 in August and 12 in September. The two size classes are (1) adult males and yearling females, and (2) adult females. There are nine categories of prey and a gut content analysis determined the amount of each, in milligrams of dry biomass, for each lizard. Table 14.9 shows these as proportions of the total. Here an interesting question concerns whether the mean niche overlap between two individuals from different size classes is the same in all four months, irrespective of whether this differs from the mean niche overlap between two individuals of the same size.

Linton *et al.*'s way of answering this question involved randomly pairing individuals from size classes (1) and (2) within each month and calculating niche overlaps from Schoener's formula to obtain the results shown in

Table 14.9 Results of stomach content analyses for short-horned lizards, *Phrynosoma douglassi brevirostre*, from near Bow Island, Alberta. The data were originally published by Linton *et al.* (1989) and come from a study by Powell and Russell (1984, 1985). The measurements for each prey category are proportions of the total, with measurements being of milligrams dry biomass. Categories are: (1) ants; (2) nonformicid Hymenoptera; (3) Homoptera; (4) Hemiptera; (5) Diptera; (6) Coleoptera; (7) Lepidoptera; (8) Orthoptera; and (9) Arachnida. The lizard size classes are (1) adult males and yearling females and (2) adult females

Month	Size	Prey category								
		1	2	3	4	5	6	7	8	9
June	1	0.33	0.00	0.00	0.12	0.00	0.28	0.00	0.26	0.02
	1	1.00	0.00	0.00	0.00	0.00	0.00	0.00	0.00	0.00
	1	0.43	0.00	0.00	0.04	0.00	0.00	0.00	0.00	0.53
	2	1.00	0.00	0.00	0.00	0.00	0.00	0.00	0.00	0.00
	2	0.14	0.00	0.00	0.00	0.00	0.00	0.00	0.00	0.86
	2	0.64	0.00	0.00	0.00	0.00	0.27	0.00	0.00	0.09
	2	0.40	0.00	0.00	0.21	0.00	0.36	0.00	0.02	0.00
	2	0.00	0.00	0.00	0.00	0.00	0.00	0.00	0.00	1.00
	2	0.47	0.07	0.00	0.05	0.00	0.42	0.00	0.00	0.00
July	1	0.44	0.00	0.00	0.00	0.00	0.00	0.00	0.00	0.56
	1	0.29	0.00	0.00	0.00	0.00	0.00	0.00	0.00	0.71
	1	0.88	0.00	0.00	0.00	0.00	0.12	0.00	0.00	0.00
	1	1.00	0.00	0.00	0.00	0.00	0.00	0.00	0.00	0.00
	1	1.00	0.00	0.00	0.00	0.00	0.00	0.00	0.00	0.00
	2	0.56	0.00	0.00	0.00	0.00	0.44	0.00	0.00	0.00
	2	0.55	0.00	0.00	0.00	0.00	0.45	0.00	0.00	0.00
	2	0.50	0.00	0.00	0.03	0.01	0.46	0.00	0.00	0.00
	2	0.24	0.00	0.00	0.00	0.00	0.36	0.00	0.40	0.00
	2	0.54	0.00	0.00	0.00	0.00	0.00	0.36	0.00	0.09
	2	0.03	0.00	0.00	0.08	0.33	0.44	0.00	0.01	0.11
August	1	0.85	0.00	0.00	0.00	0.00	0.15	0.00	0.00	0.00
	1	0.95	0.00	0.00	0.00	0.00	0.05	0.00	0.00	0.00
	1	0.59	0.00	0.00	0.00	0.11	0.31	0.00	0.00	0.00
	1	0.84	0.01	0.00	0.00	0.00	0.09	0.00	0.06	0.00
	1	0.31	0.29	0.05	0.16	0.00	0.00	0.00	0.18	0.00
	1	0.12	0.30	0.00	0.08	0.00	0.49	0.00	0.00	0.01
	1	0.11	0.00	0.00	0.08	0.00	0.03	0.00	0.77	0.00
	1	0.36	0.22	0.04	0.00	0.18	0.12	0.08	0.00	0.00
	1	0.10	0.00	0.00	0.07	0.00	0.00	0.00	0.83	0.00
	1	0.06	0.15	0.00	0.00	0.00	0.19	0.00	0.60	0.00
	2	0.72	0.00	0.00	0.00	0.00	0.28	0.00	0.00	0.00
	2	0.58	0.01	0.00	0.00	0.00	0.40	0.00	0.00	0.00
	2	0.78	0.00	0.00	0.03	0.00	0.11	0.08	0.00	0.00
September	1	0.08	0.70	0.00	0.05	0.00	0.17	0.00	0.00	0.00
	1	0.04	0.00	0.00	0.28	0.00	0.00	0.00	0.68	0.00
	1	0.03	0.12	0.00	0.00	0.00	0.16	0.37	0.32	0.00
	1	0.14	0.40	0.00	0.00	0.00	0.05	0.01	0.40	0.01
	1	1.00	0.00	0.00	0.00	0.00	0.00	0.00	0.00	0.00
	1	0.00	0.32	0.00	0.00	0.00	0.68	0.00	0.00	0.00
	2	0.06	0.12	0.00	0.00	0.00	0.00	0.00	0.82	0.00
	2	0.17	0.03	0.00	0.00	0.00	0.55	0.07	0.17	0.00
	2	0.02	0.65	0.00	0.03	0.00	0.20	0.00	0.09	0.00
	2	0.03	0.00	0.00	0.00	0.00	0.14	0.08	0.75	0.00
	2	0.10	0.12	0.00	0.08	0.00	0.00	0.00	0.71	0.00
	2	0.02	0.12	0.00	0.04	0.00	0.18	0.01	0.63	0.00

Table 14.10 Niche overlap values calculated by Linton *et al.* (1989) from samples of *Phrynosoma douglassi brevirostre*. For each month random pairs were made up with one lizard from size class 1 and one lizard from size class 2 ignoring any lizards left over

Month	Niche overlap values					
June	1.00	0.16	0.68			
July	0.55	0.24	0.54	0.14	0.68	
August	0.75	0.27	0.35			
September	0.59	0.03	0.88	0.80	0.47	0.12

Table 14.10. Any lizards that could not be paired were ignored. The niche overlap values were then subjected to a randomization one-factor analysis of variance of the type that has been discussed in section 7.1, the factor levels being months. They found no evidence of an effect.

This procedure can be criticized on the grounds that only one random pairing of individuals is considered, and it is conceivable that another pairing will give a different test result. In fact, the number of possible pairings is rather large. There are $6 \times 5 \times 4 = 120$ ways to choose the pairing for June, and 720 ways for each of July, August and September. That is, there are $120 \times 720^3 \approx 4.7 \times 10^{10}$ possible pairings, of which only one was chosen for Linton *et al.*'s randomization test.

As noted by Linton *et al.* there are 5.9×10^{13} possible randomizations of the niche overlap values to months once these have been determined by one of the 4.7×10^{10} possible random pairings of the two sizes of individuals. The total number of equally likely sets of data on the hypothesis of no differences between months is therefore the product of these two large numbers, of which only a very small fraction are allowed to be even considered by the Linton *et al.* procedure.

It must be stressed that there is nothing invalid in the Linton *et al.* test. It is simply conditional on a particular random pairing that is imposed by the data analyst. Linton *et al.* imply that this conditioning is unimportant in the sense that any random pairing gives about the same significance level. However, different pairings certainly give quite a wide variation in the F-ratios from a one-factor analysis of variance, so that this does not really seem to be true.

Pairing of individuals in the two size classes is required in order to address the question considered by Linton *et al.* However, random pairing introduces an element of arbitrariness that is not desirable. This suggests that a better test procedure is possible if a test statistic is constructed using all possible pairings (Manly, 1990b). To this end, consider all the niche overlap values that can be calculated within months for the lizard data, as shown in Table 14.11. An analysis of variance using all these values, with the months as the factor level, gives a between-months sum of squares of 0.244 and a total sum of squares of 6.853. The between-months sum of squares as a proportion of the total is therefore $Q = 0.036$, this being a statistic that makes use of all the available data.

Obviously the significance of this Q-value cannot be determined by randomization of pairs within months, since all pairs are being used in the calculation. However, bootstrapping can be carried out, where this involves regarding the lizards of each size class in each month as providing the best

Table 14.11 Niche overlaps that can be calculated for size classes (1) and (2) within months

Month	Lizard in size class (1)	Lizard in size class (2)					
		1	2	3	4	5	6
June	1	0.33	0.16	0.62	0.74	0.02	0.65
	2	1.00	0.14	0.64	0.40	0.00	0.47
	3	0.43	0.67	0.52	0.45	0.53	0.47
July	1	0.44	0.44	0.44	0.24	0.54	0.14
	2	0.29	0.29	0.29	0.24	0.38	0.14
	3	0.68	0.67	0.62	0.37	0.54	0.15
	4	0.56	0.55	0.50	0.24	0.54	0.03
	5	0.56	0.55	0.50	0.24	0.54	0.03
August	1	0.87	0.73	0.89			
	2	0.77	0.63	0.83			
	3	0.87	0.89	0.70			
	4	0.81	0.68	0.87			
	5	0.31	0.33	0.35			
	6	0.40	0.54	0.27			
	7	0.14	0.15	0.18			
	8	0.48	0.49	0.55			
	9	0.10	0.11	0.14			
	10	0.25	0.27	0.18			
September	1	0.18	0.28	0.88	0.17	0.24	0.35
	2	0.72	0.21	0.15	0.71	0.80	0.69
	3	0.47	0.47	0.40	0.57	0.47	0.63
	4	0.58	0.40	0.56	0.48	0.61	0.59
	5	0.06	0.17	0.02	0.03	0.10	0.02
	6	0.12	0.58	0.52	0.14	0.12	0.30

available information on the distribution of lizards that might have been sampled. Bootstrap samples can therefore be taken by resampling the lizards of each size class within each month, with replacement, to obtain new data for the analysis of variance. The hope is that the data obtained in this way will display the same level of variation as the data obtained from the field sampling process. If this is the case then the significance of the Q-value for field data can be determined by comparing this with the distribution of Q-values obtained from a large number of bootstrap samples.

Bootstrapping the original niche overlap values will produce new sets of data with no systematic differences between months only if the monthly means of these original values are the same. However, if observed niche overlap values used are adjusted to deviations from monthly means then the bootstrap samples will be being taken from populations with the same mean so that the null hypothesis of no differences between months becomes true, as is desirable for all bootstrap tests (Hall and Wilson, 1991). Of course, the true means of niche overlaps cannot be zero, but this is not important because the analysis of variance itself is based on deviations from means. Therefore, bootstrapping deviations from monthly means simulates sampling with the null hypothesis true, and the significance of an observed Q-value can be estimated by the proportion of bootstrap samples giving a Q-value that large, or larger.

When 1000 bootstrap samples were taken from the lizard data and the between month sums of squares as a proportion of the total sums of squares calculated, it was found that 746 of these statistics were 0.036 or more. The estimated significance level of the observed Q is therefore 74.6%. Clearly, there is no evidence of a month effect for these data.

There are two questions that will concern potential users of this test. First, is it likely to detect realistic differences between niche overlap values taken at different times? Second, what happens when the test is applied to data for which the null hypothesis of no effect is true? These questions will now be briefly addressed.

To get some idea of the power of the test it was applied to the lizard data after an alteration to make sure that the null hypothesis of no change in the level of niche overlap between months is not true. This was done by introducing a difference between the two sizes of lizard in the September data (Table 14.9) by changing the proportion of prey type 3 used by size (2) lizards from zero to P, with a corresponding proportional reduction in their use of the other types of prey. Carrying out the test on the modified data, with 1000 bootstrap samples for each value of P, gave the following significance levels:

P	0.1	0.2	0.3	0.4	0.5	0.6	0.7	0.8	0.9
Sig. level (%)	62	49	36	23	13	5	2	0.6	0.1

This does at least indicate that strong effects are likely to be detected.

To test the procedure with the null hypothesis true, 1000 sets of random data similar to the lizard data were generated. For each random set of data, the uses of the nine prey categories for each of 45 lizards were determined as independent normally distributed variables with the mean and standard deviation for each category set equal to the value found for the 45 real lizards. Negative 'uses' were set at zero. The 'random lizards' were then allocated to months and size classes with the same frequencies as for the real data. The significance level for each of the 1000 sets of random data was estimated from 100 bootstrap samples, with the expectation that the distribution of these significance levels will be approximately uniform between 0 and 1. As shown in Figure 14.2, this distribution was not quite achieved, with too many high significance levels. Nevertheless, the distribution is roughly right. It is estimated that the probability of getting a result significant at the 5% level is 0.03. It appears, therefore, that the bootstrap test tends to be slightly conservative with data of the type being considered. Whether this is a general tendency for the test

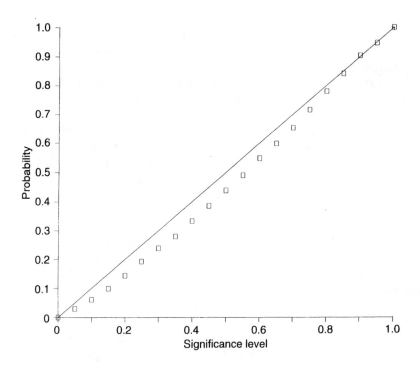

Figure 14.2 The distribution of significance levels (expressed as proportions) found for the bootstrap test applied to 1000 random sets of data. The desired distribution is shown by the diagonal line, with the probability of a significance level of p or less being equal to p. (□) simulated; (—) desired.

procedure can only be determined from a more detailed simulation study. However, the bootstrap appears to give useful results for this application.

14.5 PROBING MULTIVARIATE DATA WITH RANDOM SKEWERS

As a third example of a computer-intensive test, consider a procedure that was described by Pielou (1984) as probing multivariate data with random 'skewers'. The situation here is that samples are taken at various positions along some type of gradient, and the question of interest concerns whether there are significant trends in the sample results. Pielou used the test with samples of benthic fauna taken from the Athabasca River in Canada at sites from 40 km upstream to 320 km downstream from the town of Athabasca. Results taken on different dates were compared so as to study the effect of injecting the insecticide methoxychlor into the river either at Athabasca or 160 km downstream from the town.

The data for Pielou's test consist of values for p variables X_1 to X_p, each measured on n cases. In the Athabasca river example the variables were either absolute quantities or proportions of different invertebrate genera and the cases were from different locations, in order, along the river. However, generally the variables can measure whatever is relevant and the cases simply have to be in a meaningful order.

Pielou's 'skewers' are lines with random orientations. A large number are generated with the idea that if the n cases are approximately in order in terms of some linear combination of the X variables, then by chance some of the skewers will be in approximately the same direction. Therefore, the distribution of a statistic that measures the correspondence between the skewers and the data ordering will be different from what is expected for cases in a random order. The procedure is somewhat similar to projection pursuit (Jones and Sibson, 1987) where 'interesting' linear combinations are searched for in multivariate data sets. However, projection pursuit involves optimizing an index of 'interest' without doing random searches. An algorithm for the Pielou procedure is as follows.

Algorithm for random skewers

1. Generate p independent random numbers $a_1, a_2, \ldots, a_p$ uniformly distributed between -1 and $+1$. Scale them to $b_i = a_i/\sqrt{\sum a_i^2}$, so that $\sum b_i^2 = 1$.
2. Calculate the value of $Z = b_1 X_1 + b_2 X_2 + \ldots + b_p X_p$ for each of the n cases. This is the equation for the random skewer.

> 3. Find the rank correlation r_1 between the Z values and the case numbers for the n cases.
> 4. Repeat steps 1–3 a large number of times to determine the distribution of the rank correlations obtained by random 'skewering'.
>
> Because the scaling at step 1 will not affect the rank correlation it is not strictly necessary.

If the data are approximately in order in some direction then the distribution of correlations obtained from the above algorithm should have non-zero modes at $-\tau$ and $+\tau$, corresponding to skewers that are pointing in the opposite direction or in the direction of this order. The significance of such modes can be tested by generating distributions of rank correlations for the same number of cases but in a random order. Pielou suggests using random points for this purpose, but it seems more natural to generate distributions using the observed data in a random order so that a randomization test for significance can be used. It is then quite easy to generate the rank correlation distributions for the observed data and for some random permutations using the same random skewers.

To illustrate the procedure, consider what happens if it is applied to the lizard prey consumption data in Table 14.9. Here there are $n = 45$ lizards (ignoring the distinction between the two size classes), and $p = 9$ variables for proportions of different types of prey consumed. The 45 lizards are in an approximate time order, although within months the order is arbitrary. Obviously the ordering within months will tend to obscure time changes, but, nevertheless, any strong time trends in the consumption of different prey types should be detected. Purely for the purpose of an example it will be assumed that the lizards are in a strict order of sample time.

To begin with, 499 random permutations of the data order were generated for use at the same time as the real order. Then 500 random skewers were determined, with their corresponding rank correlations for the real data and its permutations. A suitable summary statistic for the real data is the maximum rank correlation observed (ignoring the sign) for the 500 skewers, which was 0.72. None of the permuted data sets gave such a large correlation so that the significance level is estimated as 1/500, or 0.02%. Hence there is clear evidence of time trends in prey consumption.

Figure 14.3 gives an interesting comparison between the distribution of absolute rank correlations found for the real data and the distribution found for the first random permutation of the data. As predicted by Pielou, the distributions are quite different, and the maximum correlation for the data in random order is only 0.46.

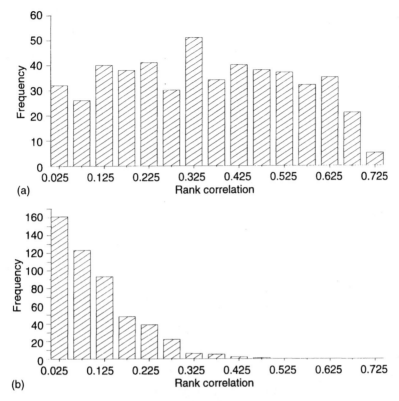

Figure 14.3 Distributions of rank correlations found with 500 random skewers for (a) the real lizard prey consumption data of Table 14.3 and (b) the lizard prey consumption data after a random reordering.

An alternative way of looking for time trends involves carrying out a multiple regression with the sample number as the dependent variable and the prey consumptions as the predictor variables. A permutation test can then be used to determine the significance of the result, as discussed in section 8.1. (Note that since the prey proportions add to 1, it is necessary to miss out one of the proportions in order to fit the regression.) This alternative approach may or may not give the same result as Pielou's method because fitting a least squares regression is not quite the same as maximizing a rank correlation. However, with the lizard data the results do agree fairly well.

This can be seen by making a comparison between the the random skewer that gives the highest rank correlation with the observed data and the fitted multiple regression. The equation of the best fitting of the 500 random skewers has the equation

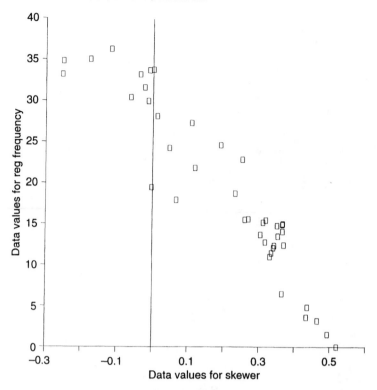

Figure 14.4 Plot of values from the multiple regression equation predicting the lizard sample number against values from the random skewer with the highest rank correlation with the sample numbers.

$$Z_1 = 0.33X_1 - 0.53X_2 - 0.01X_3 + 0.41X_4 - 0.03X_5$$
$$+ 0.41X_6 + 0.02X_7 - 0.09X_8 + 0.52X_9,$$

while a multiple regression of the sample number on the proportions of the first eight prey types gives the equation

$$Z_2 = 10.99X_1 + 43.29X_2 + 23.92X_3 - 27.60X_4 + 25.29X_5$$
$$+ 19.76X_6 + 35.372X_7 + 37.03X_8.$$

These equations seem quite different. However, plotting Z_2 against Z_1 shows that there is in fact a close relationship (Figure 14.4). Indeed, if Z_1 is coded to have the same mean and variance as Z_2, and the sign is changed so that it increases with time instead of decreasing, then the values from the two equations become almost identical. This is shown in Figure 14.5, where the coded Z_1 and Z_2 are both plotted against the lizard number in the original data.

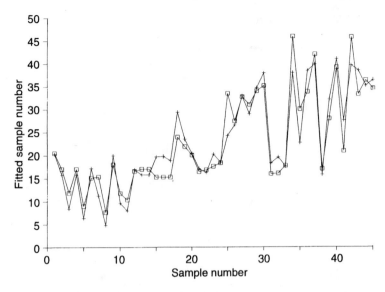

Figure 14.5 Plots of coded values of the random skewer and values from the multiple regression against the lizard order in the data. The random skewer values are coded to have the same mean and standard deviation as the multiple regression values, and to increase with the lizard number, so as to match the multiple regression as closely as possible. (□) scaled skewer; (+) regression.

Before leaving this example, yet another way of looking for a trend in data of this type can be mentioned. It is easy enough to construct a matrix of prey consumption distances between the 45 lizards, for example using equation (14.3). Another matrix can also be constructed in which the distance measure is the difference between the order of the lizards in the data. Mantel's test (section 9.2) can then be used to see if the lizards that are close in the sample order also tend to consume similar proportions of the different prey types.

14.6 ANT SPECIES SIZES IN EUROPE

A study by Cushman *et al.* (1993) of the distribution of ant species in Europe included consideration of whether the size of ant species is significantly related to the latitude at which the species occurs, because before the data were collected it was thought possible that the larger species might tend to be at higher latitudes. The question of whether the relationship between size and latitude appears to be the same for the two subfamilies Formicinae and Myrmicinae was also investigated, because

Table 14.12 The form of data on the mean size in millimetres of ant species in Europe, with samples from 107 locations, at latitudes from 50.1° to 70.0° N

Record	Latitude	Species	Size	Family
1	50.1	1	5.25	1
2	50.1	2	5.75	1
3	50.1	3	5.75	1
4	50.1	4	6.75	1
5	50.1	5	3.60	1
6	50.1	6	3.50	1
7	50.1	7	5.00	1
8	50.1	8	4.25	1
9	50.1	9	2.85	2
10	50.1	10	4.65	1
.				
.				
.				
2341	70.0	28	5.00	2

most of the ant species are in one of these subfamiles. It was assumed that the relationship between size and latitude can be approximated reasonably well by a linear regression of the form

$$E(Y) = \beta_0 + \beta_1 X, \qquad (14.4)$$

where $E(Y)$ is the expected size of a species in millimetres at latitude X.

The data available consists of 2341 records of the form shown in Table 14.12, drawn from Barrett (1979) and Collingwood (1979). There are 62 species and records for 107 different latitude stations. What makes the situation complicated is the fact that the available species records only provided one average size for each species, and that species occur at different numbers of latitudes. For example, species 13 occurs at 106 of the 107 sample stations, while species 1 only occurs 11 times. The result is that when regressions are considered of size against latitude the situation is that there are numerous instances were several different latitudes (the X variable in the regression) have exactly the same size (the Y value in the regression). This makes the analysis decidedly non-standard.

Figure 14.6 shows a plot of the species sizes against latitude, with multiple points for most species. Three regression lines are also shown. The upper line is from a regression using species from subfamily 1 (Formicinae) only, the middle line is from a regression using all species and the lower line is from a regression using species from subfamily 2 (Myrmicinae) only. On the face of it there does seem to be a difference

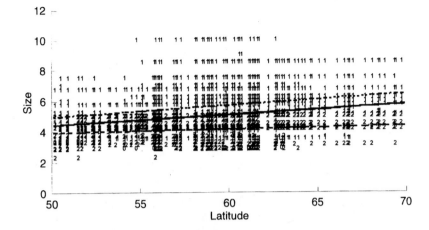

Figure 14.6 Plot of species sizes (mm) against the latitude where these species are present. Most species contribute several or many of the plotted points. Species in subfamily 1 are plotted as '1', and those in subfamily 2 plot as '2'. There are a few species plotted as '0' when they are neither of the two subfamilies. The lines shown were obtained by regressing the sizes against latitudes for species from family 1 only (top broken line), all species (middle continuous line), and species from family 2 only (lower broken line).

between the two subfamilies, with a tendency for size to increase with latitude for subfamily 1 but not for subfamily 2.

Some authors have overcome the problem of multiple observations for each species in a situation like this by relating species sizes to the mid-points of the latitudes where the species occur, as indicated in Figure 14.7. Cushman *et al.* considered this possibility, but concluded that it is a less appropriate than taking proper account of the latitudinal distribution of each species.

In order to test for relationships between species sizes and latitudes of occurrence, Cushman *et al.* allowed for the multiple records for different species using randomization tests. First, the observed regression coefficient of size on latitude for all 62 species (0.0673) was compared with the distribution obtained by randomly reallocating the sizes to the species 9999 times, keeping the latitudes of occurrence for each species equal to those for the real data. Only 7 of the 10 000 observed plus randomized values were greater than or equal to 0.0673, giving a significance level of 7/10 000, or 0.07%. This is a very highly significant result, giving clear evidence that size is related to latitude for all the species taken together. A one-sided test was used because of the initial prediction that, if anything, size would tend to increase with latitude.

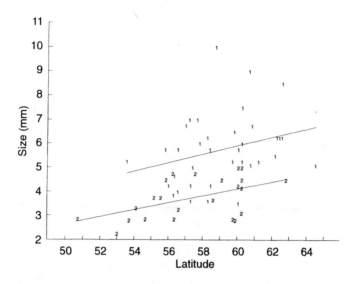

Figure 14.7 Species sizes plotted against the mid-point of the latitudes for which they are present, with one point per species. Data for subfamilies 1 and 2 are plotted as '1' and '2', respectively. The regression lines shown are based on the subfamily data, with the upper line for subfamily 1 and the lower line for subfamily 2.

Next, the same type of test was carried out just using the species in subfamily 1. In this case the regression coefficient for size on latitude (0.0755) was found to be significantly large at the 0.48% level, which is another highly significant result. (It should be noted that the p-values quoted by Cushman *et al.* for this and the previous test were multiplied by 10 by mistake. This does not, however, materially alter any conclusions.)

Finally, the same test was carried out using the species in subfamily 2. The result was that the regression coefficient for size on latitude (0.0215) was found to be significant at the 7.0% level. Hence for subfamily 2 there is no clear evidence of size tending to increase with latitude.

Having obtained these results, an obvious question concerns whether the difference between subfamily 1 and subfamily 2 in the regression coefficients for size on latitude is significantly different from zero. It is possible that this is the case, even though the coefficient is significant for subfamily 1 but not significant for subfamily 2.

It is not immediately obvious how to test for a subfamily difference in the regression coefficient of size on latitude. Apart from the multiple observations on each species, there are two further complications. The first of these is that species in subfamily 1 are generally larger than those in subfamily 2. This suggests that the constants β_0 in equation (14.4) will be different, even if the coefficients of lattitudes, β_1, are the same. The second complication is that the latitudinal distributions are not the same for the two subfamilies, with subfamily 1 tending to be at higher latitudes than subfamily 2. These two complications mean that care is needed to ensure that a test for a difference between the two regression coefficients on latitude is not confounded by the size and latitudinal distribution differences between the two subfamilies.

Cushman *et al.* considered a number of alternative randomization and bootstrap approaches for testing the equality of the regression coefficients for the two familes. Eventually they chose to use an approximate randomization test with sizes and latitudes adjusted to deviations from family means before randomizing between families, based on the following algorithm.

Algorithm for comparing size related to latitude for two subfamilies

1. The data for subfamily 1 were standardized by replacing the size for each species by the deviation of the size from the mean size for the subfamily, and latitudes were replaced with their deviations from the family mean latitude. The same standardization was carried out for subfamily 2.
2. Standardized sizes were regressed on standardized latitudes for each of the subfamilies, and the difference D_1 in the slopes was calculated.
3. The records for species were randomly reallocated to subfamilies, regressions equations were estimated for subfamilies and the difference D_2 in regression slopes was calculated. This randomization and estimation process was repeated a large number $N-1$ of times to generate statistics $D_2, \ldots D_N$.
4. The estimated significance level for D_1 was determined as the proportion of the values $D_1, D_2, \ldots, D_N$ as far or further from zero than D_1.

When the above procedure was applied with $N = 9999$ randomizations it was found that the estimated significance level of the observed slope difference ($D_1 = 0.054$) is 0.084. There is thus no clear evidence of different regression slope for the two subfamilies.

The properties of this test were examined to a limited extent as part of the Cushman *et al.* study. To begin with, the performance of the test when there is no difference in regression slope was examined by generating 1000 random sets of data with the null hypothesis true, carrying out the randomization test on each one, and seeing how many results were significant at 5% level. It was found that 37 (3.7%) of the 1000 tests gave a significant result. This is not quite significantly different from 5% (on a two-sided test at 5% level). The performance of the test therefore seems reasonable when the null hypothesis is true.

Unequal residual variation for the subspecies may be a problem with the real data because the regression of size against latitude (with multiple points for species at different latitudes) has a residual standard deviation of 1.385 when fitted for Formicinae only, and a residual standard deviation of 0.688 for Myrmicinae only. To examine the effect of this type of difference, 1000 sets of data were generated to mimic this but with the same regression slopes for subfamilies. The result was that 2.1% of slope differences were significant with a 5% level test. The test therefore seems to becomes more conservative with unequal residual variation.

The power of the test is not high relative to the observed slope difference $D_1 = 0.054$. This is demonstrated by the results of generating 1000 sets of random data, for which the average difference in regression slopes for two subfamilies was 0.11. It was found that only 45.2% of these data sets displayed a significant slope difference by the randomization test. Therefore the power of the test is not great even for detecting twice the observed difference.

14.7   CHAPTER SUMMARY

- Four examples of the use of computer-intensive methods in non-standard conditions are discussed.
- The first example concerns species presences and absences on islands, and methods for testing whether species interactions occur in the sense that some species tend to occur together, or some species seldom occur together. A Monte Carlo test for the independence of the presences and absences of different species involves comparing a test statistic calculated from observed data with the distribution obtained by simulating alternative sets of data with the null hypothesis true.
- An alternative generalized Monte Carlo test is also considered for the situation of the first example. In this case alternative test statistics are produced by a series of small changes to the initial set of data.
- The second example concerns the use of bootstrapping to test for changes with time in the niche overlap between individuals in two groups.
- The third example concerns a Monte Carlo method for searching for a linear combination of variables that changes systematically along a dimension in space. Basically, a large number of random linear

combinations ('random skewers') are generated and their correlations with the distance along a line are calculated. Meaningful linear combinations should then be associated with high correlations. A randomization test is possible for the significance of an obtained linear combination.

- The final example concerns the relationship between the size of different species of ants in Europe, and the latitude at which they occur. This can be studied using regression methods, but the complication is that often the same species (with the same size) is recorded at several or many different latitudes. The usual assumptions of regression do not apply in this case, but tests for significance based on a restricted randomization are possible. An approximate randomization test for a difference in the regression coefficient for two subfamilies of ants is also considered.

15

Bayesian methods

15.1 THE BAYESIAN APPROACH TO DATA ANALYSIS

So far, all of the methods that have been discussed in this book have been based on the classical concepts of tests of significance and confidence intervals. However, the increasing availability of computer power has had a major impact on Bayesian statistical methods as well. In particular, new computational approaches have meant that many situations that were previously not amenable to the Bayesian approach can now be handled.

The basic idea behind Bayesian statistics is to change prior probabilities for parameters taking particular numerical values to new probabilities as a result of collecting more data, with this change being achieved through the use of Bayes' theorem. The area is controversial because many statisticians do not accept the general use of prior probabilities, although Bayes' theorem itself is not in question.

As an example of the Bayesian approach, suppose that there is interest in the value of a parameter θ for a certain population, and that before any data are collected it is somehow possible to say that the θ must take one of the values $\theta_1, \theta_2, \ldots \theta_n$, and that the probability of the value being θ_i is $\pi(\theta_i)$. Suppose also that some new data are collected and the probability of observing these data is $\pi(\text{data}|\theta_i)$ if in fact $\theta = \theta_i$. Then Bayes' theorem states that the probability of θ being equal to θ_i, given the new data, is

$$\pi(\theta_i|\text{data}) = \pi(\text{data}|\theta_i)p(\theta_i)/ \sum_{j=1}^{n} \pi(\text{data}|\theta_j)\pi(\theta_j), \qquad (15.1)$$

where $\pi(\theta_i|\text{data})$ is the posterior distribution for θ. This result generalizes in a fairly obvious way for a parameter that can take any value on a continuous scale, and for situations where several parameters are involved, so that in general

$$\pi(\theta_1, \theta_2, \ldots \theta_p|\text{data}) \propto \pi(\text{data}|\theta_1, \theta_2, \ldots \theta_p)\pi(\theta_1, \theta_2, \ldots \theta_p), \qquad (15.2)$$

i.e. the posterior probability of several parameters given a set of data is proportional to the probability of the data given the parameters multiplied by the prior probability of the parameters.

There are various ways that an equation like (15.2) can be used in a computer-intensive analysis. One possibility involves generating the posterior distributions of various quantities of interest through a direct Monte Carlo simulation. This approach was used by Brown and Czeisler (1992), for example, for the analysis of data on human circadian rhythms obtained under closely controlled laboratory conditions. However, it generally requires analytic or numerical approximations to be used to calculate the posterior distribution, posing severe difficulties in situations where there are several or many parameters.

An alternative approach uses Markov chain Monte Carlo methods. The remainder of this chapter concentrates on these methods because they have attracted a great deal of interest in recent years as a means of making Bayesian methods feasible in a wide range of situations where this was not the case before.

15.2 THE GIBBS SAMPLER AND RELATED METHODS

The Gibbs sampler (Geman and Geman, 1984) is a method for approximating a multivariate distribution by taking samples only from univariate distributions. The value of this with Bayesian inference is that it makes it relatively easy to sample from a multivariate posterior distribution even when the number of parameters involved is very large.

Assume that a posterior distribution has a density function $\pi(\theta_1, \theta_2, \ldots \theta_p)$ for the p parameters $\theta_1, \theta_2, \ldots \theta_p$, and let $\pi(\theta_i|\theta_1, \ldots \theta_{i-1}, \theta_{i+1}, \ldots, \theta_p)$ denote the conditional density function for θ_i given the values of the other parameters. The problem then is to generate a large number of random samples from the posterior distribution in order to approximate both the distribution itself and the distribution of various functions of the parameters. This is done by picking arbitrary starting values $\{\theta_1(0), \theta_2(0), \ldots, \theta_p(0)\}$ for the p parameters and then changing them one by one by selecting new values as follows:

$\theta_1(1)$ is chosen from $\pi(\theta_1|\theta_2(0), \theta_3(0), \ldots, \theta_p(0))$

$\theta_2(1)$ is chosen from $\pi(\theta_2|\theta_1(1), \theta_3(0), \ldots, \theta_p(0))$

$\theta_3(1)$ is chosen from $\pi(\theta_3|\theta_1(1), \theta_2(1), \theta_4(0) \ldots, \theta_p(0))$

$$\vdots$$

$\theta_p(1)$ is chosen from $\pi(\theta_p|\theta_1(1), \theta_2(1), \ldots \theta_{p-1}(1))$.

At that stage, all of the initial starting values have been replaced, which completes one cycle of the algorithm. The process is then repeated many times to produce the sequence $\{\theta_1(1), \theta_2(1), \ldots \theta_p(1)\}$, $\{\theta_1(2), \theta_2(2), \ldots, \theta_p(2)\}$, $\ldots$, $\{\theta_1(N), \theta_2(N), \ldots \theta_p(N)\}$, which is called a

Monte Carlo chain because at each step in the algorithm the change made is dependent only on the current θ values rather than any earlier values.

Two key facts make this algorithm potentially very useful. First, it can be shown that $\{\theta_1(i), \theta_2(i), \ldots, \theta_p(i)\}$ follows the distribution with density function $\pi(\theta_1, \theta_2, \ldots, \theta_p)$ for large values of i. Second, drawing observations from the conditional distributions is often relatively straightforward, so that implementation of the Gibbs sampler is not difficult.

A complication comes in because the successive sets of generated sample values may be highly correlated, but this can be overcome by only retaining values at every rth step in the sequence, with r chosen large enough to ensure that these values have only negligible correlation. Alternatively, several different sequences can be generated with different randomly chosen starting points and only the final sets of values $\{\theta_1(N), \theta_2(N), \ldots, \theta_p(N)\}$ retained.

There are various modifications to the Gibbs sampler that have been suggested to improve its convergence properties and an alternative Hastings–Metropolis algorithm for simulation is also often used, although this has slightly more complicated transition rules (Metropolis *et al.*, 1953; Hastings, 1970).

It should also be noted that by assuming constant prior distributions for all the parameters, so that $p(\theta_1, \theta_2, \ldots, \theta_p) = 1$, the right-hand side of equation (15.2) becomes equal to the likelihood function for a set of data. Therefore Markov chain Monte Carlo methods can be used to generate and maximize a likelihood function for a non-Bayesian statistician. They can also be used for the estimation of parameters by the method of moments (Gelman, 1995).

Example 15.1: mark–recapture estimation of population size   As an example of the use of the Gibbs sampler, consider the situation of mark–recapture sampling to estimate the size of an animal population that is not changing during the sampling period. Suppose that M animals are marked and released into a population and allowed to mix freely with U unmarked animals. A sample of all the $M + U$ animals is then taken in such a way that each animal has an unknown probability p of being captured, independent of the fate of other animals. The sample contains $m = 25$ marked animals and $u = 10$ unmarked animals. The problem is to estimate U and p. See Underhill (1990) and Garthwaite *et al.* (1995) for reviews of other Bayesian approaches to this type of situation.

The data here are m and u, and the sampling scheme used means that m is an observation from a binomial distribution with probability function $\text{Prob}(m) = {}^{M}C_m p^m (1 - p)^{M-m}$, while u is an observation from an independent binomial distribution with probability function $\text{Prob}(u) = {}^{U}C_u p^u (1 - p)^{U-u}$. Therefore, the probability of the data given the unknown parameters is

$$P(m, u|U, p) = {}^{M}C_{m}p^{m}(1 - p)^{M-m} {}^{U}C_{u}p^{u}(1 - p)^{U-u}.$$

A Bayesian analysis requires that a prior distribution for U and p is now defined. It is at this point that some people have philosophical difficulties, because it is hard to imagine what this distribution is with respect to, and to decide what mathematical form it should take. On the other hand, the choice of a prior distribution seems often to be relatively unimportant for the determination of the posterior distribution, suggesting that within reason any choice may be satisfactory.

For this example the prior distribution assumed for U is the uniform distribution for the integers $1, 2, \ldots, 100$, i.e. each of these values is assumed to have been equally likely prior to collecting the data. For p a uniform prior distribution is also assumed, but this time over the interval 0 to 1. Further, the prior distributions for U and p are assumed to be independent, so that the joint prior distribution is

$$P(U, p) = C, \quad 1 \le U \le 100, \quad 0 \le p \le 1,$$

where C is a constant. Noting that the observed data are impossible for $U < 10$, it then follows from equation (15.2) that the posterior distribution of U and p takes the form

$$\begin{aligned} P(U, p|m, u) &\propto P(m, u|U, p)P(U, p) \\ &\propto {}^{M}C_{m}p^{m}(1 - p)^{M-m} {}^{U}C_{u}p^{u}(1 - p)^{U-u}, \end{aligned} \tag{15.3}$$

for $10 \le U \le 100$, $0 \le p \le 1$, with $M = 50$, $m = 25$ and $u = 10$. Because a uniform prior distribution is being assumed, the right-hand side of equation (15.3) is just the likelihood function. Therefore the calculations that are about to be described can be thought of as a means for evaluating the likelihood function instead of being a Bayesian analysis.

To generate values from the posterior distribution using the Gibbs sampler it is necessary to use the conditional distribution of U given p, and then the conditional distribution of p given U. The conditional distribution of U is obtained by treating any parts of equation (15.3) that do not involve U as being constants. Thus

$$P(U|m, u, p) \propto {}^{U}C_{u}(1 - p)^{U}. \tag{15.4}$$

Values from this distribution can be generated by selecting values of U between 10 and 100 with probabilities proportional to their values from the right-hand side of equation (15.4), remembering that $u = 10$.

The conditional distribution of p is obtained from equation (15.3) by treating any terms not involving p as constants, to give

$$P(p|m, u, U) \propto p^{m+u}(1 - p)^{M-m+U-u}. \tag{15.5}$$

For the convenience of calculations this can be approximated by a discrete

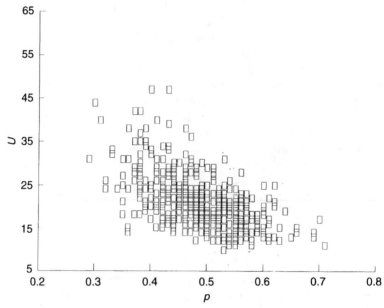

Figure 15.1 Posterior distribution of the number of unmarked animals in a population (U) and the probability of capture (p) for mark–recapture sampling, as determined using the Gibbs sampler.

distribution which gives the 101 specific values $0.00, 0.01, 0.02, \ldots, 1.00$ probabilities proportional to the right-hand side of equation (15.5), with $M = 50$, $m = 25$ and $u = 10$.

Five separate sequences $(U_1, p_1), (U_2, p_2), \ldots, (U_{5000}, p_{5000})$ were started with arbitrary different initial values U_0 and p_0 for U and p. For each sequence the process consisted of generating a value U_1 from the distribution $P(U|m, u, p_0)$, then p_1 from $P(p|m, u, U_1)$, then U_2 from $P(U|m, u, p_1)$, and so on. The sequences were then analysed for autocorrelation and it was found that the correlation was close to zero for values of (U_i, p_i) at least five steps apart. That is to say, it appears that if every fifth pair (U_i, p_i) is selected from a sequence then these are effectively independent observations from the posterior distribution for U and p. Therefore, the five sequences of length 5000 provided the equivalent of about 5000 independent observations from the posterior distribution.

There are a number of more sophisticated types of 'output analysis' that can be applied in a situation like this (Gelman and Rubin, 1992). Here it suffices to say that for the example being considered the Gibbs sampler seems to converge very quickly to stable conditions, with the five separate sequences producing very similar distributions for U and p.

Figure 15.1 shows the distribution for 1000 observations of U and p

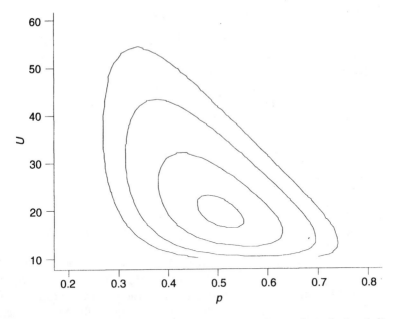

Figure 15.2 Contours for the posterior distribution of U and p calculated directly from equation (15.3). The outer incomplete contour corresponds to a density of 0.001, with the other contours, moving inwards, corresponding to densities of 0.01, 0.1 and 0.4. The outer contour is incomplete because the minimum possible value of U is 10 (the number of unmarked animals in the second sample) and the density does not fall below 0.001 with $U = 10$.

obtained by selecting every 50th pair of values from each of the five sequences, starting at 50. The fact that these observations are indeed from the true posterior distribution is confirmed by comparing this empirical distribution with contours of constant density that are calculated directly from equation (15.3), as shown in Figure 15.2.

Because there are only two parameters being considered and equation (15.3) is reasonably simple it is not difficult to sample the posterior distribution directly in the situation being considered or, indeed, to determine its properties by direct calculation. There is therefore no real need to use the Gibbs sampler at all. It must be stressed, therefore, that the purpose of this example has just been to demonstrate how the Gibbs sampler can be applied in a simple situation. The true value of the Gibbs sampler is in situations where alternative direct approaches are not feasible.

Selecting values from the distributions specified by equations (15.4) and (15.5) is quite easy and the calculations for this example were done in Lotus 1-2-3 (Lotus Development Corporation, 1991).

15.3 BIOLOGICAL APPLICATIONS

Some biological applications of Markov chain Monte Carlo methods are for analysing data on immunity following immunization for hepatitis B (Coursaget *et al.*, 1991); modelling the development of Aids (Lange *et al.*, 1992; Gilks *et al.*, 1993); mapping of diseases and relating diseases to geographically varying covariates (Gilks *et al.*, 1993; Ferrándiz *et al.*, 1995); gene mapping and other genetic analyses (Gilks *et al.*, 1993; Smith and Roberts, 1993; Guo and Thompson, 1992); the analysis of agricultural field trials (Besag and Green, 1993; Besag *et al.*, 1995); modelling the relationship between the reproductive mass and the vegetative mass of plants (Billard, 1994); mapping the ranges of animals (Heikkinen and Högmander, 1994; Högmander, 1995); the analysis of human growth (Wakefield *et al.*, 1994); the analysis of plasma concentrations of a drug at various times after the administration to patients (Wakefield *et al.*, 1994), logistic regression with unobserved covariates (Besag *et al.*, 1995), fitting proportional hazards survival models (Gelfand and Mallick, 1995), the estimation of disease prevalence (Joseph *et al.*, 1995), the maximum likelihood of the effective population size times the neutral mutation rate from molecular sequence data (Kuhner *et al.*, 1995), the fitting of models for count data (Scollnik, 1995), and the analysis of data from band recoveries of birds (Vounatsou and Smith, 1995).

15.4 FURTHER READING

The Markov chain Monte Carlo methods that have been the focus of this chapter have not yet been widely used, and the theory for their use is still being developed. In particular, there is intense interest by some statisticians in improvements to criteria for confirming convergence and in modifications to algorithms that are designed to accelerate the rate of convergence (Gelman and Rubin, 1992; Geyer, 1992; Gelfand and Sahu, 1994; Geyer and Thompson, 1995). See Smith and Roberts (1993), Besag *et al.* (1995) and Gelman *et al.* (1995) for a more detailed review of all of these matters.

The Bayesian approach to statistics is attracting increasing interest because improvements in computer technology mean that it is now possible to use this approach in many situations where this was not really feasible before. Also, other computer-intensive analyses based on Bayesian types of argument are appearing. It seems fair to say that the true value of these methods will only become apparent when more experience of their use has accumulated.

One application of a computer-intensive Bayesian type of analysis that is worth a special mention is Raftery *et al.*'s (1995) method for drawing inferences about the population dynamics of the bowhead whale, *Balaena mysticetus*. This method combines information from the input and output parameters to a deterministic population model to produce what are called postmodel distributions for quantities of interest. Here, input parameters include birth and death rates, output parameters include the current population size, and the main quantity of interest is the replacement yield. The postmodel distribution is evaluated by computer simulation.

15.5 CHAPTER SUMMARY

- The essential element of Bayesian statistical inference is the idea of using data to transform prior distributions for parameters into posterior distributions that take into account the evidence in the data. In the past the calculations involved in determining posterior distributions have been very substantial for many applications, which has restricted the use of Bayesian methods. However, recently Markov chain Monte Carlo methods have been developed that make the methods potentially much easier to use.
- The Gibbs sampler is a Markov chain Monte Carlo method that makes it possible to approximate samples from a multivariate distribution by taking samples only from univariate distributions. Essentially this means that values can be generated from a Bayesian posterior distribution for a large number of parameters in a relatively straightforward way. The Gibbs sampler is described and illustrated on an example involving population size estimation by mark–recapture methods.
- Several biological applications of Markov chain Monte Carlo methods are described, and a number of extensions and generalizations are briefly reviewed.

15.6 EXERCISE

15.1 Use a spreadsheet such as Lotus 1-2-3 (Lotus Development Corporation, 1991) to generate values from the Gibbs sampler defined by equations (15.4) and (15.5) in the manner described in Example 15.1. (This will probably require some macro programming.) Show from the generated values that the mode of the posterior distribution is at approximately $U = 20$ and $p = 0.5$, where these can be shown to be the usual maximum likelihood estimates from the model being used. Plot a suitable subset of the generated values to produce an empirical posterior distribution like that shown in Figure 15.1.

16

Conclusion and final comments

16.1 RANDOMIZATION

Randomization is a method of inference that has been available for 60 years, but has only been really practical for 20 years or so. What should have become apparent from the discussions and examples in this book is that it is mainly of value under three particular circumstances. First, it tends to have better properties than more conventional methods for standard analyses like regression and analysis of variance with extremely non-normal data. Second, randomization can sometimes be applied with situations such as testing for an association between two distance matrices where no alternative seems sensible. Third, there are cases where the observations form the entire population of interest so that randomization is the only method of inference that is possible.

What should also have become obvious is that there are many areas where more research is needed on the properties of randomization methods and the development of new methods based on randomization arguments. Examples are:

- The development of methods for comparing mean values from two or more samples when the sources of these samples may not have the same level of variation, including situations where the study design involves several factors.
- The development of small sample methods for the comparison of the amount of variation in several samples from non-normal distributions.
- Comprehensive simulation studies of the properties of randomization methods for analysis of variance, regression and multivariate tests.
- Further study of methods for comparing distance matrices in the context of distances between objects located in space where spatial autocorrelation is present.

16.2 BOOTSTRAPPING

Bootstrapping is based on the usual ideas of random sampling from populations. Since the publication of Efron's (1979a,b) classic papers the use of this method has been explosive, and it seems fair to say that many people have treated it as the complete answer to many difficult problems. Unfortunately, this is being a little too optimistic. The examples in Chapter 3 demonstrate that sometimes bootstrapping simply does not work well.

The lesson that should be learned from the examples in Chapter 3 is not that bootstrapping is ineffective. Rather, bootstrapping should be used with caution in situations where it has not been thoroughly tested out already. Generally, the theory of bootstrapping guarantees that it will work well in certain situations with large samples. With the small sample situations that have to be dealt with most commonly the properties of the method need careful study before it can be trusted, probably using simulated data.

One interesting aspect of the development of the literature on bootstrapping in the last 15 years has been the way that an initially very simple concept has led to some extremely complicated theory. This is unfortunate, and at odds with the KISS philosophy (*Keep It Simple Statistician*). Presumably it explains why it is that most uses of bootstrap confidence limits seem to be of Efron's original percentile method: users understand it, and superficially at least the results usually look reasonable. This should not be taken as an argument against the development of more theory. However, theoreticians need to realize that most people do not like to employ methods that have a mysterious basis.

No doubt the applications of bootstrapping for confidence intervals and tests of significance will continue to develop in the future. It can be hoped that the theory will be consolidated and synthesized, that a few standard methods with well-understood properties will emerge, and that these will be available for easy use in standard statistical packages.

16.3 MONTE CARLO METHODS IN GENERAL

By their nature, Monte Carlo methods as defined in this book tend to apply in rather non-standard situations. Like randomization and bootstrapping, it seems inevitable that these methods will be used more in future in biology and other areas as more people realize the power of this approach. In particular, the generalized Monte Carlo test (sections 4.2 and 14.3) seems to have many potential applications, as does the concept of the estimation of parameters using an implicit statistical model defined only by simulation (section 4.3).

At present the main problem with using these approaches to data analysis is the need to write a special computer program for each application. The computer power is already available to most potential users. It will be interesting to see the applications that develop in the next few years.

16.4 CLASSICAL VS. BAYESIAN INFERENCE

The recent upsurge of interest in Bayesian methods resulting from the realization that their use is facilitated by Markov chain Monte Carlo calculations seems likely to have a considerable impact in the near future. Initially the main uses have been in medicine, but applications in biology in general have already appeared.

Until now there have been two deterrents to the widespread use of Bayesian inference. One has been the difficulty of doing the required calculations. Markov chain Monte Carlo has largely overcome this. The other deterrent has been the philosophical difficulty of accepting the use of prior distributions for parameters. This is unchanged.

This is not the place to argue the pros and cons of Bayesian inference as an alternative to more conventional inference based on the idea of parameters as fixed unknown quantities. It is sufficient to note that a classical statistician may argue that real knowledge about prior distributions is often very minimal and allowing their use makes it possible for scientists to influence the results of an analysis with their preconceived biases. On the other hand, a Bayesian statistician may argue that all important decisions about the real world are partly subjective, and that in any cases the prior distributions used will have little effect when the data provide a reasonable amount of information.

The important thing for biologists and other scientists to realize when they see Bayesian inference being applied is that this is different from the classical inference methods based on sampling distributions. They are therefore well advised to pay careful attention to the assumptions being made. The sensitivity of results to prior distributions should often be of particular interest.

Appendix: software for computer-intensive statistics

As time goes on more software for computer-intensive methods is continually being produced, thus making these methods increasingly easy to use. Basically, four types of software are available:

(a) Spreadsheet such as Lotus 1-2-3 (Lotus Development Corporation, 1991) can be used to carry out some of the simpler randomization and bootstrap procedure. Bootstrapping is particularly easy because it often just involves setting up one line in the spreadsheet to resample the data and do the necessary calculations, and then just copying this line for each bootstrap sample required (Willemain, 1994). Randomization is more difficult because a macro has to be written to permute the data, although this is not difficult with a little experience.
(b) There are some programs available for carrying out specific types of calculation such as RT (Manly, 1996a). These are relatively easy to use in the sense that they read the data and then produce the required result. Their limitation is lack of flexibility. A list of some of these programs and their capabilities is provided below.
(c) Major statistical packages like MINITAB (Minitab Inc., 1994), SAS (SAS Institute Inc., 1990) and SPSS (SPSS Inc., 1990) incorporate programming languages that may permit some computer-intensive tests to be carried out. (See also the note below on the MULTTEST procedure in SAS.)
(d) There are various languages available that can be used to program any type of analysis. These vary from programming languages like Basic, C and Fortran, through higher level languages like MatLab (Math Works Inc., 1989) and S and S-PLUS (Becker et al., 1988), to Resampling Stats (Bruce, 1991), which has been specially designed for randomization and bootstrapping.

Each of these types of software is useful at different times. For example, in writing this book calculations were done using Lotus 1-2-3 spreadsheets, options in RT, and specially written programs in Fortran and Resampling Stats.

Some of the available programs that include computer-intensive options are:

1. BOJA (Boomsma, 1991; Dalgleish, 1995) carries out various jackknife and bootstrap analyses including regression and analysis of variance.
2. CANOCO (ter Braak, 1988) which allows randomization tests of hypotheses related to analyses on species presence and absence and abundance data.
3. MULTTEST, a procedure within SAS (SAS Institute Inc., 1990) carries out a range of the bootstrap and randomization multiple tests described by Westfall and Young (1993).
4. RT (Manly, 1996a) carries out one- and two-sample tests, analysis of variance, regression, Mantel randomization tests, tests on spatial data, time series analysis and some multivariate tests, all by sampling randomization distributions (see also below.)
5. SIMSTAT (Péladeau, 1994) carries out many standard parametric and non-parametric analyses, and incorporates the ability to repeat many of these analyses with bootstrap resampled data.
6. StatXact (Mehta and Patel, 1995) carries out a variety of tests on one, two or K samples, contingency tables, censored survival data and other situations using p-values from either sampling or enumerating full randomization distributions. A software package LogXact for exact logistic regression is available from the same company.

Further information about computer software, with some emphasis on fast algorithms for calculating p-values from full randomization distributions, are provided in the appendix to Edgington's (1995) book written by Rose Baker. Efron and Tibshirani (1993) discuss a range of functions for bootstrap calculations in the S language.

A trial version and the Fortran source code for RT are available free of charge on Western EcoSystem Technology's World-Wide Web and FTP sites. To obtain the trial version or source code over the World-Wide Web, access URL http://www.west-inc.com and follow the instructions listed there. To obtain the trial version or source code using FTP software, FTP to ftp.wyoming.com and download the file /pub/users/west-inc/RTinstr.txt for further instructions and a list of filenames in the package. An order form and pricing information for the full version of RT is available at http://www.west-inc.com or by emailing admin@west-inc.com and requesting the same. The Fortran source code is also available from http://emmy.otago.ac.nz/casm.html/rt.html.

References

Aalen, O. O. (1993) Further results on the non-parametric linear regression model in survival analysis. *Statistics in Medicine,* **12,** 1569–88.

Aastveit, A. H. (1990) Use of bootstrapping for estimation of standard deviation and confidence intervals of genetic variance and covariance components. *Biometrics Journal,* **32,** 515–27.

Abrahamson, W. G., Sattler, J. F., McCrea, K. D. and Weis, A. E. (1989) Variation in selection pressures on the goldenrod gall fly and the competitive interactions of its natural enemies. *Oecologia,* **79,** 15–22.

Abramovitch, L. and Singh, K. (1985) Edgeworth corrected pivotal statistics and the bootstrap. *Annals of Statistics,* **13,** 116–32.

Agresti, A. (1992) A survey of exact inference for contingency tables. *Statistical Science,* **7,** 131—77.

Akritas, M. G. (1986) Bootstrapping the Kaplan–Meier estimator. *Journal of the American Statistical Association,* **81,** 1032–8.

Alldredge, A. W., Deblinger, R. D. and Peterson, J. (1991) Birth and fawn bed selection by pronghorns in a sagebrush-steppe community. *Journal of Wildlife Management,* **55,** 222–7.

Alvarez-Buylla, E. R. and Slatkin, M. (1994) Finding confidence limits on population growth rates: three real examples revised. *Ecology,* **75,** 255–60.

Andersen, M. (1992) Spatial analysis of two-species interactions. *Ecology,* **91,** 134–40.

Anderson, C. S. (1995) Measuring and correcting for size selection in electrofishing mark–recapture experiments. *Transactions of the American Fisheries Society,* **124,** 663–76.

Andrade, I. and Proeçna, I. (1992) Search for a break in the Portugese GDP 1833–1985 with bootstrap methods, in *Bootstrapping and Related Techniques* (eds K. H. Jöckel, G. Rothe and W. Sender), Springer-Verlag, Berlin, pp. 133–42.

Andrén, H. (1994) Can one use nested subset pattern to reject the random sample hypothesis? Examples from boreal bird communities. *Oikos,* **70,** 489–91.

Andrews, D. F. and Herzberg, A. M. (1985) *Data,* Springer-Verlag, New York.

Archie, J. W. (1989a) A randomization test for phylogenetic information in systematic data. *Systematic Zoology,* **38,** 239–52.

Archie, J. W. (1989b) Phylogenies of plant families: a demonstration of phylogenic randomness in DNA sequence data derived from proteins. *Evolution,* **43,** 1796–1800.

Arita, H. T. (1993) Tests for morphological competitive displacement: a reassessment of parameters. *Ecology,* **74,** 627–30.

Arvesen, J. N. and Schmitz, T. H. (1970) Robust procedures for variance component problems using the jackknife. *Biometrics,* **26,** 677–86.

Babu, G. J., Rao, C. R. and Rao, M. B. (1992) Nonparametric estimation of specific occurrence/exposure rate in risk and survival analysis. *Journal of the*

American Statistical Association, **87**, 84–9.

Bacon, R. W. (1977) Some evidence of the largest squared correlation from several samples. *Econometrics*, **45**, 1997–2001.

Bailer, A. J. (1989) Testing variance for equality with randomization tests. *Journal of Statistical Computation and Simulation*, **31**, 1–8.

Baker, F. B. and Collier, R. O. (1966) Some empirical results on variance ratios under permutation in the completely randomized design. *Journal of the American Statistical Association*, **61**, 813–20.

Baker, R. (1995) Two permutation tests of equality of variances. *Statistics and Computing*, **5**, 289–96.

Baker, R. D. and Tilbury, J. B. (1993) Algorithm AS 283: rapid computation of the permutation paired and grouped *t*-tests. *Applied Statistics*, **42**, 432–41.

Barbujani, G. and Sokal, R. R. (1991) Genetic population structure of Italy. II. Physical and cultural barriers to gene flow. *American Journal of Human Genetics*, **48**, 398–411.

Barker, R. J. (1990) Paradise Shelduck band recoveries in the Wanganui district. *Notornis*, **37**, 173–81.

Barlow, W. E. and Sun, W. (1989) Bootstrapped confidence intervals for the Cox model using a linear relative risk form. *Statistics in Medicine*, **8**, 927–35.

Barnard, G. A. (1963) Discussion on Professor Bartlett's paper. *Journal of the Royal Statistical Society*, **B25**, 294.

Barrett, K. E. J. (1979) *Provisional Atlas of the Insects of the British Isles. Part 5, Hymenoptera: Formicidae*, Monks Wood Experimental Station, Huntingdon.

Bartlett, M. S. (1937) Properties of sufficiency and statistical tests. *Proceedings of the Royal Society of London*, **A160**, 268–82.

Bartlett, M. S. (1975) *The Statistical Analysis of Spatial Pattern*, Chapman & Hall, London.

Basu, D. (1980) Randomization analysis of experimental data: the Fisher randomization test. *Journal of the American Statistical Association*, **75**, 575–82.

Baumiller, T. K. and Ausich, W. I. (1992) The broken-stick model as a null hypothesis for crinoid stalk taphonomy and as a guide to the distribution of connective tissue in fossils. *Paleobiology*, **18**, 288–98.

Beaton, A. E. (1978) Salvaging experiments: interpreting least squares in non-random samples, in *Computer Science and Statistics: Tenth Annual Symposium on the Interface* (eds D. Hogben and D. Fife), U.S. Department of Commerce, Washington, pp. 137–45.

Becher, H. (1993) Bootstrap hypothesis testing. *Biometrics*, **49**, 1268–72.

Becker, R., Chambers, J. and Wilks, A. (1988) *The S Language*, Wadsworth, Belmont, California.

Bedall, F. K. and Zimmermann, H. (1976) On the generation of multivariate normal distributed random vectors by $N(0,1)$ distributed random numbers. *Biometrical Journal*, **18**, 467–71.

Behran, R. and Srivastava, M. S. (1985) Bootstrap tests and confidence regions for functions of a covariance matrix. *Annals of Statistics*, **13**, 95–115.

Bello, A. L. (1994) A bootstrap method for using imputation techniques for data with missing values. *Biometric Journal*, **36**, 453–64.

Bermingham, E. and Avise, J. C. (1986) Molecular zoogeography of freshwater fishes in the southeastern United States. *Genetics*, **113**, 939–65.

Berry, K. J. (1982) Algorithm AS179: enumeration of all permutations of multisets with fixed repetition numbers. *Applied Statistics*, **31**, 169–173.

Berry, K. J. and Mielke, P. W. (1984) Computation of exact probability values for multi-response permutation procedures (MRPP) *Communications in Statistics – Simulation and Computation*, **13**, 417–32.

Besag, J. (1978) Some methods of statistical analysis for spatial pattern. *Bulletin of the International Statistical Institute*, **47**, 77–92.

Besag, J. and Clifford, P. (1989) Generalized Monte Carlo significance tests. *Biometrika*, **76**, 633–42.

Besag, J. and Clifford, P. (1991) Sequential Monte Carlo p-values. *Biometrika*, **78**, 301–4.

Besag, J. and Diggle, P. J. (1977) Simple Monte Carlo tests for spatial pattern. *Applied Statistics*, **26**, 327–33.

Besag, J. and Green, P. J. (1993) Spatial statistics and Bayesian computation. *Journal of the Royal Statistical Society*, **B55**, 25–37.

Besag, J., Green, P., Higdon, D. and Mengersen, K. (1995) Bayesian computation and stochastic systems. *Statistical Science*, **10**, 3–66.

Besag, J. and Newell, J. (1991) The detection of clusters in rare diseases. *Journal of the Royal Statistical Society*, **A154**, 143–55.

Billard, L. (1994) The world of biometry. *Biometrics*, **50**, 899–916.

Biondini, M. E., Mielke, P. W. and Berry, K. J. (1988) Data-dependent permutation techniques for the analysis of ecological data. *Vegetatio*, **75**, 161–8.

Birks, H. J. B., Line J. M., Juggins, S., Stevenson, A. A. and ter Braak, C. J. F. (1990) Diatoms and pH reconstruction. *Philosophical Transactions of the Royal Society of London*, **B327**, 263–78.

Boik, R. J. (1987) The Fisher–Pitman permutation test: a non-robust alternative to the normal theory *F* test when variances are heterogeneous. *British Journal of Mathematical and Statistical Psychology*, **40**, 26–42.

Bookstein, F. L. (1987) Random walk and the existence of evolutionary rates. *Paleobiology*, **13**, 446–64.

Boomsma, A. (1991) *BOJA: a Program for Bootstrap and Jackknife Analysis.* IEC Progamma, P.O. Box 841, 9700 AV Groningen, The Netherlands.

Boos, D. D. and Brownie, C. (1989) Bootstrap methods for testing homogeneity of variances. *Technometrics*, **31**, 69–82.

Boos, D. D., Janssen, P. and Veraverbeke, N. (1989) Resampling from centred data in the two-sample problem. *Journal of Statistical Planning and Inference*, **21**, 327–45.

Booth, J. G., Butler, R. W. and Hall, P. (1994) Bootstrap methods for finite populations. *Journal of the American Statistical Association*, **89**, 1282–9.

Booth, J. G. and Hall, P. (1993) Bootstrap confidence regions for functional relationships in error-in-variables models. *Annals of Statistics*, **21**, 1780–91.

Borcard, D. and Legendre, P. (1994) Environmental control and spatial structure in ecological communities: an example using oribatid mites (Acari: Oribatei) *Environmental and Ecological Statistics*, **1**, 37–61.

Borcard, D., Legendre, P. and Drapeau, P. (1992) Partialling out the spatial component of ecological variation. *Ecology*, **73**, 1045–55.

Borowsky, R. (1977) Detection of the effects of selection on protein polymorphisms in natural populations by means of distance analysis. *Evolution*, **31**, 341–6. (errata in *Evolution*, **31**, 648).

Bose, A. and Babu, G. J. (1991) Accuracy of the bootstrap approximation. *Probability Theory and Related Fields*, **90**, 301–16.

Box, G. E. P. (1953) Non-normality and tests on variances. *Biometrika*, **40**, 318–35.

Bradley, J. V. (1968) *Distribution Free Statistical Methods.* Prentice-Hall, New Jersey.

Breidt, F. J., Davis, R. A. and Dunsmuir, W. T. M. (1995) Improved bootstrap prediction intervals for autoregressions. *Journal of Time Series Analysis*, **16**, 177–200.

Breiman, L. (1992) The little bootstrap and other methods for dimensionality selection in regression: X-fixed prediction error. *Journal of the American Statistical Association*, **87**, 738–54.

Breiman, L., Friedman, J., Olshen, R. and Stone, C. (1984) *Classification and Regression Trees*, Wadsworth, Belmont, California.

Breiman, L. and Spector, P. (1992) Submodel selection and evaluation in regression. The X-random case. *International Statistical Review*, **60**, 291–319.

Brey, T. (1990) Confidence limits for secondary production estimates: application of the bootstrap to the increment summation method. *Marine Biology*, **106**, 503–8.

Britten, H. B., Brussard, P. F., Murphy, D. D. and Ehrlich, P. R. (1995) A test for isolation-by-distance in central Rocky Mountain and Great Basin populations of Edith's Checkerspot butterfly (*Euphydryas editha*). *Journal of Heredity*, **86**, 204–10.

Bros, W. E. and Cowell, B. C. (1987) A technique for optimizing sample size (replication). *Journal of Experimental Marine Biology and Ecology*, **114**, 63–71.

Brown, B. M. and Maritz, J. S. (1982) Distribution-free methods in regression. *Australian Journal of Statistics*, **24**, 318–31.

Brown, E. N. and Czeisler, C. A. (1992) The statistical analysis of circadian rhythms and amplitude in constant-routine core-temperature data. *Journal of Biological Rhythms*, **7**, 177–202.

Brown, R. P. and Thorpe, R. S. (1991) Within-island microgeographic variation in body dimensions and scalation of the skink *Chalcides sexlineatus*, with testing of causal hypotheses. *Biological Journal of the Linnean Society*, **44**, 47–64.

Brown, R. P., Thorpe, R. S. and Báez, M. (1991) Parallel within-island microevolution of lizards on neighbouring islands. *Nature*, **352**, 60–2.

Brown, R. P., Thorpe, R. S. and Báez, M. (1993) Patterns and causes of morphological population differentiation in the Tenerife skink, *Chalcides viridanus*. *Biological Journal of the Linnean Society*, **50**, 313–28.

Brualdi, R. A. (1980) Matrices of zeros and ones with fixed row and column sum vectors. *Linear Algebra and its Applications*, **33**, 159–231.

Bruce, P. C. (1991) *Resampling Stats: User's Guide to IBM Version 3.0.* Resampling Stats, 612 N. Jackson Street, Arlington, VA 22201, USA.

Bryant, H. N. (1992) The role of permutation tail probability tests in phylogenetic systematics. *Systematic Biology*, **41**, 258–63.

Buckland, S. T. (1980) A modified analysis of the Jolly–Seber capture–recapture model. *Biometrics*, **36**, 419–35.

Buckland, S. T. (1982) A note on the Fourier series model for analysing line transect data. *Biometrics*, **38**, 469–77.

Buckland, S. T. (1984) Monte Carlo confidence intervals. *Biometrics*, **40**, 811–17.

Buckland, S. T., Anderson, D. R., Burnham, K. P. and Laake, J. L. (1993) *Distance Sampling: Estimating Abundance of Biological Populations*, Chapman & Hall, London.

Buckland, S. T. and Elston, D. A. (1993) Empirical models for the spatial distribution of wildlife. *Journal of Applied Ecology*, **30**, 478–95.

Buckland S. T. and Garthwaite P. H. (1990) Algorithm AS 259: Estimating confidence intervals by the Robbins–Monro search process. *Applied Statistics*, **39**, 413–24.

Buckland, S. T. and Garthwaite, P. H. (1991) Quantifying precision of mark–recapture estimates using the bootstrap and related methods. *Biometrics*, **47**, 255–68.

Burgman, M. A. (1987) An analysis of the distribution of plants on granite outcrops in southern Western Australia using Mantel tests. *Vegetatio*, **71**, 79–86.

Burke, M. D. and Yuen, K. C. (1995) Goodness-of-fit tests for the Cox model via bootstrap method. *Journal of Statistical Planning and Inference*, **47**, 237–56.

Burnham, K. P. and Overton, W. S. (1978) Estimation of the size of a closed population when capture probabilities vary among animals. *Biometrika*, **65**, 625–33.

Burnham, K. P. and Overton, W. S. (1979) Robust estimation of population size when capture probabilities vary among animals. *Ecology*, **60**, 927–36.

Burr, D. (1994) A comparison of certain bootstrap confidence intervals in the Cox model. *Journal of the American Statistical Association*, **89**, 1290–302.

Burr, D. and Doss, H. (1993) Confidence bands for the median survival time as a function of the covariates in the Cox model. *Journal of the American Statistical Association*, **88**, 1330–40.

Bycroft, C. M., Nicolaou, N. , Smith, B. and Wilson, J. B. (1993) Community structure (niche limitation and guild proportionality) in relation to the effect of spatial scale, in a *Nothofagus* forest sampled with a circular transect. *New Zealand Journal of Ecology*, **17**, 95–101.

Cade, B. S. and Richards, J. D. (1996) Least absolute deviation estimation and permutation tests as alternative regression procedures. *Biometrics*, **52**, 25–40.

Cain, A. J. and Sheppard, P. M. (1950) Selection in the polymorphic land snail *Cepaea nemoralis*. *Heredity*, **4**, 275–94.

Carpenter, J. M. (1992) Random cladistics. *Cladistics*, **8**, 147–53.

Carpenter, S. R., Frost, T. M., Heisey, D. and Kratz, T. K. (1989) Randomized intervention analysis and the interpretation of whole-ecosystem experiments. *Ecology*, **70**, 1142–52.

Carr, G. J. and Portier, C. J. (1993) An evaluation of some methods for fitting dose–response models to quantal–response developmental toxicology data. *Biometrics*, **49**, 779–91.

Case, T. J. (1983) Niche overlap and the assembly of island lizard communities. *Oikos*, **41**, 427–33.

Case, T. J. and Sidell, R. (1983) Pattern and chance in the structure of model and natural communities. *Evolution*, **37**, 832–49.

Castellano, S., Malhotra, A. and Thorpe, R. S. (1994) Within-island geographical variation of the dangerous Taiwanese snake, *Trimeresurus stejnegeri*, in relation to ecology. *Biological Journal of the Linnean Society*, **52**, 365–75.

Caton, E. L., McClelland, B. R., Patterson, D. A. and Yates, R. E. (1992) Characteristics of foraging perches used by breeding Bald Eagles in Montana. *Wilson Bulletin*, **104**, 136–42.

Chao, A. (1984) Nonparametric estimation of the number of classes in a population. *Scandinavian Journal of Statistics*, **11**, 265–70.

Chatfield, C. (1989) *The Analysis of Time Series*, 4th edn. Chapman & Hall, London.

Chen, J. and Sitter, R. R. (1993) Edgeworth expansion and the bootstrap for stratified sampling without replacement from a finite population. *Canadian Journal of Statistics*, **21**, 347–57.

Cheverud, J. M. (1988) A comparison of genetic and phenotypic correlations. *Evolution*, **42**, 958–68.

Cheverud, J. M., Wagner, G. P. and Dow, M. M. (1989) Methods for the comparative analysis of variation patterns. *Systematic Zoology*, **38**, 201–13.

Chiu, S. (1989) Detecting periodic components in a white Gaussian time series. *Journal of the Royal Statistical Society*, **B51**, 249–59.

Chung, J. H. and Fraser, D. A. S. (1958) Randomization tests for a multivariate two-sample problem. *Journal of the American Statistical Association*, **53**, 729–35.

Clarke, B. (1960) Divergent effects of natural selection on two closely related polymorphic snails. *Heredity*, **14**, 423–43.

Clarke, B. (1962) Natural selection in mixed populations of polymorphic snails. *Heredity*, **17**, 319–45.

Clarke, K. R. (1993) Non-parametric multivariate analysis of changes in community structure. *Australian Journal of Ecology*, **18**, 117–43.

Clarke, K. R. and Green, R. H. (1988) Statistical design and analysis for a 'biological effects' study. *Marine Ecology Progress Series*, **46**, 213–26.

Cole, B. J. (1981) Overlap, regularity and flowering phenologies. *American Naturalist*, **117**, 993–7.

Cole, M. J. and McDonald, J. W. (1989) Bootstrap goodness-of-link testing in generalized linear models, in *Statistical Modelling: Proceedings of GLIM 89 and the 4th International Workshop on Statistical Modelling* (ed. A. Decarli), Springer-Verlag Lecture Notes in Statistics 57, Berlin, pp. 84–94.

Cole, T. M. (1996) The use of matrix permutation tests for evaluating competing hypotheses of modern human origins. *Journal of Human Evolution*, **31**, 477–84.

Collingwood, C. A. (1979) The Formicidae (Hymenoptera) of Fennoscandinavia and Denmark. *Fauna Entomologica Scandinavica* **8**, Scandinavian Science, Klampenborg, Denmark.

Collins, M. F. (1987) A permutation test for planar regression. *Australian Journal of Statistics*, **29**, 303–8.

Connor, E. F. (1986) Time series analysis of the fossil record, in *Patterns and Processes in the History of Life* (eds D. M. Raup and D. Jablonski), Springer-Verlag, Berlin, pp. 119–47.

Connor, E. F. and Simberloff, D. (1979) The assembly of species communities: chance or competition? *Ecology*, **60**, 1132–40.

Connor, E. F. and Simberloff, D. (1983) Interspecific competition and species co-occurrence patterns on islands: null models and the evaluation of evidence. *Oikos*, **41**, 455–65.

Conover, W. J., Johnson, M. E. and Johnson, M. M. (1981) A comparative study of tests for homogeneity of variances, with applications to the outer continental shelf bidding data. *Technometrics*, **23**, 351–61.

Cook, J. R. and Stefanski, L. A. (1994) Simulation-extrapolation estimation in parametric measurement error models. *Journal of the American Statistical Association*, **89**, 1314–28.

Cope, D. A. and Lacy, M. G. (1992) Falsification of a single species hypothesis using the coefficient of variation: a simulation approach. *American Journal of Physical Anthropology*, **89**, 359–78.

Coursaget, P., Yvonnet, B., Gilks, W. R., Wang, C. C., Day, N. E., Chiron, J. P. and Diop-Mar, I. (1991) Scheduling of revaccinations against Hepatitis B virus. *Lancet*, **337**, 1180–3.

Cox, D. R. (1972) Regression model and life tables. *Journal of the Royal Statistical*

Society, **B34**, 187–220.

Cox, D. R. and Snell, E. J. (1981) *Applied Statistics: Principles and Examples*, Chapman & Hall, London.

Crivelli, A., Firinguetti, L., Montaño, R. and Muñóz, M. (1995) Confidence intervals in ridge regression by bootstrapping the dependent variable: a simulation study. *Communications in Statistics – Simulation and Computation*, **24**, 631–52.

Crowley, P. H. (1992) Density dependence, boundedness, and attraction: detecting stability in stochastic systems. *Oecologia*, **90**, 246–54.

Crowley, P. H. and Johnson, D. M. (1992) Variability and stability of a dragonfly assemblage. *Oecologia*, **90**, 260–9.

Cumming, B. F., Smol, J. P. and Birks, H. J. B. (1992) Scaled chrysophytes (Chrysophyceae and Synurophyceae) from Adirondack drainage lakes and their relationship to environmental variables. *Journal of Phycology*, **28**, 162–78.

Cushman, J. H., Lawton, J. H. and Manly, B. F. J. (1993) Latitudinal patterns in European ant assemblages: variation in species richness and body size. *Oecologia*, **95**, 30–7.

Cyr, L., Bank, H. L., Rust, P. F. and Schmehl, M. K. (1990) Comparative characteristics of estimators of the mean responding cell density in limiting dilution assays. *Communications in Statistics – Theory and Methods*, **19**, 2643–66.

Cyr, L., Rust, P. F., Peters, J. R., Schmehl, M. K. and Bank, H. L. (1993) Confidence intervals for the relative frequency of responding cells in limited dilution experiments. *Biometrics*, **49**, 491–8.

Dagosto, M. (1994) Testing positional behaviour of Malagasy lemurs: a randomization approach. *American Journal of Physical Anthropology*, **94**, 189–202.

Dale, M. R. T. and MacIsaac, D. A. (1989) New methods for the analysis of spatial pattern in vegetation. *Journal of Ecology*, **77**, 78–91.

Dalgleish, L. I. (1995) Software review: bootstrapping and jackknifing with BOJA. *Statistics and Computing*, **5**, 165–74.

Davison, A. C., Hinkley, D. V. and Schrechtman, E. (1986) Efficient bootstrap simulation. *Biometrika*, **73**, 555–66.

Davison, A. C., Hinkley, D. V. and Worton, B. J. (1992) Bootstrap likelihoods. *Biometrika*, **79**, 113–30.

Dawson, T. J., Tierney, P. J. and Ellis, B. A. (1992) The diet of the bridled nailtail wallaby (*Onychogalea fraenata*) II. Overlap in dietary niche breadth and plant preferences with the black-striped wallaby (*Macropus dorsalis*) and domestic cattle. *Wildlife Research*, **19**, 79–87.

De Angelis, D., Hall, P. and Young, G. A. (1993) Analytical and bootstrap approximations to estimator distributions in L^1 regression. *Journal of the American Statistical Association*, **88**, 1310–16.

De Beer, C. F. and Swanepoel, J. W. H. (1989) A modified Durbin–Watson test for serial correlation in multiple regression under nonnormality using the bootstrap. *Journal of Statistical Computation and Simulation*, **33**, 75–81.

Delgado, M. A. (1996) Testing serial independence using the sample distribution function. *Journal of Time Series Analysis*, **17**, 271–85.

Den Boer, P. J. (1990) On the stabilization of animal numbers. Problems of testing. 3. What do we conclude from significant test results? *Oecologia*, **83**, 38–46.

Den Boer, P. J. and Reddingius, J. (1989) On the stabilization of animal numbers. Problems of testing. 2. Confrontation with data from the field. *Oecologia*, **79**, 143–9.

Dennis, B. and Taper, M. L. (1994) Density dependence in time series observations of natural populations: estimation and testing. *Ecological Monographs*, **64**, 205–24.

Diaconis, P. and Efron, B. (1983) Computer intensive methods in statistics. *Scientific American*, **248**, 96–108.

Diamond, J. M. and Gilpin, M. E. (1982) Examination of the 'null' model of Connor and Simberloff for species co-occurrences on islands. *Oecologia*, **52**, 64–74.

DiCiccio, T. J. and Romano, J. P. (1988) A review of bootstrap confidence intervals. *Journal of the Royal Statistical Society*, **B50**, 338–70.

DiCiccio, T. J. and Romano, J. P. (1990) Nonparametric confidence limits by resampling methods and least favourable families. *International Statistical Review*, **58**, 59–76.

DiCiccio, T. J. and Tibshirani, R. (1987) Bootstrap confidence intervals and bootstrap approximations. *Journal of the American Statistical Association*, **82**, 163–70.

Dietz, E. J. (1983) Permutation tests for association between distance matrices. *Systematic Zoology*, **32**, 21–6.

Diggle, P. J. (1983) *Statistical Analysis of Spatial Point Patterns*, Academic Press, London.

Diggle, P. J. and Chetwynd, A. G. (1991) Second-order analysis of spatial clustering for inhomogenous populations. *Biometrics*, **47**, 1155–63.

Diggle, P. J. and Gratton, R. J. (1984) Monte Carlo methods of inference for implicit statistical models. *Journal of the Royal Statistical Society*, **B46**, 193–227.

Diggle, P. J. and Milne, R. K. (1983) Bivariate Cox processes: some models for bivariate spatial point patterns. *Journal of the Royal Statistical Society*, **B45**, 11–21.

Dillon, R. T. (1984) Geographical distance, environmental difference and divergence between isolated populations. *Systematic Zoology*, **33**, 69–82.

Dixit, S. S., Cumming, B. F., Birks, H. J. B., Smol, J. P., Kingston, J. C., Uutala, A. J., Charles, D. F. and Camburn, K. E. (1993) Diatom assemblages from Adirondack lakes (New York, USA) and the development of inference models for retrospective environmental assessment. *Journal of Paleolimnology*, **8**, 27–47.

Dixon, P. M. (1993) The bootstrap and the jackknife: describing the precision of ecological studies, in *Design and Analysis of Ecological Experiments* (eds S. M. Scheiner and J. Gurevitch), Chapman & Hall, New York, pp. 290–318.

Dixon, P. M., Weiner, J., Mitchell-Olds and Woodley, R. (1987) Bootstrapping the Gini coefficient of inequality. *Ecology*, **68**, 1548–51.

Dixon, W. J. (ed.) (1981) *BMDP Statistical Software,* University of California Press, Los Angeles.

Do, K. and Hall, P. (1991a) On importance resampling for the bootstrap. *Biometrika*, **78**, 161–7.

Do, K. and Hall, P. (1991b) Quasi-random resampling for the bootstrap. *Statistical Computing*, **1**, 13–22

Does, R. J. M. M., Strijbosch, L. W. G. and Albers, W. (1988) Using jackknife methods for estimating the parameter in dilution series. *Biometrics*, **44**, 1093–102.

Donegani, M. (1992) A bootstrap adaptive test for two-way analysis of variance.

Biometrical Journal, **34**, 141–6.

Donegani, M. and Unternährer, M. (1991) Estimation of the variance of an adaptive estimator by bootstrapping. *Communications in Statistics – Theory and Methods*, **20**, 3925–31.

Dopazo, J. (1994) Estimating errors and confidence intervals for branch lengths in phylogenetic trees by a bootstrap approach. *Journal of Molecular Evolution*, **38**, 300–4.

Douglas, M. E. and Endler, J. A. (1982) Quantitative matrix comparisons in ecological and evolutionary investigations. *Journal of Theoretical Biology*, **99**, 777–95.

Dow, M. M. and Cheverud, J. M. (1985) Comparison of distance matrices in studies of population structure and genetic microdifferentiation: quadratic assignment. *American Journal of Physical Anthropology*, **68**, 367–73.

Dow, M. M., Cheverud, J. M. and Friedlander, J. S. (1987) Partial correlation of distance matrices in population structure. *American Journal of Physical Anthropology*, **72**, 343–52.

Dwass, M. (1957) Modified randomization tests for non-parametric hypotheses. *Annals of Mathematical Statistics*, **28**, 181–7.

Eberhardt, L. L. (1995) Using the Lotka–Leslie model for sea otters. *Journal of Wildlife Management*, **59**, 222–7.

Edgington, E. S. (1980) *Randomization Tests*, Marcel Dekker, New York.

Edgington, E. S. (1987) *Randomization Tests*, 2nd edn, Marcel Dekker, New York.

Edgington, E. S. (1995) *Randomization Tests*, 3rd edn, Marcel Dekker, New York.

Edwards, D. (1985) Exact simulation based inference: a survey, with additions. *Journal of Statistical Computation and Simulation*, **22**, 307–26.

Efron, B. (1979a) Computers and the theory of statistics: thinking the unthinkable. *Society for Industrial and Applied Mathematics*, **21**, 460–80.

Efron, B. (1979b) Bootstrap methods: another look at the jackknife. *Annals of Statistics*, **7**, 1–26.

Efron, B. (1981a) Nonparametric standard errors and confidence intervals. *Canadian Journal of Statistics*, **9**, 139–72.

Efron, B. (1981b) Censored data and the bootstrap. *Journal of the American Statistical Association*, **76**, 312–19.

Efron, B. (1987) Better bootstrap confidence intervals. *Journal of the American Statistical Association*, **82**, 171–85.

Efron, B. (1992) Jackknife-after-bootstrap standard errors and influence functions. *Journal of the Royal Statistical Society*, **B54**, 83–127.

Efron, B. and Gong, G. (1983) A leisurely look at the bootstrap, the jackknife, and cross validation. *American Statistician*, **37**, 36–48.

Efron, B. and Tibshirani, R. (1986) Bootstrap methods for standard errors, confidence intervals, and other measures of statistical accuracy. *Statistical Science*, **1**, 54–77.

Efron, B. and Tibshirani, R. (1993) *An Introduction to the Bootstrap*, Chapman & Hall, London.

Elgar, M. A. and Harvey, P. H. (1987) Basal metabolic rates in mammals: allometry, phylogeny and ecology. *Functional Ecology*, **1**, 25–36.

Fagen, R. M. (1978) Information measures: statistical confidence limits and inference. *Journal of Theoretical Biology*, **73**, 61–79.

Faith, D. P. (1991) Cladistic permutation tests for monophyly and nonmonophyly. *Systematic Zoology*, **40**, 366–75.

Faith, D. P. and Ballard, J. W. O. (1994) Length differences and topology-dependent tests: a response to Källersjö *et al. Cladistics*, **10**, 57–64.

Faith, D. P. and Cranston, P. S. (1991) Could a cladogram have arisen by chance alone? On permutation tests for cladistic structure. *Cladistics*, **7**, 1–28.

Faith, D. P. and Cranston, P. S. (1992) Probability, parsimony and Popper. *Systematic Biology*, **41**, 252–7.

Faith, D. P. and Norris, R. H. (1989) Correlation of environmental variables with patterns of distribution and abundance of common and rare freshwater macroinvertebrates. *Biological Conservation*, **50**, 77–98.

Falck, W., Bjørnstad, O. N. and Stenseth, N. C. (1995) Bootstrap estimated uncertainty of the dominant Lyapunov exponent for Holarctic mocrotine rodents. *Proceedings of the Royal Society of London*, **B261**, 159–65.

Farris, J. S., Källersjö, M., Kluge, A. G. and Bult, C. (1994) Permutations. *Cladistics*, **10**, 65–76.

Faust, K. and Romney, A. K. (1985) The effect of skewed distributions on matrix permutation tests. *British Journal of Mathematical and Statistical Psychology*, **38**, 152–60.

Felsenstein, J. (1986) Confidence limits on phylogenies: an approach using the bootstrap. *Evolution*, **39**, 783–91.

Felsenstein, J. (1992) Estimating effective population size from samples of sequences: a bootstrap Monte Carlo integration method. *Genetical Research, Cambridge*, **60**, 209–20.

Ferrándiz, J, López, A., Llopis, A., Morales, M. and Tejerizo, M. L. (1995) Spatial interaction between neighbouring counties: cancer mortality data in Valencia (Spain) *Biometrics*, **51**, 665–78.

Ferrenberg, A. M., Landau, D. P. and Wong, Y. J. (1992) Monte Carlo simulations: hidden errors from 'good' random number generators. *Physical Review Letters*, **69**, 3382–4.

Fisher, N. I. and Hall, P. (1990) On bootstrap hypothesis tests. *Australian Journal of Statistics*, **32**, 177–90.

Fisher, N. I. and Hall, P. (1991) Bootstrap algorithms for small samples. *Journal of Statistical Planning and Inference*, **27**, 157–69.

Fisher, N. I. and Hall, P. (1992) Bootstrap methods for directional data, in *The Art of Statistical Science* (ed. K. V. Mardia), Wiley, New York, pp. 47–63.

Fisher, R. A. (1924) The influence of rainfall on the yield of wheat at Rothamsted. *Philosophical Transactions of the Royal Society of London*, **B213**, 89–142.

Fisher, R. A. (1935) *The Design of Experiments*, Oliver and Boyd, Edinburgh.

Fisher, R. A. (1936) The coefficient of racial likeness and the future of craniometry. *Journal on the Royal Anthropological Institute*, **66**, 57–63.

Flack, V. F. and Chang, P. C. (1987) Frequency of selecting noise variables in subset regression analysis: a simulation study. *American Statistician*, **41**, 84–6.

Fleming, T. H. and Partridge, B. L. (1984) On the analysis of phenological overlap. *Oecologia*, **62**, 344–50.

Flint, P. L., Pollock, K. H., Thomas, D. and Sedinger, J. S. (1995) Estimating prefledging survival: allowing for brood mixing and dependence among brood mates. *Journal of Wildlife Management*, **59**, 448–55.

Ford, E. B. (1975) *Ecological Genetics*, 4th edn, Chapman & Hall, London.

Fore, L. S., Karr, J. R. and Conquest, L. L. (1994) Statistical properties of an index of biological integrity used to evaluate water resources. *Canadian Journal of Fisheries and Aquatic Science*, **51**, 1077–87.

Fortin, M. (1994) Edge detection algorithms for two-dimensional ecological data. *Ecology*, **75**, 956–65.

Foutz, R. V., Jensen, D. R. and Anderson, G. W. (1985) Multiple comparisons in the randomization analysis of designed experiments with growth curve responses. *Biometrics*, **41**, 29–37.

Fox, B. J. (1987) Species assembly and the evolution of community structure. *Evolutionary Ecology*, **1**, 201–13.

Fox, B. J. (1989) Small-mammal community pattern in Australian heathland: a taxonomically-based rule for species assembly, in *Patterns in the Structure of Mammalian Communities* (eds D. W. Morris, Z. Abramsky, B. J. Fox, and M. R. Willig), Special Publication of the Museum of Texas Technical University 28, Lubbock, Texas, pp. 91–103.

Fox, B. J. and Brown, J. H. (1993) Assembly rules for functional groups in North American desert rodent communities. *Oikos*, **67**, 358–70.

Fox, B. J. and Brown, J. H. (1995) Reaffirming the validity of the assembly rule for functional groups or guilds: a reply to Wilson. *Oikos*, **73**, 125–32.

Fox, W. T. (1987) Harmonic analysis of periodic extinctions. *Paleobiology*, **13**, 257–71.

France, R. McQueen, D., Lynch, A. and Dennison, M. (1992) Statistical comparison of seasonal trends for autocorrelated data: a test of consumer and resource mediated trophic interactions. *Oikos*, **65**, 45–51.

Francis, I. (1974) Factor analysis: fact or fabrication. *Mathematical Chronicle*, **3**, 9–44.

Frangos, C. C. (1987) An updated bibliography on the jackknife method. *Communications in Statistics - Theory and Methods*, **16**, 1543–84.

Frangos, C. C. and Schucany, W. R. (1995) Improved bootstrap confidence intervals in certain toxicological experiments. *Communications in Statistics - Theory and Methods*, **24**, 829–44.

Frawley, W. H. (1974) Using the jackknife in testing dose responses in proportions near zero or one – revisited. *Biometrics*, **30**, 539–45.

Freedman, D. and Lane, D. (1983) A nonstochastic interpretation of reported significance levels. *Journal of Business and Economic Statistics*, **1**, 292–8.

Gabriel, K. R. and Hall, W. J. (1983) Rerandomization inference on regression and shift effects: computationally feasible methods. *Journal of the American Statistical Association*, **78**, 827–36.

Gabriel, K. R. and Hsu, C. F. (1983) Evaluation of the power of rerandomization tests with application to weather modification experiments. *Journal of the American Statistical Association*, **78**, 766–75.

Gail, M. H., Tan, W. Y. and Piantadosi, S. (1988) Tests of no treatment effect in randomized clinical trials. *Biometrika*, **75**, 57–64.

Galiano, E. F., Castro, I. and Sterling, A. (1987) A test for spatial pattern in vegetation using a Monte Carlo simulation. *Journal of Ecology*, **75**, 915–24.

Garcia-Jurado, I., González-Manteiga, W., Prada-Sánchez, J. M., Febrero-Bande, M. and Cao, R. (1995) Predicting using Box–Jenkins, nonparametric, and bootstrap techniques. *Technometrics*, **37**, 303–10.

Garthwaite, P. H. and Buckland, S. T. (1992) Generating Monte Carlo confidence intervals by the Robbins–Monro process. *Applied Statistics*, **41**, 159–71.

Garthwaite, P. H., Yu, K. and Hope, P. B. (1995) Bayesian analysis of a multiple-recapture model. *Communications in Statistics - Theory and Methods*, **24**, 2229–47.

Gates, J. (1991) Exact Monte Carlo tests using several statistics. *Journal of Statistical Computation and Simulation*, **38**, 211–18.

Gelfand, A. E. and Mallick, B. K. (1995) Bayesian analysis of proportional hazards models built from monotone functions. *Biometrics*, **51**, 843–52.

Gelfand, A. E. and Sahu, S. K. (1994) On Markov chain Monte Carlo acceleration. *Journal of Computational and Graphical Statistics*, **3**, 261–76.

Gelman, A. (1995) Method of moments using Monte Carlo simulation. *Journal of Computational and Graphical Statistics*, **4**, 36–54.

Gelman, A., Carlin, J. B., Stern, H. S. and Rubin, D. B. (1995) *Bayesian Data Analysis*. Chapman and Hall, London.

Gelman, A. and Rubin, D. B. (1992) Inference from iterative simulation using multiple sequences. *Statistical Sciences*, **7**, 457–511.

Geman, S. and Geman, D. (1984) Stochastic relaxation, Gibbs distributions and the Bayesian restoration of images. *IEEE Transactions on Pattern Recognition and Machine Intelligence*, **6**, 721–41.

George, E. O. and Mudholkar, G. S. (1990) P-values for two-sided tests. *Biometrical Journal*, **32**, 747–51.

Geyer, C. J. (1992) Practical Markov chain Monte Carlo. *Statistical Science*, **7**, 473–511.

Geyer, C. J. and Thompson, E. A. (1995) Annealing Markov chain Monte Carlo with applications to ancestral inference. *Journal of the American Statistical Association*, **90**, 909–20.

Gibbons, J. D. (1986) Randomness, tests of, in *Encyclopedia of Statistical Sciences*, **7**, 555–62, Wiley, New York.

Gibson, A. R., Baker, A. J. and Moeed, A. (1984) Morphometric variation in introduced populations of the common myna (*Acridotheres tristis*): an application of the jackknife to principal component analysis. *Systematic Zoology*, **33**, 408–21.

Gilinsky, N. L. and Bambach, R. K. (1986) The evolutionary bootstrap: a new approach to the study of taxonomic diversity. *Paleobiology*, **12**, 251–68.

Gilks, W. R., Clayton, D. G., Spiegelhalter, D. J., Best, N. G., McNeill, A. J., Sharples, L. D. and Kirby, A. J. (1993) Modelling complexity: applications of Gibbs sampling in medicine. *Journal of the Royal Statistical Society*, **B55**, 39–52.

Gilpin, M. E. and Diamond, J. M. (1982) Factors contributing to non-randomness in species co-occurrences on islands. *Oecologia*, **52**, 75–84.

Gilpin, M. E. and Diamond, J. M. (1984) Are species co-occurrences on islands non-random, and are null hypotheses useful in community ecology? In *Ecological Communities: Conceptual Issues and the Evidence* (eds D. R. Strong, D. Siberloff, L. G. Abele and A. B. Thistle), Princeton University Press, New Jersey, pp. 297–343.

Gilpin, M. E. and Diamond, J. M. (1987) Comments on Wilson's null model. *Oecologia*, **74**, 159–60.

Gladen, B. (1979) The use of the jackknife to estimate proportions from toxicological data in the presence of litter effects. *Journal of the American Statistical Association*, **74**, 278–83.

Gleason, J. R. (1988) Algorithms for balanced bootstrap simulations. *American Statistician*, **42**, 263–6.

Golbeck, A. L. (1992) Bootstrapping current life table estimators, In *Bootstrapping and Related Techniques* (eds K. H. Jöckel, G. Rothe and W. Sendler), Springer-Verlag, Berlin, pp. 197–201.

Gonzalez, M. L. and Manly, B. F. J. (1996) Analysis of variance by randomization with small data sets. *Environmetrics* (in press).

Good, P. (1994) *Permutation Tests: a Practical Guide to Resampling Methods for*

Testing Hypotheses, Springer-Verlag, New York.

Goodhall, D. W. (1974) A new method for the analysis of spatial pattern by random pairing of quadrats. *Vegetatio*, **29**, 135–46.

Gordon, A. D. (1981) *Classification*, Chapman & Hall, London.

Gray, H. L. and Schucany, W. R. (1972) *The Generalized Jackknife Statistic*, Marcel Dekker, New York.

Green, B. F. (1977) A practical interactive program for randomization tests of location. *American Statistician*, **31**, 37–9.

Green, K. (1989) Altitudinal and seasonal differences in the diets of *Antechinus swainsonii* and *A. stuartii* (Marsupialia: Dasyuridae) in relation to the availability of prey in the Snowy Mountains. *Australian Wildlife Research*, **16**, 581–92.

Greenacre, M. (1984) *The Theory and Applications of Correspondence Analysis*, Academic Press, London.

Grieg-Smith, P. (1952) The use of random and contiguous quadrats in the study of the structure of plant communities. *Annals of Botany*, **16**, 293–316.

Grieg-Smith, P. (1983) *Quantitative Plant Ecology*, Blackwell Scientific Publications, Oxford.

Grine, F. E., Demes, B., Junger, W. L. and Cole, T. M. (1993) Taxonomic affinity of the early Homo cranium from Swartkrans, South Africa. *American Journal of Physical Anthropology*, **92**, 411–26.

Guo, S. W. and Thompson, E. A. (1992) Performing the exact test of Hardy–Weinberg proportion for multiple alleles. *Biometrics*, **48**, 361–72.

Haase, P. (1995) Spatial pattern analysis in ecology based on Ripley's *K*-function: introduction and method of edge correction. *Journal of Vegetation Science*, **6**, 575–82.

Haining, R. (1990) *Spatial Data Analysis in the Social and Environmental Sciences*, Cambridge University Press, Cambridge.

Hall, P. (1986) On the number of bootstrap simulations required to construct a confidence interval. *Annals of Statistics*, **14**, 1453–62.

Hall, P. (1989) Antithetic resampling for the bootstrap. *Biometrika*, **76**, 713–24.

Hall, P. (1992a) *The Bootstrap and Edgeworth Expansion*, Springer-Verlag, New York.

Hall, P. (1992b) On the removal of skewness by transformation. *Journal of the Royal Statistical Society*, **54**, 221–8.

Hall, P. and Owen, A. B. (1993) Empirical likelihood confidence bands in density estimation. *Journal of Computational and Graphical Statistics*, **2**, 273–89.

Hall, P. and Wilson, S. (1991) Two guidelines for bootstrap hypothesis testing. *Biometrics*, **47**, 757–62.

Hall, W. J. (1985) Confidence intervals, by rerandomization, for additive and multiplicative effects. *Proceedings of the Statistical Computing Section*, American Statistical Association Conference, Las Vegas, August 1985, pp. 60–9.

Hallett, J. G. (1991) The structure and stability of small animal faunas. *Oecologia*, **88**, 383–93.

Hallin, M. and Melard, G. (1988) Rank-based tests for randomness against first-order serial dependence. *Journal of the American Statistical Association*, **83**, 1117–28.

Hamilton, M. A. and Collings, B. J. (1991) Determining the appropriate sample size for nonparametric tests for location shift. *Technometrics*, **33**, 327–37.

Harkness, R. D. and Isham, V. (1983) A bivariate spatial point pattern of ants' nests. *Applied Statistics*, **32**, 293–303.

Harper, J. C. (1978) Groupings by locality in community ecology and paleoecology: tests of significance. *Lethaia*, **11**, 251–7.

Harris, R. J. (1985) *A Primer on Multivariate Statistics*, Academic Press, Orlando.

Harris, W. F. (1986) The breeding ecology of the South Island Fernbird in Otago wetlands, PhD Thesis, University of Otago, Dunedin, New Zealand.

Harshman, J. (1994) The effects of irrelevant characters on bootstrap values. *Systematic Biology*, **43**, 419–24.

Harvey, L. E. (1994) Spatial patterns of inter-island plant and bird species movements in the Galápagos Islands. *Journal of the Royal Society of New Zealand*, **24**, 45–63.

Hasegawa, M. and Kishino, H. (1994) Accuracies of the simple methods for estimating the bootstrap probability of a maximum-likelihood tree. *Molecular Biology and Evolution*, **11**, 142–5.

Haslett, S. J. and Wear, R. G. (1985) Biomass estimation of *Artemia* at Lake Grassmere, Marlborough, New Zealand. *Australian Journal of Marine and Freshwater Research*, **36**, 537–57.

Hastings, W. K. (1970) Monte Carlo sampling methods using Markov chains and their applications. *Biometrika*, **57**, 97–109.

Haukka, J. K. (1995) Correction for covariate measurement error in generalized linear models – a bootstrap approach. *Biometrics*, **51**, 1127–32.

Hayes, D. L. (1995) *Recovery Monitoring of Pigeon Guillemot Populations in Prince William Sound, Alaska. Exxon Valdez* Oil Spill Restoration Report (Restoration Project 94173), U.S. Fish and Wildlife Service, Anchorage, Alaska.

Heikkinen, J. and Högmander, H. (1994) Fully Bayesian approach to image restoration with an application in biogeography. *Applied Statistics*, **43**, 569–82.

Heltshe, J. F. (1988) Jackknife estimate of the matching coefficient of similarity. *Biometrics*, **44**, 447–60. (correction in *Biometrics*, **45**, 1347).

Heltshe, J. F. and Forrester, N. E. (1983a) Estimating species richness using the jackknife procedure. *Biometrics*, **39**, 1–11.

Heltshe, J. F. and Forrester, N. E. (1983b) Estimating diversity using quadrat sampling. *Biometrics*, **39**, 1073–6.

Heltshe, J. F. and Forrester, N. E. (1985) Statistical evaluation of the jackknife estimate of diversity when using quadrat samples. *Ecology*, **66**, 107–11.

Hemelrijk, C. K. (1990) A matrix partial correlation test used in investigations of reciprocity and other social interaction patterns at group level. *Journal of Theoretical Biology*, **143**, 405–20.

Higgins, S. S., Bendel, R. B. and Mack, R. N. (1984) Assessing competition among skewed distributions of plant biomass: an application of the jackknife. *Biometrics*, **40**, 131–7.

Higham, C. F. W., Kijngam, A. and Manly, B. F. J. (1980) An analysis of prehistoric canid remains from Thailand. *Journal of Archaeological Science*, **7**, 149–165.

Hillis, D. M. and Bull, J. J. (1993) An empirical test of bootstrapping as a method for assessing confidence in phylogenetic analysis. *Systematic Biology*, **42**, 182–92.

Hinch, S. G., Somers, K. M. and Collins, N. C. (1994) Spatial autocorrelation and assessment of habitat–abundance relationships in littoral zone fish. *Canadian Journal of Fisheries and Aquatic Science*, **51**, 701–12.

Hinkley, D. V. (1980) Discussion on Basu's paper. *Journal of the American Statistical Association*, **75**, 582–4.

Hinkley, D. V. (1983) Jackknife methods. *Encyclopedia of Statistical Sciences*, **4**, 280–7.

Hinkley, D. V. (1988) Bootstrap methods. *Journal of the Royal Statistical Society*, **B50**, 321–37.

Hinkley, D. V. and Shi, S. (1989) Importance sampling and the nested bootstrap. *Biometrika*, **76**, 435–46.

Hjorth, J. S. U. (1994) *Computer Intensive Statistical Methods: Validation, Model Selection and Bootstrap*. Chapman & Hall, London.

Hoeffding, W. (1952) The large sample power of tests based on permutations of observations. *Annals of Mathematical Statistics*, **23**, 169–92.

Hoffman, A. (1985) Patterns of family extinction depend on definition and geologic timescale. *Nature*, **315**, 659–62.

Hogg, S. E., Murray, D. L. and Manly, B. F. J. (1978) Methods of estimating throughfall under a forest. *New Zealand Journal of Science*, **21**, 129–36.

Högmander, H. (1995) *Methods of Spatial Statistics in Monitoring of Wildlife Populations*. Jyäskylä Studies in Computer Science, Economics and Statistics, University of Jyäskylä.

Holyoak, M. (1993) The frequency of detection of density dependence in insect orders. *Ecological Entomology*, **18**, 339–47.

Holyoak, M. (1994) Identifying delayed density-dependence in time series data. *Oikos*, **70**, 296–304.

Holyoak, M. and Crowley, P. H. (1993) Avoiding erroneously high levels of detection in combinations of semi-independent tests. *Oecologia*, **95**, 103–14.

Holyoak, M. and Lawton, J. H. (1992) Detection of density dependence from annual censuses of bracken-feeding insects. *Oecologia*, **91**, 425–30.

Hopkins, W. G., Wilson, N. C. and Russell, D. G. (1991) Validation of the physical activity instrument for the Life in New Zealand national survey. *American Journal of Epidemiology*, **133**, 73–82.

Howe, M. A., Geissler, P. H. and Harrington, B. A. (1989) Population trends of North American shorebirds based on the International Shorebird Survey. *Biological Conservation*, **49**, 185–99.

Hu, F. and Zidek, J. V. (1995) A bootstrap based on the estimating equations of the linear model. *Biometrika*, **82**, 263–75.

Huang, J. S. (1991) Efficient computation of the performance of bootstrap and jackknife estimators of the variance of L-statistics. *Journal of Statistical Computation and Simulation*, **38**, 45–66.

Hubbard, A. E. and Gilinsky, N. L. (1992) Mass extinctions as statistical phenomena: an examination of the evidence using χ^2 tests and bootstrapping. *Paleobiology*, **18**, 148–60.

Hubert, L. J. (1985) Combinatorial data analysis: association and partial association. *Psychometrika*, **4**, 449–67.

Hubert, L. J. and Schultz, J. (1976) Quadratic assignment as a general data analysis strategy. *British Journal of Mathematical and Statistical Psychology*, **29**, 190–241.

Huggins, R. M. (1989) On the statistical analysis of capture experiments. *Biometrika*, **76**, 133–40.

Hurlbert, S. H. (1978) The measurement of niche overlap and some relatives. *Ecology*, **59**, 67–77.

Hurvich, C. M., Simonoff, J. S. and Zeger, S. L. (1991) Variance estimation for sample autocovariances: direct and resampling approaches. *Australian Journal of Statistics*, **33**, 23–42.

Hurvich, C. M. and Tsai, C. (1990) The impact of model selection on inference in linear regression. *American Statistician*, **44**, 214–17.

Hutchings, M. J. (1979) Standing crop and pattern in pure stands of *Merculialis perennis* and *Rubus fruticosus* in mixed deciduous woodland. *Oikos*, **31**, 351–7.

Jackson, D. A. (1993) Stopping rules in principal components analysis: a comparison of heuristical and statistical approaches. *Ecology*, **74**, 2204–14.

Jackson, D. A. (1995) Bootstrapped principal components – reply to Mehlman *et al. Ecology*, **76**, 644–5.

Jackson, D. A. and Somers, K. M. (1989) Are probability estimates from the permutation model of Mantel's test stable? *Canadian Journal of Zoology*, **67**, 766–9.

Jackson, D. A., Somers, K. M. and Harvey, H. H. (1992) Null models and fish communities: evidence of nonrandom patterns. *American Naturalist*, **139**, 930–51.

Jacobs, J. A. (1986) From the core or from the skies? *Nature*, **323**, 296–7.

Jain, A. K. and Dubes, R. C. (1988) *Algorithms for Clustering Data*, Prentice-Hall, New York.

Jaksić, F. M. and Medel, R. G. (1990) Objective recognition of guilds: testing for statistically significant species clusters. *Oecologia*, **82**, 87–92.

James, G. S. (1954) Tests of linear hypotheses in univariate and multivariate analysis when the ratios of the population variances are unknown. *Biometrika*, **41**, 19–43.

Jöckel, K. H. (1986) Finite sample properties and asymptotic efficiency of Monte Carlo tests. *Journal of Statistical Computation and Simulation*, **14**, 336–47.

Joern, A. and Lawlor, L. R. (1980) Food and microhabitat utilization by grasshoppers from arid grasslands: comparison with neutral models. *Ecology*, **61**, 591–9.

Joern, A. and Lawlor, L. R. (1981) Guild structure in grasshopper assemblages based on food and microhabitat resources. *Oikos*, **37**, 93–104.

John, R. D. and Robinson, J. (1983) Significance levels and confidence intervals for permutation tests. *Journal of Statistical Computation and Simulation*, **16**, 161–73.

Johns, M. V. (1988) Importance sampling for bootstrap confidence intervals. *Journal of the American Statistical Association*, **83**, 709–14.

Johnson, S. G. and Johnston, R. F. (1989) A multifactorial study of variation in interclutch interval and annual reproduction success in the feral pigeon, *Columba livia. Oecologia*, **80**, 87–92.

Jones, M. C. and Sibson, R. (1987) What is projection pursuit? *Journal of the Royal Statistical Society*, **A150**, 1–36.

Joseph, L., Gyorkos, T. W. and Coupal, L. (1995) Bayesian estimation of disease prevalence and the parameters of diagnostic tests in the absence of a gold standard. *American Journal of Epidemiology*, **141**, 263–72.

Kadmon, R. (1995) Nested species subsets and geographical isolation: a case study. *Ecology*, **76**, 458–65.

Kadmon, R. and Pulliam, H. R. (1993) Island biogeography: effect of geographical isolation on species composition. *Ecology*, **74**, 977–81.

Källersjö, M., Farris, J. S., Kluge, A. G. and Bult, C. (1992) Skewness and permutation. *Cladistics*, **8**, 275–87.

Kaplan, E. L. and Meier, P. (1958) Nonparametric estimation from incomplete observations. *Journal of the American Statistical Association*, **53**, 457–81.

Karr, J. R. and Martin, T. E. (1981) Random numbers and principal components:

further searches for the unicorn, in *The Use of Multivariate Statistics in Studies of Wildlife Habitat* (ed. D. E. Capen), United States Department of Agriculture, general technical report, RM-87, pp. 20-4.

Kemp, W. P. and Dennis, B. (1993) Density dependence in rangeland grasshoppers (Orthoptera: Acrididae) *Oecologia*, **96**, 1-8.

Kempthorne, O. (1952) *The Design and Analysis of Experiments*, Wiley, New York.

Kempthorne, O. (1955) The randomization theory of statistical inference. *Journal of the American Statistical Association*, **50**, 946-67.

Kempthorne, O. and Doerfler, T. E. (1969) The behaviour of some significance tests under experimental randomization. *Biometrika*, **56**, 231-48.

Kennedy, P. E. (1995) Randomization tests in econometrics. *Journal of Business and Economic Statistics*, **13**, 85-94.

Kennedy, P. E. and Cade, B. S. (1996) Randomization tests for multiple regression. *Communications in Statistics – Computation and Simulation* (in press).

Kim, Y. B., Haddock, J. and Willemain, T. R. (1993) The binary bootstrap: inference with autocorrelated binary data. *Communications in Statistics – Simulation and Computation*, **22**, 205-16.

Kincaid, W. B. and Bryant, E. H. (1983) A geometric method for evaluating the null hypothesis of random habitat utilization. *Ecology*, **64**, 1463-70.

Kitchell, J. A., Estabrook, G. and MacLeod, N. (1987) Testing for equality of rates of evolution. *Paleobiology*, **13**, 272-85.

Kirby, J. M. (1991a) Multiple functional regression – I. Function minimization technique. *Computers and Geosciences*, **17**, 537-47.

Kirby, J. M. (1991b) Multiple functional regression – II. Rotation followed by classical regression techniques. *Computers and Geosciences*, **17**, 895-905.

Kirkpatrick, M. and Slatkin, M. (1993) Searching for evolutionary patterns in the shape of a phylogenetic tree. *Evolution*, **47**, 1171-81.

Knapp, S. J. and Bridges, W. C. (1988) Parametric and jackknife confidence interval estimators for two-factor mating design genetic variance ratios. *Theoretical and Applied Genetics*, **76**, 385-92.

Knapp, S. J., Bridges, W. C. and Yang, M. (1989) Nonparametric confidence interval estimators for heritability and expected selection response. *Genetics*, **121**, 891-8.

Knox, G. (1964) Epidemiology of childhood leukaemia in Northumberland and Durham. *British Journal of Preventive and Social Medicine*, **18**, 17-24.

Knox, R. G. (1989) Effects of detrending and rescaling on correspondence analysis: solution stability and accuracy. *Vegetatio*, **83**, 129-36.

Knox, R. G. and Peet, R. K. (1989) Bootstrapped ordination: a method for estimating sampling effects in indirect gradient analysis. *Vegetatio*, **80**, 153-65.

Knuth, D. E. (1981) *The Art of Computer Programming. Volume 2, Semi-numerical Algorithms*, Addison-Wesley, Reading, Massachusetts.

Koch, G. G., Elashoff, J. D. and Amara, I. A. (1988) Repeated measurements – design and analysis. *Encyclopedia of Statistical Sciences*, **8**, 46-73.

Konigsberg, L. W., Kramer, A., Donnelly, S. M., Relethford, J. H. and Blangero, J. (1994) Modern human origins. *Nature*, **372**, 228-9.

Kramer, A., Donnelly, S. M., Kidder, J. H., Ousley, S. D. and Olah, S. M. (1995) Craniometric variation in large-bodied hominoids: testing the single-species hypothesis for *Homo habilis*. *Journal of Human Evolution*, **29**, 443-62.

Krzanowski, W. J. (1993) Permutational tests for correlation matrices. *Statistics and Computing*, **3**, 37-44.

Kuhner, M. K., Yamato, J. and Felsenstein, J. (1995) Estimating effective population size and mutation rate from sequence data using Metropolis–Hastings sampling. *Genetics*, **140**, 1421–30.

Lake, J. A. (1995) Calculating the probability of multitaxon evolutionary trees: bootstrappers gambit. *Proceedings of the National Academy of Sciences, USA* **92**, 9662–6.

Lambert, D. (1985) Robust two-sample permutation tests. *Annals of Statistics*, **13**, 606–25.

Lange, N., Carlin, B. P. and Gelfand, A. E. (1992) Hierarchical Bayes models for the progression of HIV infection using longitudinal $CD4^+$ counts. *Journal of the American Statistical Association*, **87**, 615–32.

Lapointe, F., Kirsch, J. A. W. and Bleiweiss, R. (1994) Jackknifing of weighted trees: validation of phylogenies reconstructed from distance matrices. *Molecular Phylogenetics and Evolution*, **3**, 256–67.

Lapointe, F. and Legendre, P. (1990) A statistical framework to test the consensus of two nested classifications. *Systematic Zoology*, **39**, 1–13.

Lapointe, F. and Legendre, P. (1991) The generation of untrametric matrices representing dendrograms. *Journal of Classification*, **8**, 177–200.

Lapointe, F. and Legendre, P. (1992a) A statistical framework to test the consensus among additive trees (cladograms) *Systematic Biology*, **41**, 158–71.

Lapointe, F. and Legendre, P. (1992b) Statistical significance of the matrix correlation coefficient for comparing independent phylogenetic trees. *Systematic Biology*, **41**, 378–84.

Larkin, R. P. (1992) Comparing actual and random distributions of nearest neighbours. *Auk*, **109**, 202–3.

Lawlor, L. R. (1980) Structure and stability in natural and randomly constructed communities. *American Naturalist*, **116**, 394–408.

LeBlanc, M. and Crowley, J. (1993) Survival trees by goodness of split. *Journal of the American Statistical Association*, **88**, 457–67.

Lecointre, G., Philippe, H., Vân Lê, H. L. and Le Guyader, H. (1993) Species sampling has a major impact on phylogenetic inference. *Molecular Phylogenetics and Evolution*, **2**, 205–24.

Legendre, P. (1993) Spatial autocorrelation: trouble or new paradigm? *Ecology*, **74**, 1659–73.

Legendre, P. and Fortin, M. J. (1989) Spatial pattern and ecological analysis. *Vegetatio*, **80**, 107–38.

Legendre, P., Lapointe, F. and Casgrain, P. (1994) Modelling brain evolution from behaviour: a permutational regression approach. *Evolution*, **48**, 1487–99.

Legendre, P., Oden, N. L., Sokal, R. R., Vaudor, A. and Kim, J. (1990) Approximate analysis of variance of spatially autocorrelated regional data. *Journal of Classification*, **7**, 53–75.

Legendre, P. and Troussellier, M. (1988) Aquatic heterotrophic bacteria: modelling in the presence of spatial autocorrelation. *Limnology and Oceanography*, **33**, 1055–67.

Léger, C., Politis, D. N. and Romano, J. P. (1992) Bootstrap technology and applications. *Technometrics*, **34**, 378–98.

Lenski, R. E. and Service, P. M. (1982) The statistical analysis of population growth rates calculated from schedules of survivorship and fecundity. *Ecology*, **63**, 655–62.

Levene, H. (1960) Robust tests for equality of variance, in *Contributions to Probability and Statistics*, (eds I. Olkin, S. G. Ghurye, W. Hoeffding, W. G.

Madow and H. B. Mann), Stanford University Press, Stanford, pp. 278–92.

Levin, B. and Robbins, H. (1983) Urn models for regression analysis with applications to employment discrimination. *Law and Contemporary Problems*, **46**, 247–67.

Lin, J. (1989) Approximating the normal tail probability and its inverse for use on a pocket calculator. *Applied Statistics*, **38**, 69–70.

Link, W. A. and Sauer, J. R. (1995) Estimation and confidence intervals for empirical mixing distributions. *Biometrics*, **51**, 810–21.

Linton, L. R., Edgington, E. S. and Davies, R. W. (1989) A view of niche overlap amenable to statistical analysis. *Canadian Journal of Zoology*, **67**, 55–60.

Livshits, G., Sokal, R. R. and Kobyliansky, E. (1991) Genetic affinities of Jewish populations. *American Journal of Human Genetics*, **49**, 131–46.

Lo, S. K. (1991) On the analysis and application of measures of linkage disequilibrium. *Australian Journal of Statistics*, **33**, 249–59.

Lock, R. H. (1986) Using the computer to approximate permutation tests. *Proceedings of the Statistical Computing Section*, American Statistical Association Conference, Chicago, August 1986, pp. 349–52.

Lokki, H. and Saurola, P. (1987) Bootstrap methods for two-sample location and scatter problems. *Acta Ornithologica*, **23**, 133–47.

Long, A. D. and Singh, R. S. (1995) Molecules versus morphology: the detection of selection acting on morphological characters along a cline in *Drosophila melanogaster*. *Heredity*, **74**, 569–81.

Lotter, A. F. and Birks, H. J. B. (1993) The impact of the Laacher See Tephra on terrestrial and aquatic ecosystems in the Black Forest, southern Germany. *Journal of Quaternary Science*, **8**, 263–76.

Lotus Development Corporation (1991) *User's Guide: Lotus 1-2-3, Release 2.3*, Lotus Development Corporation, Cambridge, Massachusetts.

Lotwick, H. W. and Silverman, B. W. (1982) Methods for analysing spatial processes of several types of point. *Journal of the Royal Statistical Society*, **B44**, 406–13.

Ludwig, J. A. and Goodhall, D. W. (1978) A comparison of paired with blocked quadrat variance methods for the analysis of spatial pattern. *Vegetatio*, **38**, 49–59.

Luo, J. and Fox, B. J. (1994) Diet of the eastern mouse (*Pseudomys gracilicaudatus*) II. Seasonal and successional patterns. *Wildlife Research*, **21**, 419–31.

Luo, J. and Fox, B. J. (1996) A review of the Mantel test in dietary studies: effect of sample size and inequality of sample sizes. *Wildlife Research*, **23**, 267–88.

Lutz, T. M. (1985) The magnetic record is not periodic. *Nature*, **317**, 404–7.

Lyons, N. I. and Hutcheson, K. (1986) Estimation of Simpson's diversity when counts follow a Poisson distribution. *Biometrics*, **42**, 171–6.

MacArthur, R. H. and Levins, R. (1967) The limiting similarity, convergence and divergence of coexisting species. *American Naturalist*, **101**, 377–85.

Madansky, A. (1988) *Prescriptions for Working Statisticians*, Springer-Verlag, New York.

Manly, B. F. J. (1977) A simulation experiment on the application of the jackknife with Jolly's method for the analysis of capture–recapture data. *Acta Theriologica*, **22**, 215–23.

Manly, B. F. J. (1983) Analysis of polymorphic variation in different types of habitat. *Biometrics*, **39**, 13–27.

Manly, B. F. J. (1985) *The Statistics of Natural Selection on Animal Populations*, Chapman & Hall, London.

Manly, B. F. J. (1986) Randomization and regression methods for testing for associations with geographical, environmental and biological distances between populations. *Researches on Population Ecology*, **28**, 201–18.

Manly, B. F. J. (1988) The comparison and scaling of student assessment marks in several subjects. *Applied Statistics*, **37**, 385–95.

Manly, B. F. J. (1990a) *Stage-Structured Populations: Sampling, Analysis and Simulation*, Chapman & Hall, London.

Manly, B. F. J. (1990b) On the statistical analysis of niche overlap data. *Canadian Journal of Zoology*, **68**, 1420–2.

Manly, B. F. J. (1992) Bootstrapping for determining sample sizes in biological studies. *Journal of Experimental Marine Biology and Ecology*, **158**, 189–96.

Manly, B. F. J. (1993) A review of computer-intensive multivariate methods in ecology, in *Multivariate Environmental Statistics* (eds G. P. Patil and C. R. Rao), Elsevier Science Publishers, Amsterdam, pp. 307–46.

Manly, B. F. J. (1994a) *Multivariate Statistical Methods: A Primer*, 2nd edn, Chapman & Hall, London.

Manly, B. F. J. (1994b) CUSUM methods for detecting changes in monitored environmental variables, in *Statistics in Ecology and Environmental Monitoring* (eds D. J. Fletcher and B. F. J. Manly), University of Otago Press, Dunedin, pp. 225–38.

Manly, B. F. J. (1995a) A note on the analysis of species co-occurrences. *Ecology*, **76**, 1109–15.

Manly, B. F. J. (1995b) Randomization tests to compare means with unequal variation. *Sankhya*, **57B**, 200–22.

Manly, B. F. J. (1996a) *RT, A Program for Randomization Testing, Version 2.0*, Centre for Applications of Statistics and Mathematics, University of Otago.

Manly, B. F. J. (1996b) Are there bumps in body-size distributions? *Ecology*, **77**, 81–6.

Manly, B. F. J. (1996c) On comparing sample means by randomization, with unequal variances, in *Proceedings of the A. C. Aitken Centenary Conference* (eds L. Kavalieris, F. C. Lam, L. Roberts and J. Shanks), Otago University Press, Dunedin, pp. 183–92.

Manly, B. F. J. (1996d) The statistical analysis of artefacts in graves: presence and absence data. *Journal of Archaeological Science*, **23**, 473–84.

Manly, B. F. J. and Gonzalez, M. L. (1997) Analysis of variance by randomization: problems and solutions. Submitted for publication.

Manly, B. F. J., McAlevey, L. and Stevens, D. (1986) A randomization procedure for comparing group means on multiple measurements. *British Journal of Mathematical and Statistical Psychology*, **39**, 183–9.

Manly, B. F. J. and McAlevey, L. (1987) A randomization alternative to the Bonferroni inequality with multiple *F* tests. *Proceedings of the Second International Tampere Conference in Statistics* (eds T. Pukkila and S. Puntanen), Department of Mathematical Sciences, University of Tampere, Finland, pp. 567–73.

Manly, B. F. J. and Patterson, G. B. (1984) The use of Weibull curves to measure niche overlap. *New Zealand Journal of Zoology*, **11**, 337–42.

Mantel, N. (1967) The detection of disease clustering and a generalized regression approach. *Cancer Research*, **27**, 209–20.

Mantel, N., Tukey, J. W., Ciminera, J. L. and Heyse, J. F. (1982) Tumorgenicity assays including the use of the jackknife. *Biometrical Journal*, **24**, 579–96.

Mantel, N. and Varland, R. S. (1970) A technique of nonparametric multivariate analysis. *Biometrics*, **26**, 547–58.

Mardia, (1971) The effect of nonnormality on some multivariate tests and robustness to nonnormality in the linear model. *Biometrika*, **58**, 105–121.

Maritz, J. S. (1981) *Distribution-Free Statistical Methods*, Chapman & Hall, London.

Maritz, J. S. (1995) A permutation test allowing for missing values. *Australian Journal of Statistics*, **37**, 153–9.

Markus, M. T. and Visser, R. A. (1992) Applying the bootstrap to generate confidence regions in multiple correspondence analysis, in *Bootstrapping and Related Techniques* (eds K. H. Jöckel, G. Rothe and W. Sendler), Springer-Verlag, Berlin, pp. 71–5.

Marriott, F. H. C. (1979) Barnard's Monte Carlo tests: how many simulations? *Applied Statistics*, **28**, 75–7.

Marshall, R. J. (1989) Statistics, geography and disease. *New Zealand Statistician*, **24**, 11–16.

Math Works Inc. (1989) *Matlab User's Guide*, The Math Works Inc., 21 Eliot Street, South Natick, MA 01760, USA.

Mayfield, H. (1961) Nesting success calculated from exposure. *Wilson Bulletin*, **73**, 255–61.

Mayfield, H. (1975) Suggestions for calculating nesting success. *Wilson Bulletin*, **87**, 456–66.

McCullagh, P. and Nelder, J. A. (1989) *Generalized Linear Models*, 2nd edn, Chapman & Hall, London.

McDonald, J. W. and Smith, P. W. F. (1995) Exact conditional tests of quasi-independence for triangular contingency tables: estimating attained significance levels. *Applied Statistics*, **44**, 143–51.

McKechnie, S. W., Ehrlich, P. R. and White, R. R. (1975) Population genetics of *Euphydryas* butterflies. I. Genetic variation and the neutrality hypothesis. *Genetics*, **81**, 571–94.

McLachlan, G. J. (1980) The efficiency of Efron's 'bootstrap' approach applied to error rate estimation in discriminant analysis. *Journal of Statistical Computation and Simulation*, **11**, 273–9.

McLeod, A. (1985) A remark on algorithm AS 183. *Applied Statistics*, **34**, 198–200.

McPeek, M. S. and Speed, T. P. (1995) Modelling interference in genetic recombination. *Genetics*, **139**, 1031–44.

Mead, R. (1974) A test for spatial pattern at several scales using data from a grid of contiguous quadrats. *Biometrics*, **30**, 295–307.

Mehlman, D. W., Shepherd, U. L. and Kelt, D. A. (1995) Bootstrapping principal components – a comment. *Ecology*, **76**, 640–3.

Mehta, C. R. and Patel, N. R. (1995) *StatXact 3 for Windows*, Cytel Software Corporation, 675 Massachusetts Avenue, Cambridge, MA 02139, USA.

Mehta, C. R., Patel, N. R. and Senchaudhuri, P. (1988a) Importance sampling for estimating exact probabilities in permutational inference. *Journal of the American Statistical Association*, **83**, 999–1005.

Mehta, C. R., Patel, N. R. and Wei, L. J. (1988b) Constructing exact significance tests with restricted randomization rules. *Biometrika*, **75**, 295–302.

Metropolis, N., Rosenbluth, A. W., Rosenbluth, M. N., Teller, A. H. and Teller, E. (1953) Equations of state calculations by fast computing machines. *Journal of Chemical Physics*, **21**, 1087–92.

Meyer, J. S., Ingersoll, C. G., McDonald, L. L. and Boyce, M. (1986) Estimating uncertainty in population growth rates: jackknife vs. bootstrap techniques.

Ecology, **67**, 1156–66.

Mielke, P. W. (1978) Clarification and appropriate inference for Mantel and Valand's nonparametric multivariate analysis technique. *Biometrics*, **34**, 277–82.

Mielke, P. W. (1984) Meteorological applications of permutation techniques based on distance functions, in *Handbook of Statistics 4: Nonparametric Methods* (eds P. R. Krishnaiah and P. K. Sen), North-Holland, Amsterdam, pp. 813–30.

Mielke, P. W., Berry, K. J. and Johnson, E. S. (1976) Multi-response permutation procedures for *a priori* classifications. *Communications in Statistics*, **A5**, 1409–24.

Milan, L. and Whittaker, J. (1995) Application of the parametric bootstrap to models that incorporate a singular value decomposition. *Applied Statistics*, **44**, 31–49.

Miller, R. G. (1968) Jackknifing variances. *Annals of Mathematical Statistics*, **39**, 567–82.

Miller, R. G. (1974) The jackknife – a review. *Biometrika*, **61**, 1–15.

Mingoti, S. A. and Meeden, G. (1992) Estimating the total number of species using presence and absence data. *Biometrics*, **48**, 863–75.

Minitab Inc. (1994) *MINITAB User's Guide, Release 10 for Windows*, Minitab Inc., 3081 Enterprise Drive, State College, Pennsylvania 16801-3008.

Minta, S. and Mangel, M. (1989) A simple population estimate based on simulation for capture-recapture and capture–resight data. *Ecology*, **70**, 1738–51.

Mitchell, E. A., Scragg, R. and Stewart, A. W. (1991) Results from the first year of the New Zealand cot death study. *New Zealand Medical Journal*, **104**, 71–6.

Mitchell, E. A., Taylor, B. J. and Ford, R. P. K. (1992) Four modifiable and other major risk factors for cot death: the New Zealand study. *Journal of Paediatrics and Child Health*, **28** (Supplement), 53–8.

Mitchell-Olds, T. and Bergelson, J. (1990) Statistical genetics of an annual plant, *Impatiens capensis*. I. Genetic basis of quantitative variation. *Genetics*, **124**, 407–15.

Montgomery, D. C. (1984) *Design and Analysis of Experiments*, Wiley, New York.

Montgomery, D. C. and Peck, E. A. (1982) *Introduction to Linear Regression Analysis*, Wiley, New York.

Morgan, B. J. T. (1984) *Elements of Simulation*, Chapman & Hall, London.

Moulton, L. H. and Zeger, S. L. (1989) Analysing repeated measures on generalized linear models via the bootstrap. *Biometrics*, **45**, 381–94.

Moulton, L. H. and Zeger, S. L. (1991) Bootstrapping generalized linear models. *Computational Statistics and Data Analysis*, **11**, 53–63.

Mueller, L. D. (1979) A comparison of two methods for making statistical inferences on Nei's measure of genetic distance. *Biometrics*, **35**, 757–63.

Mueller, L. D. and Altenberg, L. (1985) Statistical inference on measures of niche overlap. *Ecology*, **66**, 1204–10.

Mueller, L. D. and Ayala, F. J. (1981) Dynamics of single-species population growth: experimental and statistical analysis. *Theoretical Population Biology*, **20**, 101–17.

Mueller, L. D. and Ayala, F. J. (1982) Estimation and interpretation of genetic distance in empirical studies. *Genetical Research, Cambridge*, **40**, 127–37.

Muller, H. G. and Wang, J. L. (1990) Bootstrap confidence intervals for effective doses in the probit model for dose–response data. *Biometrical Journal*, **32**, 529–44.

Myklestad, Å. and Birks, H. J. B. (1993) A numerical analysis of the distribution patterns of *Salix L.* species in Europe. *Journal of Biogeography,* **20,** 1–32.

Nantel, P. and Neumann, P. (1992) Ecology of ectomycorrhizal-basidiomycete communities on a local vegetation gradient. *Ecology,* **73,** 99–117.

Navidi, W. C. (1995) Bootstrapping a method of phylogenetic inference. *Journal of Statistical Planning and Inference,* **43,** 169–84.

Nemec, A. F. L. and Brinkhurst, R. O. (1988a) Using the bootstrap to assess statistical significance in the cluster analysis of species abundance data. *Canadian Journal of Fisheries and Aquatic Science,* **45,** 965–70.

Nemec, A. F. L. and Brinkhurst, R. O. (1988b) The Fowlkes–Mallow statistic and the comparison of two independently determined dendrograms. *Canadian Journal of Fisheries and Aquatic Science,* **45,** 971–5.

Neter, J., Wasserman, W. and Kutner, M. H. (1983) *Applied Linear Regression Models,* Irwin, Homewood, Illinois.

O'Brien, P. C. (1984) Procedures for comparing samples with multiple endpoints. *Biometrics,* **40,** 1079–87.

Oden, N. L. (1991) Allocation of effort in Monte Carlo simulation for power of permutation tests. *Journal of the American Statistical Association,* **86,** 1074–6.

Oden, N. L. and Sokal, R. R. (1992) An investigation of three-matrix permutation tests. *Journal of Classification,* **9,** 275–90.

Oja, H. (1987) On permutation tests in multiple regression and analysis of covariance problems. *Australian Journal of Statistics,* **29,** 91–100.

Openshaw, S. (1994) Two exploratory space–time–attribute pattern analysers relevant to GIS, in *Spatial Analysis and GIS* (eds S. Fotheringham and P. Rogerson), Taylor & Francis, London, pp. 83–104.

O'Quigley, J. and Pessione, F. (1991) The problem of covariate-time interaction in a survival study. *Biometrics,* **47,** 101–15.

Ord, J. K. (1985) Periodogram analysis. *Encyclopedia of Statistical Sciences,* **6,** 679–82.

Ord, J. K. (1988) Time series. *Encyclopedia of Statistical Sciences,* **9,** 245–55.

Orlowski, L. A., Grundy, W. D. and Mielke, P. W. (1991) An empirical coverage test for the *g*-sample problem. *Mathematical Geology,* **23,** 583–9.

Owen, A. B. (1988) Empirical likelihood ratio confidence intervals for a single functional. *Biometrika,* **75,** 237–49.

Owen, A. B. (1990) Empirical likelihood ratio confidence regions. *Annals of Statistics,* **18,** 90–120.

Pagano, M. and Tritchler, D. (1983) On obtaining permutation distributions in polynomial time. *Journal of the American Statistical Association,* **78,** 435–40.

Page, R. D. M. (1988) Quantitative cladistic biogeography: constructing and comparing area cladograms. *Systematic Zoology,* **37,** 254–70.

Pagel, M. D. and Harvey, P. H. (1988) Recent developments in the analysis of comparative data. *Quarterly Review of Biology,* **63,** 413–40.

Palmer, M. W. (1990) The estimation of species richness by extrapolation. *Ecology,* **71,** 1195–8.

Palmer, M. W. (1991) Estimating species richness: the second-order jackknife reconsidered. *Ecology,* **72,** 1512–3.

Parr, W. C. and Schucany, W. R. (1980) The jackknife: a bibliography. *International Statistical Review,* **48,** 73–8.

Patterson, B. D. (1990) On the temporal development of nested subsets of species composition. *Oikos,* **59,** 330–42.

Patterson, B. D. and Atmar, W. (1986) Nested subsets and the structure of insular mammalian faunas and archipelagos. *Biological Journal of the Linnean Society*, **28**, 65–82.

Patterson, B. D. and Brown, J. H. (1991) Regionally nested patterns of species composition in granivorous rodent assemblages. *Journal of Biogeography*, **18**, 395–402.

Patterson, C. and Smith, A. B. (1987) Is the periodicity of extinctions a taxonomic artefact? *Nature*, **330**, 248–52.

Patterson, G. B. (1986) A statistical method for testing for dietary differences. *New Zealand Journal of Zoology*, **13**, 113–5.

Patterson, G. B. (1992) The ecology of a New Zealand grassland lizard guild. *Journal of the Royal Society of New Zealand*, **22**, 91–106.

Pearson, E. S. (1937) Some aspects of the problem of randomization. *Biometrika*, **29**, 53–64.

Pearson, S. M. (1991) Food patches and the spacing of individual foragers. *Auk*, **108**, 355–62.

Pearson, S. M. (1992) Reply to Larkin. *Auk*, **109**, 203–4.

Peck, R., Fisher, L. and Ness, J. V. (1989) Approximate confidence intervals for the number of clusters. *Journal of the American Statistical Association*, **84**, 184–91.

Péladeau, N. (1994) *SIMSTAT User's Guide*, Provalis Research, 500 Adam Street, Montreal, QC, Canada, H1V 1W5.

Penttinen, A., Stoyan, D. and Henttonen, H. M. (1992) Marked point processes in forest statistics. *Forest Science*, **38**, 806–24.

Perry, J. N. (1995a) Spatial aspects of animal and plant distribution in patchy farmland habitats, in *Ecology and Integrated Farming Systems* (eds D. M. Glen, M. P. Greaves and H. M. Anderson), Wiley, London, pp. 221–42.

Perry, J. N. (1995b) Spatial analysis by distance indices. *Journal of Animal Ecology*, **64**, 303–14.

Perry, J. N. and Hewitt, M. (1991) A new index of aggregation for animal counts. *Biometrics*, **47**, 1505–18.

Perry, R. I. and Smith, S. J. (1994) Identifying habitat associations of marine fishes using survey data: an application to the northwest Atlantic. *Canadian Journal of Fisheries and Aquatic Science*, **51**, 589–602.

Pielou, E. C. (1984) Probing multivariate data with random skewers: a preliminary to direct gradient analysis. *Oikos*, **42**, 161–5.

Pierce, G. J., Thorpe, R. S., Hastie, L. C., Brierley, A. S., Guerra, A., Boyle, P. R., Jamieson, R. and Avila, P. (1994) Geographical variation in *Loligo forbesi* in the northeast Atlantic Ocean: analysis of morphometric data and tests of causal hypotheses. *Marine Biology*, **119**, 541–7.

Pitman, E. J. G. (1937a) Significance tests which may be applied to samples from any populations. *Journal of the Royal Statistical Society*, **B4**, 119–30.

Pitman, E. J. G. (1937b) Significance tests which may be applied to samples from any populations. II. The correlation coefficient test. *Journal of the Royal Statistical Society*, **B4**, 225–32.

Pitman, E. J. G. (1937c) Significance tests which may be applied to samples from any populations. III. The analysis of variance test. *Biometrika*, **29**, 322–35.

Pleasants, J. M. (1990) Null-model tests for competitive displacement: the fallacy of not focusing on the whole community. *Ecology*, **71**, 1078–84.

Pleasants, J. M. (1994) A comparison of test statistics used to detect competitive displacement in body size. *Ecology*, **75**, 847–50.

Plotnick, R. E. (1989) Application of bootstrap methods to reduced major axis line fitting. *Systematic Zoology*, **38**, 144–53.

Pocock, S. J., Geller, N. L. and Tsiatis, A. (1987) The analysis of multiple endpoints in clinical trials. *Biometrics*, **43**, 487–98.

Politis, D. N. and Romano, J. P. (1994) The stationary bootstrap. *Journal of the American Statistical Association*, **89**, 1303–13.

Pollard, E., Lakhani, K. H. and Rothery, P. (1987) The detection of density-dependence from a series of annual censuses. *Ecology*, **68**, 2046–55.

Pollock, K. H. and Otto, M. C. (1983) Robust estimation of population size in closed animal populations from capture–recapture experiments. *Biometrics*, **39**, 1035–49.

Poole, R. W. and Rathcke, B. J. (1979) Regularity, randomness, and aggregation in flowering phenologies. *Science*, **203**, 470–1.

Popham, E. J. and Manly, B. F. J. (1969) Geographical distribution of the Dermaptera and the continental drift hypothesis. *Nature*, **222**, 981–2.

Powell, G. L. and Russell, A. P. (1984) The diet of the eastern short-horned lizard (*Phrynosoma douglassi brevirostre*) in Alberta and its relationship to sexual size dimorphism. *Canadian Journal of Zoology*, **62**, 428–40.

Powell, G. L. and Russell, A. P. (1985) Growth and sexual size dimorphism in Alberta populations of the eastern short-horned lizard, *Phrynosoma douglassi brevirostre*. *Canadian Journal of Zoology*, **63**, 139–54.

Prager, M. H. and Hoenig, J. M. (1989) Superposed epoch analysis: a randomization test of environmental effects on recruitment with application to chub mackerel. *Transactions of the American Fisheries Society*, **118**, 608–18.

Prager, M. H. and Hoenig, J. M. (1992) Can we determine the significance of key-event effects on a recruitment time series? – A power study of superposed epoch analysis. *Transactions of the American Fisheries Society*, **121**, 123–31.

Quang, P. X. (1990) Confidence intervals for densities in line transect sampling. *Biometrics*, **46**, 459–72.

Quang, P. X. and Lanctot, R. B. (1991) A line transect model for aerial surveys. *Biometrics*, **47**, 1089–102.

Quenouille, M. H. (1956) Notes on bias in estimation. *Biometrika*, **43**, 353–60.

Quinn, J. F. (1987) On the statistical detection of cycles in extinctions in the marine fossil record. *Paleobiology*, **13**, 465–78.

Raftery, A. E., Givens, G. H. and Zeh, J. E. (1995) Inference from a deterministic population dynamics model for bowhead whales. *Journal of the American Statistical Association*, **90**, 402–16.

Rampino, M. R. and Stothers, R. B. (1984a) Terrestrial mass extinctions, cometary impacts and the sun's motion perpendicular to the galactic plane. *Nature*, **308**, 709–12.

Rampino, M. R. and Stothers, R. B. (1984b) Geological rythms and cometary impacts. *Science*, **226**, 1427–31.

Raz, J. (1989) Analysis of repeated measurements using nonparametric smoothers and randomization tests. *Biometrics*, **45**, 851–71.

Raz, J. and Fein, G. (1992) Testing for heterogeneity of evoked potential signals using an approximation to an exact permutation tests. *Biometrics*, **48**, 1069–80.

Raup, D. M. (1985a) Magnetic reversals and mass extinctions. *Nature*, **314**, 341–3.

Raup, D. M. (1985b) Rise and fall of periodicity. *Nature*, **317**, 384–5.

Raup, D. M. (1987) Mass extinctions: a commentary. *Palaeontology*, **30**, 1–13.

Raup, D. M. and Boyajian G. E. (1988) Patterns of generic extinction in the fossil record. *Paleobiology*, **14**, 109–25.

Raup, D. M. and Jablonski, D. (eds) (1986) *Patterns and Processes in the History of Life*, Springer-Verlag, Berlin.

Raup, D. M. and Sepkoski, J. J. (1984) Periodicity of extinctions in the geologic past. *Proceedings of the National Academy of Sciences*, **81**, 801–5.

Raup, D. M. and Sepkoski, J. J. (1988) Testing for periodicity of extinction. *Science*, **241**, 94–6.

Raybould, S. (1990) Childhood leukaemia in the Tyne and Wear: investigating a viral hypothesis. *Statistician*, **39**, 349–63.

Reddingius, J. and den Boer, P. J. (1989) On the stabilization of animal numbers. Problems of testing. 1. Power estimates and estimation errors. *Oecologia*, **78**, 1–8.

Reed, W. J. (1983) Confidence estimation of ecological aggregation indices based on counts – a robust procedure. *Biometrics*, **39**, 987–98.

Reid, N. (1981) Estimating the median survival time. *Biometrika*, **68**, 601–8.

Rejmánkova, E., Savage, H. M., Rejmánek, M., Arredondo-Jimenez, J. I. and Roberts, D. R. (1991) Multivariate analysis of relationships between habitats, environmental factors and occurrence of anopheline mosquito larvae *Anopheles albimanus* and *A. pseudopunctipennis* in southern Chiapas, Mexico. *Journal of Applied Ecology*, **28**, 827–41.

Rencher, A. C. and Pun, F. C. (1980) Inflation of R^2 in best subset regression. *Technometrics*, **22**, 49–53.

Reyment, R. A. (1982) Phenotypic evolution in a Cretaceous foraminifer. *Evolution*, **36**, 1182–99.

Ripley, B. D. (1979) Simulating spatial patterns: dependent samples from a multivariate density. *Applied Statistics*, **28**, 109–12.

Ripley, B. D. (1981) *Spatial Statistics*, Wiley, New York.

Ripley, B. D. (1990) Thoughts on pseudorandom number generators. *Journal of Computational and Applied Mathematics*, **31**, 153–63.

Riska, B. (1985) Group size factors and geographical variation of morphometric correlation. *Evolution*, **39**, 792–803.

Roberts, A. and Stone, L. (1990) Island-sharing by archipelago species. *Oecologia*, **83**, 560–7.

Robertson, P. K. (1990) Controlling for time-varying population distributions in disease clustering studies. *American Journal of Epidemiology*, **132** (Supplement), 131–5.

Robertson, P. K. and Fisher, L. (1983) Lack of robustness in time–space clustering. *Communications in Statistics – Simulation and Computation*, **12**, 11–22.

Robinson, J. (1973) The large-sample power of permutation tests for randomization models. *Annals of Statistics*, **1**, 291–6.

Robinson, J. (1982) Saddlepoint approximations for permutation tests and confidence intervals. *Journal of the Royal Statistical Society*, **B44**, 91–101.

Rodrigo, A. G., Kelly-Borges, M., Bergquist, P. R. and Bergquist, P. L. (1993) A randomization test of the null hypothesis that two cladograms are sample estimates of a parametric phylogenetic tree. *New Zealand Journal of Botany*, **31**, 257–68.

Roff, D. A. (1995) The estimation of genetic correlations from phenotypic correlations: a test of Cheverud's conjecture. *Heredity*, **74**, 481–90.

Roff, D. A. and Bentzen, P. (1989) The statistical analysis of mitochondrial DNA polymorphisms: χ^2 and the problem of small samples. *Molecular Biology and Evolution*, **6**, 539–45.

Roff, D. A. and Preziosi, R. (1994) The estimation of the genetic correlation: the use of the jackknife. *Heredity*, **73**, 544–8.

Rohlf, F. J. (1965) A randomization test of the nonspecificity hypotheses in numerical taxonomy. *Taxon*, **14**, 262–7.

Romano, J. P. (1989) Bootstrap and randomization tests of some nonparametric hypotheses. *The Annals of Statistics*, **17**, 141–59

Romesburg, H. C. (1985) Exploring, confirming and randomization tests. *Computers and Geosciences*, **11**, 19–37.

Romesburg, H. C. (1989) Zorro: a randomization test for spatial pattern. *Computers and Geosciences*, **15**, 1011–17.

Rosenbaum, P. R. (1988) Permutation tests for matched pairs with adjustments for covariates. *Applied Statistics*, **37**, 401–11.

Rosenbaum, P. R. (1991) Discussing hidden bias in observational studies. *Annals of Internal Medicine*, **115**, 901–5.

Rosenbaum, P. R. (1994) Coherence in observational studies. *Biometrics*, **50**, 368–74.

Rosenbaum, P. R. and Krieger, A. M. (1990) Sensitivity analysis for the two-sample permutation inferences in observational studies. *Journal of the American Statistical Association*, **85**, 493–8.

Rosenberg, D. K., Overton, W. S. and Anthony, R. G. (1995) Estimation of animal abundance when capture probabilities are low. *Journal of Wildlife Management*, **59**, 252–61.

Routledge, R. D. (1980) Bias in estimating the diversity of large, uncensused communities. *Ecology*, **61**, 276–81.

Routledge, R. D. (1984) Estimating ecological components of diversity. *Oikos*, **42**, 23–9.

Ryti, R. T. and Gilpin, M. E. (1987) The comparative analysis of species occurrence patterns on archipelagos. *Oecologia*, **73**, 282–7.

Sale, P. F. (1974) Overlap in resource use, and interspecific competition. *Oecologia*, **17**, 245–56.

Salsburg, D. (1971) Testing dose responses on proportions near zero or one with the jackknife. *Biometrics*, **27**, 1035–41.

Sanderson, M. J. (1989) Confidence limits on phylogenies: the bootstrap revisited. *Cladistics*, **5**, 113–29.

Sanderson, M. J. (1995) Objections to bootstrapping phylogenies: a critique. *Systematic Biology*, **44**, 299–320.

SAS Institute Inc. (1990) *SAS Language Reference, Version 6*, SAS Institute, Cary, North Carolina 27513.

Sayers, B. M., Mansourian, B. G., Phan T. T. and Bogel, K. (1977) A pattern analysis study of a wild-life rabies epizootic. *Medical Informatics*, **2**, 11–34.

Schenker, N. (1985) Qualms about bootstrap confidence intervals. *Journal of the American Statistical Association*, **80**, 360–1.

Schluter, D. (1988) Estimating the form of natural selection on a quantitative trait. *Evolution*, **42**, 849–61.

Schluter, D. and Smith, J. N. M. (1986) Natural selection on beak and body size in the song sparrow. *Evolution*, **40**, 221–31.

Schmoyer, R. L. (1994) Permutation tests for correlation in regression errors. *Journal of the American Statistical Association*, **89**, 1507–16.

Schnell, G. D., Watt, D. J. and Douglas, M. E. (1985) Statistical comparison of proximity matrices: applications in animal behaviour. *Animal Behaviour*, **33**, 239–53.

Schoener, T. W. (1968) The *Anolis* lizards of Bimini: resource partitioning in a complex fauna. *Ecology*, **49**, 704–26.

Schork, N. (1993) Combining Monte Carlo and Cox tests of non-nested hypotheses. *Communications in Statistics – Simulation and Computation*, **22**, 939–54.

Scollnik, D. P. M. (1995) Bayesian analysis of overdispersed Poisson models. *Biometrics*, **51**, 1117–26.

Seber, G. A. F. (1979) Transects of random length, in *Sampling Biological Populations* (eds R. M. Cormack, G. P. Patil and D. S. Robson), International Cooperative Publishing House, Maryland, pp. 183–92.

Seber, G. A. F. (1984) *Multivariate Observations*. Wiley, New York.

Sepkoski, J. J. (1986) Phanerozoic overview of mass extinctions, in *Patterns and Processes in the History of Life* (eds D. M. Raup and D. Jablonski), Springer-Verlag, Berlin, pp. 277–95.

Sepkoski, J. J. and Raup, D. M. (1986) Was there 26-myr periodicity of extinctions? *Nature*, **321**, 533.

Shipley, B. (1993) A null model for competitive hierarchies in competition matrices. *Ecology*, **74**, 1693–9.

Simberloff, D. (1987) Calculating probabilities that cladograms match: a method of biogeographical inference. *Systematic Zoology*, **36**, 175–95.

Simberloff, D. and Connor, E. F. (1981) Missing species combinations. *American Naturalist*, **118**, 215–39.

Simon, J. L. and Bruce, P. C. (1991) *RESAMPLING STATS, User's Guide IBM Version 3.0*, Resampling Stats, Arlington, Virginia 22201.

Simpson, G. G., Roe, A. and Lewontin, R. C. (1960) *Quantitative Zoology*, Harcourt, Brace and World, New York.

Sitnikova, T., Rzhetsky, A. and Nei, M. (1995) Interior-branch and bootstrap tests of phylogenetic trees. *Molecular Biology and Evolution*, **12**, 319–33.

Sitter, R. R. (1992) Comparing three bootstrap methods for survey data. *Canadian Journal of Statistics*, **20**, 135–54.

Skibinski, D. O. F., Woodwark, M. and Ward, R. D. (1993) A quantitative test of the neutral theory using pooled allozyme data. *Genetics*, **135**, 233–48.

Smith, A. F. M. and Roberts, G. O. (1993) Bayesian computation via the Gibbs sampler and related Monte Carlo methods. *Journal of the Royal Statistical Society*, **B55**, 3–23.

Smith, E. P. (1985) Estimating the reliability of diet overlap measures. *Environmental Biology of Fishes*, **13**, 125–38.

Smith, E. P. and van Belle, G. (1984) Nonparametric estimation of species richness. *Biometrics*, **40**, 119–29.

Smith, S. J. (1980) Comparison of two methods of estimating the variance of the estimate of catch per unit effort. *Canadian Journal of Fisheries and Aquatic Science*, **37**, 2346–51.

Smith, T. B. (1993) Disruptive selection and the genetic basis of bill size polymorphism in the African finch *Pyrenestes*. *Nature*, **363**, 618–20.

Smith, W. (1989) ANOVA-like similarity analysis using expected species shared. *Biometrics*, **45**, 873–81.

Smith, W., Kravitz, D. and Grassle, J. F. (1979) Confidence intervals for similarity measures using the two-sample jackknife, in *Multivariate Methods in Ecological Work* (eds L. Orloci, C. R. Rao and W. M. Stiteler), International Cooperative Publishing House, Maryland, pp. 253–62.

Smouse, P. E., Long, J. C. and Sokal, R. R. (1986) Multiple regression and correlation extensions of the Mantel test of matrix correspondence. *Systematic*

Zoology, **35**, 727–32.

Snijders, T. A. B. (1991) Enumeration and simulation methods for 0–1 matrices with given marginals. *Psychometrika*, **56**, 397–417.

Sokal, R. R. (1979) Testing statistical significance of geographical variation patterns. *Systematic Zoology*, **28**, 227–32.

Sokal, R. R., Lengyel, I. A., Derish, P. A., Wooten, M. C. and Oden, N. L. (1987) Spatial autocorrelation of ABO serotypes in mediaeval cemeteries as an indicator of ethnic and familial structure. *Journal of Archaeological Science*, **14**, 615–33.

Sokal, R. R., Oden, N. L., Legendre, P., Fortin, M., Junhyong, K. and Vaudor, A. (1989) Genetic differences among language families in Europe. *American Journal of Physical Anthropology*, **79**, 489–502.

Sokal, R. R., Oden, N. L., Legendre, P., Fortin, M., Kim, J., Thomson, B. A., Vaudor, A, Harding, R. M. and Barbujani, G. (1990) Genetics and language in European populations. *American Naturalist*, **135**, 157–75.

Sokal, R. R., Oden, N. L. and Thomson, B. A. (1988) Genetic changes across language boundaries in Europe. *American Journal of Physical Anthropology*, **76**, 337–61.

Sokal, R. R., Smouse, P. E. and Neel, J. V. (1986) The genetic structure of a tribal population, the Yanomama Indians. XV. Patterns inferred by autocorrelation analysis. *Genetics*, **114**, 259–87.

Solow, A. R. (1989a) Bootstrapping sparsely sampled spatial point patterns. *Ecology*, **70**, 379–82.

Solow, A. R. (1989b) A randomization test for independence of animal locations. *Ecology*, **70**, 1546–9.

Solow, A. R. (1990) A randomization test for misclassification probability in discriminant analysis. *Ecology*, **71**, 2379–82.

Solow, A. R. (1993) A simple test for change in community structure. *Journal of Animal Ecology*, **62**, 191–3.

Solow, A. R. and Gaines, A. G. (1993) Mapping water quality by local scoring. *Canadian Journal of Statistics*, **21**, 123–30.

Solow, A. R. and Smith, W. (1991) Detecting cluster in a heterogenous community sampled by quadrats. *Biometrics*, **47**, 311–17.

Somers, K. M. and Green, R. H. (1993) Seasonal patterns in trap catches of the crayfish *Cambarus bartoni* and *Orconectes virilis* in six south-central Ontario lakes. *Canadian Journal of Zoology*, **71**, 1136–45.

Somers, K. M. and Jackson, D. A. (1993) Adjusting mercury concentration for fish-size covariation: a multivariate alternative to bivariate regression. *Canadian Journal of Fisheries and Aquatic Science*, **50**, 2388–96.

Soms, A. P. and Torbeck, L. D. (1982) Randomization tests for *K* sample binomial data. *Journal of Quality Technology*, **14**, 220–5.

Spielman, R. S. (1973) Differences among Yanomama Indian villages: do the patterns of allele frequencies, anthropometrics and map locations correspond? *American Journal of Physical Anthropology*, **39**, 461–80.

Spino, C. and Pagano, M. (1991) Efficient calculation of the permutation distribution of trimmed means. *Journal of the American Statistical Association*, **86**, 729–37.

SPSS Inc. (1990) *SPSS Reference Guide*, SPSS Inc., 444 N. Michigan Avenue, Chicago, Illinois 60611.

Squires, J. R., Anderson, S. H. and Lockman, D. C. (1992) Habitat selection of nesting and wintering trumpeter swans, in *Wildlife* 2001 (eds D. R. McCullogh and R. H. Barnett), Elsevier Applied Science, London, pp. 665–75.

Stauffer, D. F., Garton, E. O. and Steinhorst, R. K. (1985) A comparison of principal components from real and random data. *Ecology*, **66**, 1693–8.

Steel, R. G. D. and Torrie, J. H. (1980) *Principles and Procedures of Statistics*, McGraw-Hill, New York.

Stigler, S. M. and Wagner, M. J. (1987) A substantial bias in nonparametric tests for periodicity in geophysical data. *Science*, **238**, 940–5.

Stigler, S. M. and Wagner, M. J. (1988) Response to Raup and Sepkoski. *Science*, **241**, 96–9.

Still, A. W. and White, A. P. (1981) The approximate randomization test as an alternative to the *F* test in analysis of variance. *British Journal of Mathematical and Statistical Psychology*, **34**, 243–52.

Stone, L. and Roberts, A. (1990) The checkerboard score and species distributions. *Oecologia*, **85**, 74–9.

Stone, L. and Roberts, A. (1992) Competitive exclusion, or species aggregation: an aid in deciding. *Oecologia*, **91**, 419–24.

Stothers, R. (1979) Solar activity cycle during classical antiquity. *Astronomy and Astrophysics*, **77**, 121–7.

Stothers, R. (1989) Structure and dating errors in the geologic time scale and periodicity in mass extinctions. *Geophysical Research Letters*, **16**, 119–22.

Strand, L. (1972) A model for stand growth. *IUFRO Third Conference Advisory Group of Forest Statisticians*, INRA, Paris, pp. 207–16.

Strauss, E. R. (1982) Statistical significance of species clusters in association analysis. *Ecology*, **63**, 634–9.

Strong, D. R., Szyska, L. A. and Simberloff, D. S. (1979) Tests of community-wide character displacement against null hypotheses. *Evolution*, **33**, 897–913.

Swanepoel, C. J. and Frangos, C. C. (1994) Bootstrap confidence intervals for the slope parameter of a logistic model. *Communications in Statistics – Simulation and Computation*, **23**, 1115–26.

Swanepoel, J. W. H. and van Wyk, J. W. J. (1986) The comparison of two spectral density functions using the bootstrap. *Journal of Statistical Computation and Simulation*, **24**, 271–82.

Swartzman, G., Huang, C. and Kaluzny, S. (1992) Spatial analysis of Bering Sea groundfish survey data using generalized additive models. *Canadian Journal of Fisheries and Aquatic Science*, **49**, 1366–78.

Swed, F. S. and Eisenhart, C. (1943) Tables for testing for randomness of grouping in a sequence of alternatives. *Annals of Mathematical Statistics*, **14**, 83–6.

ter Braak, C. J. F. (1988) *CANOCO – A Fortran Program for Canonical Community Ordination by Partial Detrended Canonical Correlation Analysis, Principal Components Analysis and RedundancyAnalysis, Version 2.1*, Technical Report LWA-88-02, GLW, Postbus 100, 6700 AC Wageningen, Holland.

ter Braak, C. J. F. (1992) Permutation versus bootstrap significance tests in multiple regression and ANOVA, in *Bootstrapping and Related Techniques* (ed. K. H. Jöckel), Springer-Verlag, Berlin, pp. 79–86.

Thombs, L. A. and Schucany, W. R. (1990) Bootstrap prediction intervals for autoregression. *Journal of the American Statistical Association*, **85**, 486–92.

Thorpe, R. S. and Brown, R. P. (1991) Microgeographical clines in the size of mature male *Gallotia galloti* (Squamata: Lacertidae) on Tenerife: causal hypotheses. *Herpetologia*, **47**, 28–37.

Tilley, S. G., Verrell, P. A. and Arnold, S. J. (1990) Correspondence between sexual isolation and allozyme differentiation: a test in the salamander

Desmognathus ochrophaeus. Proceedings of the National Academy of Science, **87,** 2715–19.

Tokeshi, M. (1986) Resource utilization, overlap and temporal community dynamics: a null model analysis of a epiphytic chironomid community. *Journal of Animal Ecology,* **55,** 491–506.

Tonkyn, D. W. and Cole, B. J. (1986) The statistical analysis of size ratios. *American Naturalist,* **128,** 66–81.

Tritchler, D. (1984) On inverting permutation tests. *Journal of the American Statistical Association,* **79,** 200–7.

Tu, D. and Gross, A. J. (1995) Accurate confidence intervals for the ratio of specific occurrence/exposure rates in risk and survival analysis. *Biometrical Journal,* **37,** 611–26.

Tukey, J. W. (1958) Bias and confidence in not quite large samples (Abstract) *Annals of Mathematical Statistics,* **29,** 614.

Underhill, L. G. (1990) Bayesian estimation of the size of closed populations. *Ring,* **13,** 235–54.

Underhill, L. G. and Prŷs-Jones, R. P. (1994) Index numbers for waterbird populations. I. Review and methodology. *Journal of Applied Ecology,* **31,** 463–80.

Upton, G. J. G. (1984) On Mead's test for pattern. *Biometrics,* **40,** 759–66.

Van der Maarel, E., Noerst, V. and Palmer, M. W. (1995) Variation in species richness on small grassland quadrats: niche structure or small-scale plant mobility? *Journal of Vegetation Science,* **6,** 741–52.

Van Dongen, S. (1995) How should we bootstrap allozyme data? *Heredity,* **74,** 445–7.

Van Dongen, S. and Backeljau, T. (1995) One- and two-sample tests for single-locus inbreeding coefficients using the bootstrap. *Heredity,* **74,** 129–35.

Vassiliou, A., Ignatiades, L. and Karydis, M. (1989) Clustering of transect phytoplankton collections with a quick randomization algorithm. *Journal of Experimental Marine Biology and Ecology,* **130,** 135–45.

Veall, M. R. (1987) Bootstrapping and forecast uncertainty: a Monte Carlo analysis, in *Time Series and Econometric Modelling* (eds I. B. MacNeill and G. J. Umphrey), D. Reidel Publishing Company, Dordrecht, pp. 373–84.

Verdonschot, P. F. M. and ter Braak, C. J. F. (1994) An experimental manipulation of oligochaete in mesocosms treated with chlorpyrifos or nutrient additions: multivariate analyses with Monte Carlo permutation tests. *Hydrobiologia,* **278,** 251–66.

Verdú, M. and García-Fayos, P. (1994) Correlations between the abundances of fruits and frugivorous birds: the effect of temporal autocorrelation. *Acta Œcologia,* **15,** 791–6.

Vickery, W. L. (1991) An evaluation of bias in *k*-factor analysis. *Oecologia,* **85,** 413–8.

Vounatsou, P. and Smith, A. F. M. (1995) Bayesian analysis of ring-recovery data via Markov chain Monte Carlo simulation. *Biometrics,* **51,** 687–708.

Vuilleumier, F. (1970) Insular biogeography in continental regions. I. The northern Andes of South America. *American Naturalist,* **104,** 373–88

Waddle, D. (1994) Matrix correlation tests support a single origin for modern humans. *Nature,* **368,** 452–4.

Waddle, D. (1996) Interpreting results of matrix correlation tests: a response to Cole. *Journal of Human Evolution,* **31,** 485–8.

Wahrendorf, J., Becher, H. and Brown, C. C. (1987) Bootstrap comparison of non-nested generalized linear models: applications in survival analysis and epidemiology. *Applied Statistics*, **36**, 72–81.

Wakefield, J. C., Smith, A. F. M., Racine-Poon, A. and Gelfand, A. E. (1994) Bayesian analysis of linear and non-linear population models by using the Gibbs sampler. *Applied Statistics*, **43**, 201–21.

Wald, A. (1947) *Sequential Analysis*, Wiley, New York.

Ward, P. S. (1985) Taxonomic congruence and disparity in an insular ant fauna: *Rhytidoponera* in New Caledonia. *Systematic Zoology*, **34**, 140–51.

Watkins, A. and Wilson, J. B. (1992) Fine-scale community structure of lawns. *Journal of Ecology*, **80**, 15–24.

Watkins, A. and Wilson, J. B. (1994) Plant community structure, and its relation to the vertical complexity of communities: dominance/diversity and spatial rank consistency. *Oikos*, **70**, 91–8.

Weatherburn, C. E. (1962) *A First Course in Mathematical Statistics*, Cambridge University Press, Cambridge.

Weber, N. C. (1986) On the jackknife and bootstrap techniques for regression models, in *Pacific Statistical Congress* (eds I. S. Francis, B. F. J. Manly and F. C. Lam), Elsevier, The Netherlands, pp. 51–5.

Weinberg, S. L., Carroll, J. D. and Cohen, H. S. (1984) Confidence regions for IDSCAL using the jackknife and bootstrap techniques. *Psychometrika*, **49**, 475–91.

Welch, B. L. (1937) On the *z*-test in randomized blocks and Latin squares. *Biometrika*, **29**, 21–52.

Welch, W. J. (1987) Rerandomizing the median in matched-pairs designs. *Biometrika*, **74**, 609–14.

Welch, W. J. and Gutierrez, L. G. (1988) Robust permutation tests for matched-pairs designs. *Journal of the American Statistical Association*, **83**, 450–5.

Westfall, P. H. and Young, S. S. (1993) *Resampling-Based Multiple Testing: Examples and Methods for p-Value Adjustment*, Wiley, New York.

White, G. C. and Garrott, R. M. (1990) *Analysis of Wildlife Radio-Tracking Data*, Academic Press, New York.

Wichmann, B. A. and Hill, I. D. (1982) Algorithm AS 183: an efficient and portable pseudo-random number generator. *Applied Statistics*, **31**, 188–90 (correction in *Applied Statistics*, **33**, 123 (1984)).

Willemain, T. R. (1994) Bootstrapping on a shoestring: resampling using spreadsheets. *American Statistician*, **48**, 40–42.

Williams, M. R. (1995) Critical values of a statistic to detect competitive displacement. *Ecology*, **76**, 646–7.

Williams, P. H. (1996) Mapping variations in the strength and breadth of biogeographical transition zones using species turnover. *Proceedings of the Royal Society of London*, **B263**, 579–88.

Williams, S. M., Scragg, R., Mitchell, E. A., Taylor, B. J., Allen, E. M., Becroft, D. M. O., Ford, R. P. K., Hassall, I. B., and Stewart, A. W. (1996) Growth and the sudden infant death syndrome. *Acta Paediatrica* (in press).

Williamson, M. (1985) Apparent systematic effects on species-area curves under isolation and evolution, in *Statistics in Ornithology* (eds B. J. T. Morgan and P. M. North), Springer-Verlag Lecture Notes in Statistics 29, Berlin, pp. 171–8.

Wilson, J. B. (1987) Methods for detecting non-randomness in species co-occurrences: a contribution. *Oecologia*, **73**, 579–82.

Wilson, J. B. (1988) Community structure in the flora of islands in Lake Manapouri, New Zealand. *Journal of Ecology*, **76**, 1030–42.

Wilson, J. B. (1989) A null model of guild proportionality, applied to stratification of a New Zealand temperate rain forest. *Oecologia*, **80**, 263–7.

Wilson, J. B. (1995) Null models for assembly rules: the Jack Horner effect is more insidious than the narcissus effect. *Oikos*, **72**, 139–44.

Wilson, J. B., Agnew, A. D. Q. and Partridge, T. R. (1994) Carr texture in Britain and New Zealand: community convergence compared with a null model. *Journal of Vegetation Science*, **5**, 109–16.

Wilson, J. B., Allen, R. B. and Lee, W. G. (1995b) An assembly rule in the ground and herbaceous strata of a New Zealand rain forest. *Functional Ecology*, **9**, 61–4.

Wilson, J. B. and Gitay, H. (1995) Limitations to species coexistence: evidence for competition from field observations, using a patch model. *Journal of Vegetation Science*, **6**, 369–76.

Wilson, J. B., Gitay, H. and Agnew, A. D. Q. (1987) Does niche limitation exist? *Functional Ecology*, **1**, 391–7.

Wilson, J. B., James, R. E., Newman, J. E. and Myers, T. E. (1992a) Rock pool algae: species composition determined by chance? *Oecologia*, **91**, 150–2.

Wilson, J. B., Peet, R. K. and Sykes, M. T. (1995d) What constitutes evidence of community structure? A reply to van der Maarel, Noest and Palmer. *Journal of Vegetation Science*, **6**, 753–8.

Wilson, J. B. and Roxburgh, S. H. (1994) A demonstration of guild-based assembly rules for a plant community, and determination of intrinsic guilds. *Oikos*, **69**, 267–76.

Wilson, J. B., Roxburgh, S. H. and Watkins, A. J. (1992b) Limitations to plant species coexistence at a point: a study in a New Zealand lawn. *Journal of Vegetation Science*, **3**, 711–14.

Wilson, J. B., Steel, J. B., Newman, J. E. and Tangney, R. S. (1995a) Are bryophyte communities different? *Journal of Bryology*, **18**, 689–705.

Wilson, J. B., Sykes, M. T. and Peet, R. K. (1995c) Time and space in the community structure of a species-rich limestone grassland. *Journal of Vegetation Science*, **6**, 729–40.

Wilson, J. B. and Watkins, A. J. (1994) Guilds and assembly rules in lawn communities. *Journal of Vegetation Science*, **5**, 591–600.

Wilson, J. B. and Whittaker, R. J. (1995) Assembly rules demonstrated in a saltmarsh community. *Journal of Ecology*, **83**, 801–7.

Winston, M. R. (1995) Co-occurrences of morphologically similar species of stream fishes. *American Naturalist*, **145**, 527–45.

Wolda, H. and Dennis, B. (1993) Density dependence tests, are they? *Oecologia*, **95**, 581–91.

Wolfe, D. (1976) On testing equality of related correlation coefficients. *Biometrika*, **63**, 214–15.

Wolfe, D. (1977) A distribution free test for related correlations. *Technometrics*, **19**, 507–9.

Womble, W. H. (1951) Differential systematics. *Science*, **114**, 315–22.

Wormald, N. C. (1984) Generating random regular graphs. *Journal of Algorithms*, **5**, 247–80.

Wright, D. H. and Reeves, J. H. (1992) On the meaning and measurement of nestedness of species assemblages. *Oecologia*, **92**, 416–28.

Wright, S. J. and Biehl, C. C. (1982) Island biogeographic distributions: testing for random, regular and aggregated patterns of species occurrence. *American Naturalist*, **119**, 345–57.

Younger, M. S. (1985) *A First Course in Linear Regression*, Duxbury Press, Boston.

Yule, G. U. and Kendall, M. G. (1965) *An Introduction to the Theory of Statistics*, Griffin, London.

Zahl, S. (1977) Jackknifing an index of diversity. *Ecology*, **58**, 907–13.

Zeisel, H. (1986) A remark on algorithm AS 183. *Applied Statistics*, **35**, 89.

Zerbe, G. O. (1979a) Randomization analysis of the completely randomized design extended to growth and response curves. *Journal of the American Statistical Association*, **74**, 215–21.

Zerbe, G. O. (1979b) Randomization analysis of randomized block experiments extended to growth and response curves. *Communications in Statistics, Theory and Methods*, **8**, 191–205.

Zerbe, G. O. and Murphy, J. R. (1986) On multiple comparisons in the randomization analysis of growth and response curves. *Biometrics*, **42**, 795–804.

Zerbe, G. O. and Walker, S. H. (1977) A randomization test for comparison of groups of growth curves with different polynomial design matrices, *Biometrics*, **33**, 653–7.

Zick, W. (1956) The influence of various factors upon the effectiveness of maleic hydrazide in controlling quack grass, *Agropyron repens*. PhD Thesis, University of Wisconsin, Madison.

Zhang, J. and Boos, D. D. (1992) Bootstrap critical values for testing homogeneity of covariance matrices. *Journal of the American Statistical Association*, **87**, 425–9.

Zhang, J. and Boos, D. D. (1993) Testing hypotheses about covariance matrices using bootstrap methods. *Communications in Statistics – Theory and Methods*, **22**, 723–39.

Zimmerman, G. M., Goetz, H. and Mielke, P. W. (1985) Use of an improved statistical method for group comparisons to study effects of praire fire. *Ecology*, **66**, 606–11.

Zobel, K. and Zobel, M. (1988) A new null hypothesis for measuring the degree of plant community organization. *Vegetatio*, **75**, 17–25.

Zobel, K., Zobel, M. and Peet, R. K. (1993) Change in pattern diversity during secondary succession in Estonian forests. *Journal of Vegetation Science*, **4**, 489–98.

Author index

Subject index

Page numbers appearing in **bold** refer to figures and page numbers appearing in *italic* refer to tables.